Decision Analysis

APPLICATIONS OF MODERN TECHNOLOGY IN BUSINESS

Series Editor: Eric Clemons
The Wharton School University of Pennsylvania Philadelphia, Pennsylvania

Founding Editor: Howard L. Morgan,
University of Pennsylvania

COMPUTERS AND BANKING:
Electronic Funds Transfer Systems and Public Policy
Edited by Kent W. Colton and Kenneth L. Kraemer

DATA COMMUNICATIONS:
An Introduction to Concepts and Designs
Robert Techo

DATA BASE ADMINISTRATION
Jay-Louise Weldon

DECISION ANALYSIS
Geoffrey Gregory

OFFICE AUTOMATION:
A User-Driven Method
Don Tapscott

Decision Analysis

Geoffrey Gregory

Professor of Management Sciences
and
Head of the Department of Management Studies
Loughborough University of Technology

PLENUM PRESS • NEW YORK AND LONDON

Plenum Press, New York
A Division of Plenum Publishing Corporation
233 Spring Street, New York, N.Y. 10013
ISBN 0-306-42854-7
Library of Congress Catalog Card Number
87-042 964

First published in Great Britain 1988 by Pitman

British Library Cataloguing in Publication Data
Gregory, Geoffrey
Decision analysis.
1. Decision-making
I. Title
658.4'0354 HD30.23

Printed and bound in Great Britain

Contents

Preface

This book is based on a series of lectures which I gave to students on the MBA course at the Australian Graduate School of Management in the University of New South Wales. Giving these lectures provided me with the task of taking the general problems of decision-making and showing the insight which a quantitative approach afforded. The general theme is therefore that of the mathematical model, a theme which I borrow unashamedly from operational research. As with operational research the mathematical model is only a convenient way of presenting evidence in a medium which can be understood by the appropriately trained mind. It is a clear way of stating the assumptions relating the strands of the argument to the real world. It is not a total representation of the real world, as indeed no representation could ever be. Inevitably also the decision-maker himself has to live with the outcomes of his decision, and nobody would wish to deny him (or her) the right to exercise judgement. In this book I put forward some of the quantitative techniques which could clarify the issues involved.

In deciding what does or does not go into this book I have been influenced by a number of factors. I felt that it was necessary to include basic material on probability theory, decision criteria, decision trees and utility. This is the core of what is generally accepted as decision analysis. Other topics such as linear programming, dynamic programming, branch-and-bound, goal programming, game theory and simulation are included for two reasons. Firstly they concern quantitative approaches to decision-making and are therefore in by right. Secondly I wanted to make the book a text which would appeal to final-year undergraduates or postgraduates on business or management courses and also to people taking professional examinations in management or accountancy. To do this it was appropriate to include a range of topics which not only were naturally linked, but also covered a substantial part of the normal syllabus for a quantitative methods course at this level. Topics such as network analysis and inventory control have been left out either because they did not fit easily into the context I had set for the book or because I felt that I could add little to the material already available in most standard texts on management science or operational research. A knowledge of advanced mathematics is not required although some facility in handling equations and plotting graphs is expected. Calculus is used only rarely, and then only for the benefit of the more enlightened purist; others may safely skip over such interludes but be prepared to accept the results derived.

The book ends with a chapter on risk engineering. This is the work of Dr Dale F. Cooper of Spicer and Pegler Associates and Professor C. B. Chapman of the Department of Accounting and Management Science, University of Southampton. It is a revised version of two papers, one published in *Underground Space,* **9** (1c), 1985, and the other in *Further Developments in Operational Research,* G. K. Rand and R. W. Eglese (eds) (1985). I am grateful to the authors and to the publishers, Pergamon Press, for permission to publish.

Each chapter ends with further suggested reading and also with a number of problems and practical exercises. Some of the problems are open-ended in the sense that they are intended to promote discussion and there is no unequivocal answer. When there is a numerical element to a question, the answer is given in Appendix 1.

Coming closer to home I must acknowledge a great debt to all who have helped with this work. Not quite so close are the students on the MBA course in Sydney whose reactions to what was really a first draft of the book were most helpful. Other students on a variety of courses here at Loughborough and before that at the University of Lancaster must have guided me in the way the material has been presented. My colleagues at Loughborough are a constant source of helpful criticism, and particular mention must be made of Dr Peter Stratfold whose programming dexterity produced the results in the example of Chapter 13.

It is customary to save the last but most important acknowledgement of gratitude to the secretarial assistance received. Here I have been most fortunate. The early chapters were typed in first draft by Yvonne Marshall and it is a tribute to her undoubted competence that when the task was handed over to Lisa Thirlby continuity was maintained. To Lisa Thirlby I am enormously indebted, not only for the remarkable quality of her work and for the initiatives which she regularly displayed, but also for the way she invariably responded with an unruffled calmness when the pressure was on. Finally a thank you to my wife Brenda and a promise now to mend the toaster.

Loughborough *Geoffrey Gregory*
October 1986

1

Models of decision analysis

OBJECTIVES

This chapter is intended to set the scene. The reader is given a taste of the analytical approach to decision-making, introducing concepts which will be followed up in subsequent chapters. Decision-making involves more than a form of mathematical analysis and indeed the two aspects of the process – objective and subjective – should not be separated. Notwithstanding this last comment, the emphasis in the book is on quantitative methods, and it is important that the reader should study carefully the brief survey of the psychological aspects included in this chapter.

1.1 Decisions, large and small

Everybody takes decisions, and they do this at regular intervals throughout their conscious hours. You, the reader, have just taken the decision to read this section of this book, and you are now taking a sequence of decisions which will cause you either to continue reading or to take up some alternative occupation. Your actions are influenced by the benefits you anticipate from the decisions you are taking and also from the alternative decisions which you see as being available to you. Most of the decisions which we take are of a minor nature in terms of the consequences and of the facility we have to rectify the situation if the actual outcome does not live up to our expectation. You can put this book down and resume your knitting or snooker if you so wish, and you have lost only your own time. On the other hand, personal decisions like changing a job, moving house or deciding the location of your annual holiday would merit considerably more attention. Take, for example, the decision on your holiday. Before making a booking most people would consider it worthwhile to examine a number of alternative locations and times, having filtered out most of the possibilities on the grounds of cost, or of having already visited, or some personal prejudice. For the alternatives remaining you then need to gather information on characteristics such as basic living costs, travel costs, entertainment costs, proximity to mountains (or beaches), weather patterns, popularity with other visitors, standards of comfort in the hotel, etc. Some of this information, such as living and travel costs, will be readily available through the travel agent. Other information, like the weather at the time of year or the local geography, will require some research effort. A third category of information typified by the description 'standards of comfort' is not only difficult to determine without some form of survey of previous visitors, it is also difficult to define. The extent to which you are prepared to go to find this information depends on the value you put on being satisfied or disappointed with your selected holiday. There is moreover a sequential aspect to the decision problem. You can make your initial decision in a way that leaves options open so that, for example, a touring holiday of the South of France can, according to taste, be converted into a gentle enjoyment of the seaside if the weather is hot or more energetic trips to the mountains if it is cooler. Although these decisions do not have to be taken until later, their availabilities form an integral part of the initial decision.

The above example illustrates many of the features of a decision problem. There is the need to list all the alternatives to be considered, there is the sequential nature of decisions whereby one decision leads to outcomes from which further decisions can be made, and there is the need for information. Not all decisions warrant the elaborate structure we shall develop in this book, which is reassuring. Exactly the same argument holds for the commercial organization where decisions are being taken on behalf of and for the benefit of the corporate body. It may be a good idea for a firm to buy its paper clips from the cheapest supplier giving a satisfactory quality, but the savings in comparison with buying from a recognized supplier would hardly justify the effort of the investigation. Decisions concerned with opening a new branch or launching a new product could have far-reaching effects on the organization's viability, and here there is no doubt that some contemplation of the issues involved is justified.

1.2 Structure of a decision

To make any progress in the structure of a more analytical form of decision-making, we need to look more closely at the constituent parts of a decision. A decision is defined as the selection process leading to a particular action being taken. It is sometimes difficult to distinguish without being pedantic between an action and a decision, but strictly an action is the realization of a decision. The intention in this book is to make the distinction wherever possible, but not to let matters of definition obstruct the flow of an argument. Another issue, which in this case we shall leave to the more philosophically minded, is that if we have a decision to make, we should first decide whether or not we shall use the approach which follows in this book or rely on 'hunch' or intuition. In turn we must before this decide whether to decide. . . .

Put in its basic form, a decision has the following components:

(a) A list of alternative actions to be considered in the decision sequence. Before amplifying this item, take into account component (b).
(b) A list of all possible outcomes which can result from actions under (a). The decision is broken down into chronological stages. At each stage a list of alternative actions is compiled, and for each action we compile a list of what can then happen. Each of these may take us to the next decision stage where we have a further set of actions from which to choose. The procedure continues with the decision-maker's selections alternated by outcome possibilities which are outside his control.
(c) Data on the consequences of all feasible combinations of actions and outcomes. This may range from an actual profit or loss and therefore be measured in some precise way, to a more subjective consequence such as public esteem, quality of life or customer goodwill.
(d) Assessments of the likelihoods of the various outcome possibilities listed in (b). Sometimes this can be done by examining data from previous similar experiences. Records of sales of our product over previous weeks should give a strong indication of what the stock level should be for next week. Sometimes we have to rely on expert opinion backed by the results of sampling from the market. Sometimes we have only the views of experts and they themselves might have difficulty in rationalizing the way in which they came to their opinions.
(e) A decision criterion. Components (a)–(d) have described the structure of the decision, but they have not told us what to do.

The action to take may by now be obvious, but it is more likely that, because of the outcomes beyond our control, all we shall be able to do is to confine the consequences to one of a set whose relative likelihoods are known. A set of such consequences corresponds to combinations of each action at each stage. Some sets may have consequences which have little variation but are generally at a relatively low level. These correspond to 'low risk' strategies. Other sets may offer the opportunity of very attractive returns, but also include some which would create serious problems for the decision-maker. If he is a gambler he may be prepared to take such actions. There are, of course, other possibilities, but in order to make his mind up, the decision-maker is being influenced by some outlook or attitude towards what could happen when he decides which actions to take.

1.3 An illustration

The above description of the general structure of a decision has necessarily been rather abstract, so to illustrate the points being made we return to our would-be holiday-maker. In consultation with his family he has resolved the problem to one of three possibilities:

A. motor tour
B. book hotel at the seaside
C. hire a boat for a canal tour.

All possibilities are of two weeks' duration.

He then looks further at each initial action. Under action A, the weather may be good or indifferent. Dependent on this he can decide either to spend the time on the coast, or to move inland to the mountains. Finally for A, the family's enjoyment will depend on how crowded the chosen resorts are.

Action B is a little more risky as he is booking an hotel which is completely unknown to him. It may be either good or bad. He would book the hotel for the first week only and then decide whether to continue at the seaside or to spend the second week at a capital city hotel. If the hotel at the seaside turns out to be poor, he is confident that he could find a satisfactory one for the second week by local exploration. At the seaside, or in the capital city, the family's enjoyment of the second week would depend on the weather, although rainy weather in the city would not be as serious.

Finally, with action C he has a similar apprehension to that experienced with B. The family may not take to life on a canal boat. After a week, though, he has the option of spending the second week in the rustic setting of the open country or of turning the boat towards a more urban region where his family will not be so dependent on the boat for their pleasures. Again the total enjoyment will be dependent on the weather.

His options may be summarized as in Table 1.1. In this particular case, after the initial action has been taken, each subsequent action can be combined with all the following actions. There are therefore $3 \times 2 \times 2 \times 2 = 24$ possible outcomes to the decision.

Table 1.1

	(*a*)	(*b*)	(*a*)	(*b*)
A.	Motor tour	weather good weather indifferent	coast mountains	crowded comfortable
B.	Seaside hotel	hotel good hotel bad	stay capital city	weather good weather poor
C.	Canal boat	enjoy boat boat dismal	go rural go urban	weather good weather poor

Turning now to component (c) of the structure he has to allocate consequences to all of the 24 outcomes. Here he has a problem, since the real measure of the success of the action taken is the pleasure experienced in the holiday. He has, in fact, set aside a certain amount of money to pay for the holiday, and therefore the cost of each action will be approximately the same. He cannot therefore measure the consequences directly in monetary values. To convert a subjective scale of pleasure into something more quantitative, he asks himself the following question:

> 'How much would I be prepared to pay for a holiday where I took particular actions open to me and where certain particular outcomes occurred?'

To simplify his thoughts, he measures the answer in terms of a premium that he would be prepared to pay to guarantee the outcomes,

Table 1.2

(*a*) *actions*	(*b*) *outcomes*	*probabilities*	(*a*) *actions*	(*b*) *outcomes*	*probabilities*	*consequences*
A Motor tour	weather good	0.8	coast	crowded	0.8	100
				comfortable	0.2	400
			mountains	crowded	0.3	0
				comfortable	0.7	200
	weather indifferent	0.2	coast	crowded	0.4	−400
				comfortable	0.6	−300
			mountains	crowded	0.5	−200
				comfortable	0.5	0
B Seaside hotel	hotel good	0.7	stay	weather good	0.6	300
				weather poor	0.4	100
			central city	weather good	0.5	200
				weather poor	0.5	100
	hotel bad	0.3	stay	weather good	0.6	100
				weather poor	0.4	−100
			central city	weather good	0.5	0
				weather poor	0.5	−500

C Canal boat	enjoy boat	0.5	go rural	weather good	0.6	200
				weather poor	0.4	0
			go urban	weather good	0.5	−100
				weather poor	0.5	−200
	boat dismal	0.5	go rural	weather good	0.6	−200
				weather poor	0.4	−300
			go urban	weather good	0.5	−100
				weather poor	0.5	−100

or alternatively a refund that he would regard as fairly compensating for them. For example, suppose he went on the motor tour and the weather was good; if then he went to the coast and the region was not crowded he would be prepared to pay an extra £400 for his holiday. On the other hand, suppose he chose an hotel at the seaside which turned out to be bad; if then he went to the city and the weather was poor he would feel that a refund of £500 would just about compensate him for his misfortunes. In this way after much deliberation he is able to allocate consequences to all eventualities.

He then turns to component (d). Here he is able to turn to weather records for the various regions and to formulate some notions of the chances of the various outcomes. His knowledge of the quality of canal barges, seaside hotels and touring resort popularities is a little vague, but by asking around his more experienced acquaintances, he is able to put values on the probabilities. He is aware that some of these figures may be unreliable.

A summary of the structure is shown in Table 1.2. Note that we now have to separate out each stage of the decision process, even though the outcomes or actions are the same.

He now takes a decision of the type: Motor tour. If the weather is good, go to the coast; if the weather is indifferent go to the mountains.

This is a complete specification of his actions for any intermediate outcomes. There are twelve such specifications and because of the uncontrollable nature of the intermediate and final outcomes, each of them in this case has four possible consequences. In the decision quoted he will end up with consequences of 100, 400, −200 or 0. He can assess the probabilities of each consequence – in this case they are 0.64, 0.16, 0.10 and 0.10 respectively.

If all twelve specifications of actions are analysed in this way, it will be seen that some can be dismissed on the grounds that better consequences can be achieved by an alternative action, no matter what the subsequent outcome is. For example, if he is touring and the weather is indifferent, going to the coast can never give a larger consequence value than going to the mountains. Two other specifications of actions can be eliminated in this way (see if you can find them) leaving nine to be considered. To decide between these he needs a decision criterion. Several of these have been proposed and will be described in more detail in Chapter 4. Broadly they amount to considerations of the risk which the decision-maker is prepared to accept. Will he accept the risk of a very low return or consequence if there is a chance of a correspondingly high consequence? Does he want to play safe, and guarantee that he always receives the best minimum return? Does he want to look back after the decision has been taken and feel that in view of what actually happened his actions turned out reasonably well? Attitudes like these of the decision-maker can be translated into specific decision criteria which in turn lead to a particular set of outcomes being taken. Different people, therefore, faced with the same problem could plausibly take different actions because they have different attitudes, particularly towards the important factor of risk. This theme will be developed further throughout the book.

1.4 The aims of decision analysis

Enough has been said in this chapter about decision analysis to show why the procedures will only be used in decision problems of some importance. No doubt a subconscious use of the approach will be made in problems of less significance, but the depth and detail of the analysis can only be justified by a measurable return. What then is this return? Most people would be convinced of its value, if it guaranteed taking the right action. This the approach cannot do. The risks inherent in the problem will still be present and nothing can be done to eliminate them. What the analysis should do is firstly recognize their existence and secondly point out their consequences for actions taken. Even with the most precise analysis the decision-maker can, if he is unlucky, make what appears in hindsight to be a very bad decision. The adjective 'bad' (and by implication its opposite 'good') is being used in the unfair sense of comparing what a decision-maker did with what he would have done had he known the outcome of the decision problem. In tossing a coin is it really a bad decision to call 'heads' and then to find that the coin falls 'tails'? It cannot be, if the coin is fair. Similarly, a nation may have the problem of deciding on the primary form of energy for its future power stations. Evidence may simply suggest that coal will be the cheapest form of fuel for the operating life of the proposed power stations, and on the evidence the decision is taken to design and locate the stations accordingly. At a late stage of the construction a new and extensive off-shore oil and gas field is found, very much against expectations. The important point is not that this possibility was omitted from consideration. It may have been looked into carefully but all the evidence and expert opinion indicated a low probability of such a discovery. At the time, therefore, the correct action was taken. Events subsequently proved that the action was not the one that would have turned out best. Clearly there is a case for taking robust actions which can be modified in the light of developments. This can be expensive, and it can appear in the disguise of procrastination, which can also be the worst possible action.

There will, therefore, be an element of chance (or luck) about decision-making. The argument for decision analysis is that in the long run good decisions will be taken more often than they would otherwise be. Critics are asked to take a long-term view in their assessment of a decision-maker's performance. This is understandable if the decisions are of a repetitive nature, as for example would be the case in deciding how much stock of a product to hold for each week's trading. The view is not so easily taken when we are looking at what are essentially 'one-off' situations, such as opening a new branch, launching a new product or changing an investment policy. Some decisions are bound to be more far-reaching than others, and although the decision-maker may have a commendable 'batting average', his successes may not be quite as important as his failures. Perhaps more importantly his successes may not be witnessed quite so vividly by his peers as he might wish. More will be said about this aspect of decision-making in the following section. Decision analysis does not remove the chance element in decision-making. It can reduce it and it certainly attempts to quantify it, but no decision-maker can hope to make the right decision always. Any emphasis in this direction will lead to the avoidance of all situations except those offering the clearest choices and those of least importance.

1.5 Behavioural aspects

Decision analysis, as described in this book, is essentially an approach which proposes a *mathematical model* of a decision problem from which, by means of statistical analysis, a course of action is proposed. The use of the mathematical model is the link with Operational Research (O.R.), where mathematical modelling is the distinctive theme of this widely accepted management discipline. A mathematical model is simply a representation of the salient features of a problem by mathematical relationships, from which a solution may be derived. Often the mathematics is very elementary, but a number of fairly elaborate techniques have been evolved (linear programming, theory of queues, etc.) by O.R. workers to solve mathematical models which have cropped up in their work. It is unfortunate that these techniques have, in some quarters, been seen as Operational Research itself, whereas the truth is that their development is almost an historic accident. They were needed by O.R. workers and this need provided the impetus for research in these fruitful areas. Established statistical techniques have been used extensively by O.R. workers without the same associative label. The lesson for decision analysis from O.R. is that although the mathematical model leads to some quite precise mathematical analysis, there is still a strong subjective element in the approach, namely in the formulation of the model. Any mathematical model is an approximation to reality and decision analysis is no exception. For example, in deciding whether or not to launch a new product, assumptions have to be made on the following points:

(a) The timing of the launch. Are there only a limited number of occasions when the product should appear on the market? (e.g. for Spring sales, in anticipation of competitors).
(b) The impact on the company's other products. To what extent does a new design of motor vehicle detract from the manufacturer's existing range?
(c) Manufacturing implications. How versatile are the company's production facilities?
(d) The product life cycle. How can this be assessed by early sales?
(e) Publicity for the launch. How much should be spent on advertising the launch, and what is the relationship between advertising expenditure and actual sales?
(f) Passing information to dealers (retailers). How can the effectiveness be measured?

These and other factors pose questions which cannot be answered directly, but which can be assessed by means of assumptions in a mathematical model. Solving the mathematical model is then an exercise which not only provides an answer to the problem but, perhaps more importantly, provides an insight into the important features of the problem. Put simply, it is a question of which assumptions really matter in the sense that a minor change could lead to a different decision being taken, and for which assumptions is the solution insensitive?

Decision analysis is, therefore, not a means whereby the decision-maker is to be replaced by an automatic procedure. Considerable skill will have to be exercised in formulating the decision model. Moreover the output from the model – the recommended action – will only be as reliable as the model and its attendant data. The output will be evidence used by the decision-maker to make up his mind, and the weight he puts on the evidence should depend on his knowledge of the reliability of the assumptions he has had to make. The strength of the approach is that these assumptions are clearly stated.

This is only one aspect of the subjective nature of the decision-making process. At the final stage of coming to a decision other considerations will make their presence felt. These are mainly of a psychological nature. They are clearly very important, but as this is an account of the analysis of the decision model, mention will only be made of some of them, without any real attempt to discuss how they are recognized and resolved. The importance of this aspect of decision-making cannot be over-emphasized, and there is a vast literature on the subject (see for example Janis and Mann (1977)). Business games, for example, provide a simulation of business decision-making situations, but it is very difficult for a business game to create the form of stress which a real-life decision-maker faces when he has an important decision to take. The analogy between business games and decision analysis is by no means precise, but there is an element of detachment about decision analysis which has to be remembered. Care must be taken, particularly if the analysis is carried out for the decision-maker by a specialist, but it is the decision-maker who has to live with the consequences of his action.

Perhaps one of the main psychological factors is the personal motivation of the decision-maker. As a member of a commercial organization he is acting in what he believes to be the best interests of the corporate body. Nevertheless, he knows that if the action he takes turns out to be a bad one, he may face social disapproval, his reputation as a man who knows the market will suffer, and his career prospects may be damaged. Even if his associates are sufficiently enlightened to understand the risks involved, he may perceive these personal interests as being at stake. On the positive side, there is, of course, the self esteem and peer status of having taken a correct action. The influence of these personal issues depends very much on how the individual views his position within the organization.

Two aspects of the stress in decision-making can apparently have opposite effects. One is to evade responsibility either by passing the responsibility to another person or by procrastination, hoping that the decision will not have to be taken. The other aspect is to take the decision prematurely so that the source of the tension is removed. Both of these aspects are at variance with the decision analysis approach, which assumes that information will be gathered through a painstaking canvass of alternative actions, their probabilities and their consequences, and that a decision will be taken then and only then.

A phenomenon which runs counter to the completely objective view of decision-making is that of bolstering. When some of the consequences of a decision are of a rather subjective nature, the decision-maker may initially view these various outcomes dispassionately. Having taken his decision, however, he tends to magnify the attractiveness of his chosen action and to play down its disadvantages. Moreover, he spreads the alternatives by viewing his (now rejected) alternatives in the opposite light. Bolstering is a means of persuading himself that he has taken the right action. It is commonly met when people make a selection from a choice of similar products, such as motor vehicles or computer systems. Unless there is strong evidence to the contrary, as for example would arise with serious breakdown problems, the purchaser justifies his choice to himself and to associates using the arguments mentioned. It is particularly likely to happen when the decision has been based on several criteria such as initial cost, running cost, reliability, versatility, ease of use. etc., and the decision-maker changes the emphasis or weight he puts on each criterion to confirm the wisdom of his choice.

In some cases decisions may have to be taken where the appropriate data and information generally would be costly and time-consuming to collect. Alternatively, it may simply not exist. Developments in rapidly advancing technologies such as information processing or robotics are currently of this nature, where a firm may have to take decisions concerning its future office management or manufacturing design, anticipating equipment or systems not yet commercially available. There are experts who understand these developments, and the firm should make use of their expertise in some collective way. Inevitably there will be differences of opinion and the outcome of the deliberations will have a significant subjective element. An effective way of bringing together the views of experts is to use what is known as the Delphi method.

Briefly the method is as follows. A number of experts are enlisted. They are sent a questionnaire which has been drawn up by a small monitor group. The monitor group then summarizes the results of the questionnaire, and then sends out a second questionnaire. This questionnaire allows the respondents to modify their opinions based on the collective reaction. This feedback may go through several cycles before a consensus is reached (if indeed it ever is). In the process a skilful monitoring group will have brought out further relevant points from the respondents; they will try to understand and to feed back the real issues on which disagreements exist between respondents; and finally they must feed back a satisfactory evaluation. The whole process is carried out without face-to-face exchange. Not only is this cheaper, but it permits a much larger group to be used. Also it avoids the dominance that a more forceful or articulate individual can obtain in committee meetings.

The Delphi method is essentially a procedure for structuring a communication process amongst a group of individuals. In areas like the application of decision analysis to policy formulation it is possibly more useful in generating alternative actions and scenarios than in taking an actual decision. It is rather like the popular concept of brainstorming, but carried out by correspondence. A further account can be found in Linstone and Turoff (1975).

Finally, mention must be made of a concept introduced by Herbert Simon (1976). Simon argues that human beings have limited information-processing capabilities, and because of this they are prepared to sacrifice the benefit of a comprehensive survey of the alternatives with all their information requirements and analyses. Instead they look for an alternative which is better than their current method of operation and settle for that. This involves consideration of the alternative in some depth, but it avoids consideration of all the alternatives, all but one of which will, of course, be discarded. Simon calls this *satisficing*. It constitutes an improvement, but it is not the best. It can also mean that changes are made in modest steps if carried out sequentially, rather than in a more radical, disruptive and riskier switch. Clearly there will be times when the gentler approach of satisficing is more appropriate than straight optimization, and there will be times when opportunities will be missed by using this outlook. A firm can introduce refining improvements to its production line or it can redesign the whole process for computer-aided manufacture. One may be objectively superior to the other, but the success or failure of the latter will depend very much on attitudes within the firm. Satisficing can also be attractive to groups accepting responsibility for a decision where individuals within the group have irreconcilable views. It is then being used as a form of compromise, presenting a general improvement but avoiding the issue of determining the best action.

1.6 Further reading

Mathematical Modelling

Rivett, B. H. P., *Model Building for Decision Analysis*, Wiley, Chichester (1980).

Behavioural Aspects

Cooke, S. and Slack, N., *Making Management Decisions*, Prentice-Hall, London (1984).
Janis, I. L. and Mann, L., *Decision Making*, Macmillan, New York (1977).
Linstone, H. A. and Turoff, M. (eds) *The Delphi Method: Techniques and Applications*, Addison-Wesley, Reading Mass. (1975).
Radford, K. J., *Modern Managerial Decision Making*, Reston, Va. (1981).
Simon, H., *Administrative Behaviour: A Study of Decision-Making Processes in Administrative Application*, Macmillan, New York (1976).

1.7 Practical exercises

1 Brittel Parts plc is a manufacturer of component parts for the motor industry. It has supplied 4000 ratchet manifolds out of a contract of 6000 agreed with a motor manufacturer. The manifold includes a casting specific to the contract and Brittel buys these in batches of 2000 at a price of £3000. The supplier of castings now offers Brittel a batch of 6000 castings at a total price of £7000. It is an offer which will not be repeated (according to the supplier) and the price of £3000 for batches of 2000 castings still holds.

The motor manufacturer is in the process of considering bids to supply his next batch of ratchet manifolds and will announce his choice shortly, but not before he expects to be receiving a substantial part of the last 2000 of the existing contract with Brittel. Normally the contracts are for 6000 manifolds. Any castings which Brittel buys but does not use are of no value.

What options does Brittel Parts have? What information would they like to possess, and how much effort do you feel they ought to expend to obtain it?

2 You are engaged by a small firm to advise on the feasibility of running their own canteen service. The firm normally has approximately 100 employees (of whom 70 are women) on the premises between 8.30 a.m. and 4.30 p.m., so the main need occurs during the lunch break. A room is available and provision could be made for simple cooking facilities. There is also a meals delivery service in the area, for which the firm would need to provide standard heating ovens.

At present some of the employees bring their own lunches and eat them either in their place of work or, if the weather is fine, in a nearby park. There is a fish and chip shop and a sandwich bar about a quarter of a mile away, and some employees use these on occasions. A public house a mile away provides good quality bar snacks. The local market is also a mile distant from the firm and on Thursdays, a number of the female employees visit the market to do a substantial part of their week's shopping. Since the local bus service runs at inconvenient times, they do this by car.

How would you set about advising the firm? What information would you need?

3 Evaluate any of the following decisions as might be appropriate to your own circumstances. Structure the decision in the form of the five components outlined in Section 1.2, indicating where you might have difficulty in obtaining data.

(a) Buying a new car
(b) Emigrating to Australia
(c) Applying for a new job
(d) Building a tunnel to the Isle of Wight
(e) It is not raining, but do I take my umbrella when I walk to work (1 mile) today?

4 Consider a national issue such as the introduction of a 60 miles per hour speed limit on all roads. What evidence is needed and which bodies are likely to be influential in the making of such a decision?

2

Probability and random variables

OBJECTIVES

By the end of this chapter the reader should have an understanding of

(a) the concepts of probability and conditional probability,
(b) statistical distributions,
(c) populations and random variables,
(d) measures of centrality and variability,
(e) the concept of expected value,

and an ability to

(a) operate the laws of probability,
(b) use the binomial, Poisson, normal and beta distributions,
(c) estimate the mean and standard deviation from data.

2.1 Objective and subjective probabilities

Most people have a good understanding of the concept of probability. For example, when a standard coin is tossed, we say that there is a chance of 0.5 that it will fall heads, 0.5 tails. We base this argument on the physical attributes of the coin (its basic symmetry) and on the fairness of the way in which the experiment – tossing the coin – will be carried out. Moreover, we are prepared to verify our assertion by carrying out a large number of trials with the coin. In these trials we would not expect exactly half of the tosses to come down heads and half tails, but provided that the frequencies were reasonably close to fifty–fifty we would feel that our assertion was justified. In fact, had the coin been in any way suspect, we could have based our assignment of the probability number entirely on the proportions of heads and tails found in this way. Probabilities derived in this way are said to be *objective* or *empirical*. The method assumes that the assigned value can be derived either by arguing from the symmetries of the experimental mechanism or by actual experimentation. In the first case if we are prepared to assume that the coin is a flat disc the probabilities assigned by any reasonable individual would be 0.5 for either face of the coin. In the second case the proportions falling heads or tails would be assigned as probabilities, again by our reasonable individual. There is a distinction between the two methods, since in the latter case two reasonable individuals carrying out the same experiment of tossing the coin 500 times are likely to produce different answers since the coin will not fall heads, say, the same number of times for each individual. It is the method of assigning the probability which is objective, not the actual value assigned.

The argument is contrary to the philosophy of determinism, which says that all outcomes can be determined precisely, and it is only our ignorance of the laws of nature which prevent this happening. Instead we resort to arguing in terms of probabilities. Determinism is not taken seriously nowadays by the practising statistician.

Returning now to the concept of objective probabilities, it is not difficult to find instances where the approach does not meet our needs. Consider the following statements:

(a) The Sterling–US dollar exchange rate will fall tomorrow.
(b) Richard III murdered the Princes in the Tower.

Statement (a) concerns a mechanism with outcomes – fall, steady, rise – which at present are unknown. It is similar to tossing a coin, but there is no means whereby we can argue objectively from the mechanism, nor can we reproduce tomorrow a large number of times. Nevertheless, it is an outcome on which much could depend, and certainly the likelihood of such an outcome could influence a decision to transfer currency. An equivalent concept to that of objective probability is needed.

Statement (b), which does not have the commercial impact of (a), brings out a further point for consideration. The same arguments about the experimental mechanism and the inability to replicate hold, but in this case there is the added fact that the event is in the past. Richard either did it or he did not! What then has this got to do with probabilities? Probability in this sense refers to an individual's knowledge or ignorance. How strong is the evidence and what does each of us individually make of it? We certainly need (particularly for statement (a)) the concept of probability to progress with our decision-making. The approach used is to form an equivalent measure which acts like an objective probability, but which depends on an individual's experience coupled with whatever information appears relevant. It is therefore quite reasonable to discover that two people acting knowledgeably and sensibly will propose different values for the same outcome. The values should not be very different and presumably it should be possible for the people to reconcile their differences and produce a single figure. Such a figure is known as a *subjective probability*.

There are two approaches to the measurement of subjective probability. The direct method, as its name implies, simply requires the subject to assign an appropriate number to the outcome in question. If the subject is sufficiently experienced in these matters, he or she will be making comparisons in his or her mind with readily assessed objective probabilities. For the less experienced it may be necessary to specify a mechanism such as a freely rotating pointer which comes to rest at some point on the circumscribing circle. The probability that it comes to rest within a particular arc is equal to the ratio of the length of this arc to the total circumference. Thus, by adjusting the length of the arc, the equivalence of this outcome can be made to any other outcome and the subjective probability thereby defined. Taken literally, such a procedure would involve refining the comparisons by trial and error until an equivalence is reached. Even this may be difficult, since there is always the danger that the subject may not be able to retain his standards in his mind and could produce inconsistencies where he has said that one arc is less likely than the outcome in question but then that a shorter arc is more likely. This could easily happen in the later stages of the comparisons, particularly if some time elapses between the questions.

In practice no such elaborate procedure takes place. Subjective probabilities use a fairly coarse scale of measurement, being satisfied with one decimal place in the middle of the zero-one range and refining beyond this level towards the extremes. It is usually possible to determine how sensitive is the action taken to the subjective probability assigned, so that, if necessary, further thought can be given to the value used.

As an illustration of one way in which the mind can be focused on the issue in question so that a subjective probability is produced, consider question 1 at the end of the previous chapter. If the contract was to be renewed, Brittel Parts would buy the batch of 6000 at

Table 2.1

	Contract renewed	*Contract not renewed*
Buy 6000	£10 000	£7000
Do not buy 6000	£12 000	£3000
	Cost of castings	

£7000 plus an extra 2000 at the normal price of £3000, yielding a total cost of £10 000. Conversely, if the contract were not renewed, Brittel Parts would be better off if they had just bought their needed 2000 castings at the normal price of £3000. Their dilemma can be summarized as in Table 2.1. Clearly their action depends on π, the probability that the contract will be renewed. If π is close to 1, they will be tempted to take advantage of the cheaper batch, but if π is close to zero they buy only the 2000 castings of their immediate needs.

Suppose that Brittel are comparing the offer of the cheaper batch with the normal cost of the castings. If they accept the offer Brittel will be £2000 better off if the contract is renewed but £4000 worse off if it is not. Considering the offer is therefore equivalent to considering a gamble with a probability π of winning £2000 and $1 - \pi$ of losing £4000. How much is this gamble worth to the firm? If they believe π is close to 1, they would be prepared to pay up to £2000 for the gamble, but if π is close to zero, they would expect to be paid close to £4000 before the gamble is taken. We ask, therefore, what price £C would Brittel be prepared to pay for the opportunity of this offer of the cheaper castings, where it is understood that a negative value of C means that Brittel would be paid the amount $-C$ to take the gamble.

The relationship between C and π is found in the following way. Suppose that the purchase situation is repeated a large number N, of times. Brittel will expect to win £2000 $N\pi$ times and to lose £4000 $N(1 - \pi)$ times. The total gain will therefore be

$$2000N\pi - 4000N(1 - \pi)$$

For equivalence this will equal the total amount paid, CN. Hence for a given value of C, cancelling out the N value, we have

$$C = 2000\pi - 4000(1 - \pi)$$
$$\pi = \frac{4000 + C}{6000}$$

If, for example, Brittel is willing to pay £500 for this opportunity, the implication is that a subjective probability of 0.75 is attached to gaining the renewed contract. On the other hand, if Brittel requires a payment of £1000 to accept the gamble, the subjective probability is 0.5.

Psychological considerations come into the assessment of subjective probabilities. People generally have difficulties in making clear judgements, as inevitably their minds are unable to arrange their experiences in a clear and unbiased form. Tversky and Kahneman (1974) put these difficulties under three broad headings. There is *representativeness* where, for example, a firm receiving raw materials from a number of suppliers may wish to assess the probability that poor quality in his output is the result of a bad batch from a particular supplier. His assessment should take into account not only the quality reputation of this supplier but also the fraction of the input raw material which comes from this source. Other manifestations are the tendencies for people to believe that runs of outcomes must average out. For example, people argue that if we have had two good winters, we must be due for a bad one next year. The second heading is that of *availability*. Broadly this refers to the bias caused by the ease with which people tend to recollect positive instances and ignore the negative ones. Two or three business failures in a particular industry tend to give the impression that the probability of failure in this industry is higher than it really is. The human mind retains certain information and possibly even exaggerates its numerical impact. The third heading concerns situations where the assessments are made by starting from some bench mark and adjusting to meet the circumstances of the outcome in question. This is known as *anchoring*. For example, the 0.5 probability of a coin falling tails might be taken as a starting point for comparison. This then could be refined to more complex coin-tossing outcomes, but it appears that the assessment finally reached depends on the starting point taken. It is as if the person involved believes that he or she should not depart too far from the starting point. There is also a tendency for people to underestimate extremes. Several investigations have demonstrated this phenomenon. For example it is useful to know, for a new product, what sales level is to be exceeded with probability 0.1 (the ninetieth percentile). Very often this value is underestimated. The purpose of this discussion is to arrive at a concept of probability which can be used for all eventualities. As we have seen, this often requires a strong subjective element, which can be accommodated, but in so doing we should be aware of a number of sources of bias.

2.2 The laws of probability

In this section we shall speak of an *experiment*, meaning some mechanism which generates outcomes to which we shall attach probabilities. The experiment may be a very simple one such as tossing a coin, where there are just the two reasonable outcomes, heads or tails. We ignore possibilities such as the coin ending up balanced on its edge (it is to land on a flat rigid surface) or the coin disintegrating in mid-air. There will always be the need to specify the outcomes and therefore to exercise judgement in eliminating outcomes from consideration. The manufacture of a bobbin of a particular yarn is also in our terminology an 'experiment', generating a number of outcomes, such as the weight of yarn on the bobbin, the breaking strength of the yarn and the number of faults in the yarn. A third illustration where the mechanism of the experiment is rather more abstract is the means whereby a firm's customers place their orders for a particular product in a particular week. The outcome as expressed by the total demand is easily recognizable, but the mechanism which governs the decision to order and then the amount ordered within the population of customers and potential customers is very complex indeed. Fortunately, we can usually recognize some pattern in the demand from week to week and can then compress a highly complicated generating system into relatively simple probability statements about what will happen in any week.

Take any experiment and denote its outcomes by $E_1, E_2, E_3, \ldots$. Suppose, following the arguments of the first section of this chapter, we attach by some means the probabilities $\Pr(E_1), \Pr(E_2), \Pr(E_3), \ldots$ to these outcomes. Care has to be exercised in the way these probabilities are attached, and in particular they must follow certain laws and conventions. Denoting the general outcome by E, we must have

$$0 \leqslant \Pr(E) \leqslant 1$$

To progress further we need some definitions.

Mutually exclusive. Outcomes E_i and E_j are said to be mutually exclusive if, when E_i occurs, E_j cannot occur, and vice versa. They cannot occur together.

Suppose, for example, our experiment consists of cutting a standard pack of cards, and suppose that we define

E_i : card is an ace
E_j : card is a king

Then clearly E_i and E_j are mutually exclusive. [Note that it may be that neither event happens.] Now consider the further outcome

E_k : card is a club

Then E_i and E_k are not mutually exclusive, since both will occur if the cut card is the ace of clubs.

Exhaustive. Outcomes E_i and E_j are said to be exhaustive if whenever the experiment is carried out either E_i or E_j or both E_i and E_j must occur.

Again, using the example of cutting a pack of cards, define

E_i : card is a heart or an ace
E_j : card is a spade, diamond or club

Then E_i and E_j are exhaustive. They overlap with the aces of spades, diamonds and clubs and are therefore in this case not mutually exclusive.

When we can specify a set of outcomes which are mutually exclusive and exhaustive, this consists of a list of outcomes of which one, but only one, outcome will occur at any run of the experiment. For any given experiment such a list is not unique. In the card-cutting experiment, the *elementary* list of mutually exclusive and exhaustive outcomes is a list of all of the 52 cards. Alternatively we can measure the outcome simply by the suit of the card cut, and the list is then the four outcomes – spades, hearts, diamonds, clubs. This list has the same properties. A further list is the 13 card values, ignoring the suit.

We must now clarify two items of terminology. When we speak of (E_1 and E_2) we speak of an outcome which occurs when and only when both E_1 and E_2 occur. It follows that (E_1 and E_2) is the same as (E_2 and E_1). Also if E_1 and E_2 are mutually exclusive, the joint outcome (E_1 and E_2) is empty – it cannot occur. We shall find it useful to denote this empty outcome by the symbol ø.

We shall also speak of an outcome (E_1 or E_2). This refers to the outcome when either E_1 or E_2 or both E_1 and E_2 occur. A further piece of useful notation is to denote by $\bar{E}$ the outcome complementary to E, that is $\bar{E}$ occurs whenever E does not, and vice versa.

We are now in a position to state the first law of probability: Suppose E_1 and E_2 are two mutually exclusive outcomes. Then

$$\Pr(E_1 \text{ or } E_2) = \Pr(E_1) + \Pr(E_2). \tag{2.1}$$

This is known as the *addition law*.

Note that it extends to any number of mutually exclusive outcomes.

$$\Pr(E_1 \text{ or } E_2 \text{ or } \ldots \text{ or } E_n) = \Pr(E_1) + \Pr(E_2) + \ldots + \Pr(E_n)$$

Furthermore, if the outcomes $E_1, E_2, \ldots, E_n$ are also exhaustive the outcome E_1 or E_2 or . . . or E_n must occur and therefore has the probability of 1. In this case, therefore

$$\Pr(E_1) + \Pr(E_2) + \ldots + \Pr(E_n) = 1.$$

It is often helpful to think of the experiment as the selection of a point within a space conveniently depicted as a square. Outcomes E_1, E_2 etc. are regions within the square, so that, for example, if the point selected lies within E_1, the event E_1 occurs. If therefore E_1 and E_2 are mutually exclusive, their regions will not overlap. If they are exhaustive, they will together cover the whole space. Furthermore if we contrive the areas of the regions to be their probabilities, the addition law then becomes obvious. An implication of this probability/area relationship is that the total area of the space shall be unity. Such a diagram is known as a Venn diagram. The addition law is illustrated in Fig. 2.1.

One use of the addition law is the rule that since for any outcome E, the outcomes E and $\bar{E}$ are mutually exclusive and exhaustive and therefore

$$\Pr(E) + \Pr(\bar{E}) = 1.$$

To state the next law we need to introduce the concept of *conditional probability*. This refers to the modification we make to the probability of an outcome when we know that some other outcome has occurred. Suppose, for example, a firm produces carpets in three qualities –

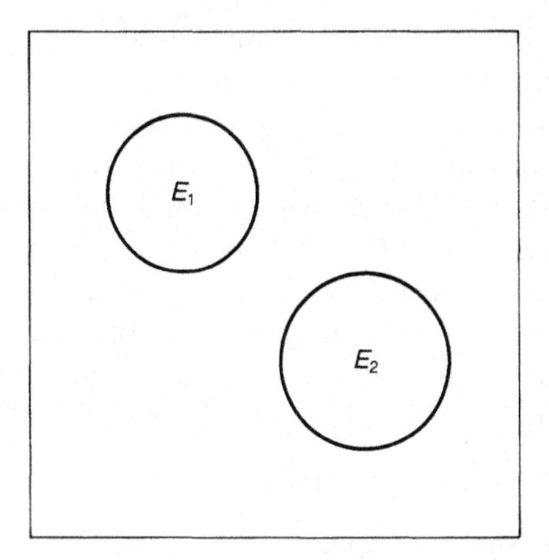

Fig. 2.1 Illustration of addition law by Venn diagram

regular, special and de-luxe. The firm's customers consist of 100 dealers, and for the sake of this illustration, we shall assume that any dealer only handles one quality. The dealers are either in the home country or export and the pattern of their practices has been established (Table 2.2).

If all the dealers send in regular monthly orders, the probability that a randomly selected order is for the regular quality is $(30 + 5)/100 = 0.35$. If, however, we know that the order has come from overseas, it can only have come from one of the 40 dealers, of whom 5 order regular quality. The probability is therefore $5/40 = 0.125$. Denoting the separate outcomes of quality and location by their initial letters (R, S, D, H, E) we have assigned the probability

$$\Pr(R) = 0.35$$

To denote the probability of the outcome R conditional on E having happened, we write

$$\Pr(R|E) = 0.125,$$

Table 2.2

	Quality required		
	Regular	*Special*	*De-luxe*
Home	30	20	10
Export	5	15	20

the vertical line separating the outcome whose probability is required from that which we know to have happened.

When the unconditional probability $\Pr(R)$ is being compared with the conditional probability $\Pr(R|E)$ we refer to the former as a *prior* probability and to the latter as a *posterior* probability. The implication of these descriptions is that the conditional probability includes further information on R – namely that E has happened – and that we have therefore modified our probability in the light of this information.

The second law of probability relates joint probabilities to conditional probabilities in the form

$$\Pr(E_1 \text{ and } E_2) = \Pr(E_1|E_2)\Pr(E_2). \tag{2.2}$$

It is also known as the *multiplication law.*

In the data of the example we have

$$\begin{aligned} \Pr(R \text{ and } E) &= \tfrac{5}{100} \\ \Pr(R|E) &= \tfrac{5}{40} \\ \Pr(E) &= \tfrac{40}{100}. \end{aligned}$$

The result can also be verified from a Venn diagram (Figure 2.2). The outcome $E_1|E_2$ requires that a point occurs in the doubly hatched region given that it is in the singly hatched region.

The multiplication law extends to the joint occurrence of several outcomes in the form:

$$\begin{aligned} \Pr(E_1 \text{ and } E_2 \text{ and } E_3) &= \Pr(E_1|E_2 \text{ and } E_3)\Pr(E_2 \text{ and } E_3) \\ &= \Pr(E_1|E_2 \text{ and } E_3)\Pr(E_2|E_3)\Pr(E_3). \end{aligned}$$

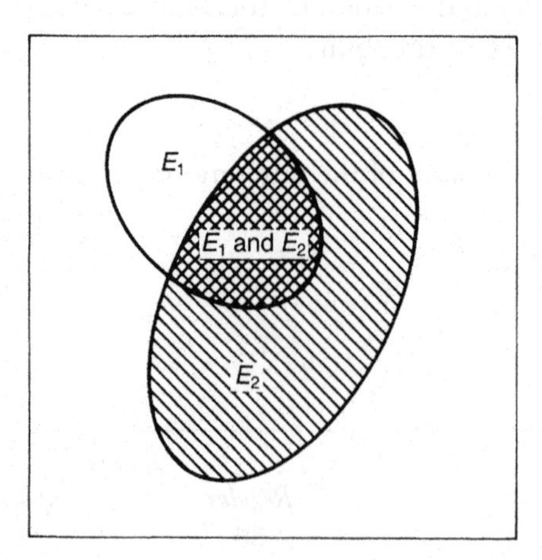

Fig. 2.2 Illustration of multiplication law by Venn diagram

Now suppose that we apply the multiplication law to the experiment of cutting a standard pack of cards. Let A be the outcome that the card cut is an ace and H be the outcome that it is a heart.

$$\text{Then } \Pr(A) = \tfrac{1}{13}, \quad \Pr(H) = \tfrac{1}{4}, \qquad \Pr(A \text{ and } H) = \tfrac{1}{52}.$$

The conditional outcome $A|H$ is the outcome that the card is an ace, given that it is a heart. From the multiplication law

$$\Pr(A|H) = \Pr(A \text{ and } H)/\Pr(H) = \tfrac{1}{52}/\tfrac{1}{4} = \tfrac{1}{13}.$$

In this case $\Pr(A|H) = \Pr(A)$, which only states in statistical language that your chance of cutting an ace is unaltered if you are told that the cut card is a heart. The outcome A is said to be independent of H. In this case it follows that

$$\Pr(A \text{ and } H) = \Pr(A|H)\ \Pr(H) = \Pr(A)\ \Pr(H), \text{ and}$$
$$\Pr(A \text{ and } H) = \Pr(H \text{ and } A) = \Pr(H|A)\ \Pr(A),$$

so that $\Pr(H|A) = \Pr(H)$. The property is symmetrical in the two outcomes, and we can therefore say that the outcomes are independent. Usually we can argue from the experiment that two of its outcomes, E_1 and E_2, will be independent and from this it follows that

$$\Pr(E_1 \text{ and } E_2) = \Pr(E_1)\ \Pr(E_2).$$

The concept extends to several variables, resulting in

$$\Pr(E_1 \text{ and } E_2 \text{ and } \ldots \text{ and } E_n) = \Pr(E_1)\ \Pr(E_2) \ldots \Pr(E_n).$$

There is also a more general form of the addition law which applies to the case where the outcomes are not mutually exclusive. The law then becomes

$$\Pr(E_1 \text{ or } E_2) = \Pr(E_1) + \Pr(E_2) - \Pr(E_1 \text{ and } E_2).$$

This can be proved rigorously from the addition and multiplication laws as stated, and it can be readily demonstrated from the Venn diagram. The reader should try both approaches.

The third law which, as we shall show, can be proved from the other two, concerns outcomes $E_1, E_2, \ldots, E_n$ which are mutually exclusive and exhaustive, and any other outcome A. If, therefore, A occurs, it must occur with one and only one of the E_i.

$$A = (A \text{ and } E_1) \text{ or } (A \text{ and } E_2) \text{ or } \ldots \text{ or } (A \text{ and } E_n).$$

Since the E_i are mutually exclusive, the outcomes (A and E_i) must also be mutually exclusive.

Hence $\Pr(A) = \Pr(A \text{ and } E_1) + \Pr(A \text{ and } E_2) + \ldots + \Pr(A \text{ and } E_n)$ by the addition law, equation (2.1).

By the multiplication law (equation (2.2)),

$$\begin{aligned} \Pr(A \text{ and } E_i) &= \Pr(A|E_i)\Pr(E_i) \text{ and hence} \\ \Pr(A) &= \Pr(A|E_1)\Pr(E_1) + \Pr(A|E_2)\Pr(E_2) + \ldots \\ &\quad + \Pr(A|E_n)\Pr(E_n). \end{aligned} \tag{2.3}$$

This is a surprisingly useful formula. For example, we have two machines E and F, which make printed circuits. Machine E produces 2 per cent defective circuits, whereas machine F produces 4 per cent defectives. The outputs of the two machines are combined, with machine F making 3 printed circuits for every 2 made by machine E. What is the probability that a circuit randomly selected from the mixture is defective?

If E and F are the outcomes that the circuit was made by machines E and F respectively and if D is the outcome that the circuit is defective, we have

$$\begin{aligned} \Pr(E) &= 0.4 \qquad & \Pr(F) &= 0.6 \\ \Pr(D|E) &= 0.02 \qquad & \Pr(D|F) &= 0.04. \end{aligned}$$

Also E and F are mutually exclusive and exhaustive, and hence by the third law (equation (2.3))

$$\begin{aligned} \Pr(D) &= \Pr(D|E)\Pr(E) + \Pr(D|F)\Pr(F) \\ &= (0.02)(0.4) + (0.04)(0.6) = 0.032. \end{aligned}$$

Finally in this section we can now derive an important result known as Bayes' formula, named after Thomas Bayes, an eighteenth-century English clergyman. Again, suppose that $E_1, E_2, \ldots, E_n$ are mutually exclusive and exhaustive outcomes and that A is any other outcome.

Then for any of the outcomes, say E_i,

$$\Pr(E_i|A)\Pr(A) = \Pr(E_i \text{ and } A) = \Pr(A|E_i)\Pr(E_i)$$

using the multiplication law (equation (2.2)) twice. Dividing both sides by $\Pr(A)$ and using the third law (equation (2.3)) we have

$$\Pr(E_i|A) = \frac{\Pr(A|E_i)\Pr(E_i)}{\Pr(A|E_1)\Pr(E_1) + \Pr(A|E_2)\Pr(E_2) + \ldots + \Pr(A|E_n)\Pr(E_n)}. \tag{2.4}$$

Bayes' formula, stated above, reverses the order of the conditional probabilities. Alternatively, it may be seen as a way of relating the posterior probability $\Pr(E_i|A)$ to the prior probability $\Pr(E_i)$.

Returning to the example of the manufacturer of printed circuits, suppose that he has selected a circuit from the mixed output of the two machines and that it is defective. What is the probability that it came from machine E? The prior probability that any circuit comes from machine E is $\Pr(E)$, which equals 0.4. We now require the posterior probability $\Pr(E|D)$.

The outcomes E and F are mutually exclusive and exhaustive. Hence, by equation (2.4), Bayes' formula,

$$\begin{aligned}\Pr(E|D) &= \Pr(D|E)\Pr(E)/[\Pr(D|E)\Pr(E) + \Pr(D|F)\Pr(F)] \\ &= (0.02)(0.4)/[(0.02)(0.4) + (0.04)(0.6)] \\ &= 0.008/0.032 = 0.25.\end{aligned}$$

The information that the circuit is defective has reduced the prior probability of 0.4 that it came from machine E to the posterior probability of 0.25. It follows that

$$\Pr(F|D) = 1 - \Pr(E|D) = 0.75$$

2.3 Statistical distributions

2.3.1 The binomial distribution

We turn now to ways in which we can handle experiments where the outcomes are measurements and we wish to make statements or inferences from the outcomes about the generating mechanism of the experiment. To be more specific, suppose that we have a production line whose finished product is a radio amplifier unit. The final stage of inspection is simply to test to see if it works, but a small sample of the output is subjected to a more rigorous form of testing. Suppose, therefore, that batches of 20 amplifiers are taken, and denote by X the number of amplifiers within a batch which fail the test. Then X is the outcome of an experiment, and any realization of X is an integer between 0 and 20. Following the arguments of the previous section we could assign probabilities to all these values, subject to the usual constraints that they are positive values and that they sum to unity. In this instance, however, it is possible to argue for a form of the probabilities because of the mechanism of the experiment generating the values.

The statement of the generating mechanism is simple but the implications for this case and for further developments are very important. Here we simply assume that any amplifier produced on this line has a probability π of being defective. The fate of each amplifier is independent of its neighbours, which means that finding a defective amplifier does not mean that we are entitled to a sequence of non-defectives to follow, making up the proportion π in the surrounding group. We are just as likely to obtain a defective following a defective as we are anywhere in the output sequences. The probability π corresponds to the proportion of the output only in the long run. This means that in our sample of 20, taken in random fashion from the output, we can obtain X values between 0 and 20, although clearly some values of X are more likely than others. It is the explicit statement of these probabilities that we can now attempt.

Consider first the value 0 for X. If X is to be 0, then all amplifiers must be non-defective, and by the multiplication law, since the testing of each amplifier is independent of any other, the probability must be $(1 - \pi)^{20}$. We write

$$\Pr(0) = (1 - \pi)^{20}.$$

A similar argument follows for any value of X, the only complication being that when X is greater than 0, allowance must be made for the number of ways in which the defectives can occur. In the case of X being equal to 1, the single defective can occur first, second, etc. to twentieth. Each of these outcomes has the probability $\pi(1 - \pi)^{19}$, and since there are 20 of them, by the addition law

$$\Pr(1) = 20\pi(1 - \pi)^{19}.$$

With $X = 2$, the number of outcomes is the number of ways of selecting the two positions out of the twenty available. There are 190 such outcomes so that

$$\Pr(2) = 190\pi^2 (1 - \pi)^{18}.$$

Fortunately, we do not have to work out these multiples by a complete listing. There is a simple formula for the number of ways of selecting x items from a group of n where the order of selection does not matter. 'Items' here refers to the position in the sample for the defective amplifiers. In the mathematical literature this is referred to as the number of combinations of n items, taking x at a time, although the word combinations appears to be unfortunate. Writing $\binom{n}{x}$ for this number, the appropriate formula is

$$\binom{n}{x} = \frac{n!}{x!(n-x)!}$$

where $a! = a(a-1)(a-2)\ldots 3 \,.\, 2 \,.\, 1$, the product of all integers less than or equal to a. $n!$ is defined as given for positive integer n, and $0! = 1$ by definition. We state this result without proof.

The general result for the sample of amplifiers is

$$\Pr(x) = \frac{20!}{x!(20-x)!}\pi^x (1-\pi)^{20-x},$$

valid for $x = 0, 1, 2, \ldots, 20$.

We are using the small letter x to denote general values taken by the variable X. The capital letter is simply a way of avoiding having to repeat the definition of the variable – number of defective amplifiers in a sample of twenty – every time reference has to be made to it.

Taking an even more general view, if a sample of size n is taken and if each item has a probability π of being defective, the probability that the sample contains x defectives is

$$\Pr(x) = \frac{n!}{x!(n-x)!}\pi^x(1-\pi)^{n-x}, \qquad x = 0, 1, 2, \ldots, n. \tag{2.5}$$

This is the formula for individual probabilities when the variable X has the *binomial distribution.* The generating mechanism for such a distribution is that there should be n independent trials (items in the sample) each of which has a probability π of success (being defective). The number of successes (defectives) then has the binomial distribution. Note that our interpretation of the mechanism could have been applied to non-defective amplifiers, replacing π by $1 - \pi$. It can also be shown that the probabilities so found sum to unity over the range of x.

In appearance the distribution is skewed unless $\pi = 0.5$. Figure 2.3 shows a plot of the probabilities when $n = 20$ and $\pi = 0.1$. Figure 2.4 has $n = 20$, $\pi = 0.04$ showing how the modal value is at $x = 0$ and the probabilities approach zero for relatively small values of x. Finally, Figure 2.5 demonstrates the symmetrical case $n = 20$, $\pi = 0.5$, showing the bell-shaped configuration.

Values of the probabilities are tabulated quite extensively. For

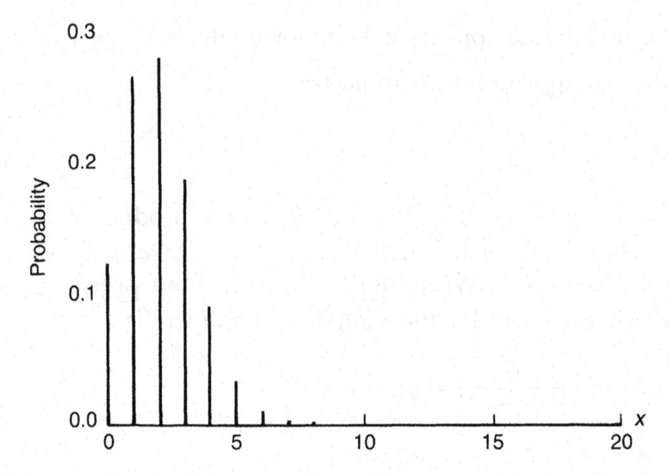

Fig. 2.3 Binomial probabilities $n = 20$, $\pi = 0.1$

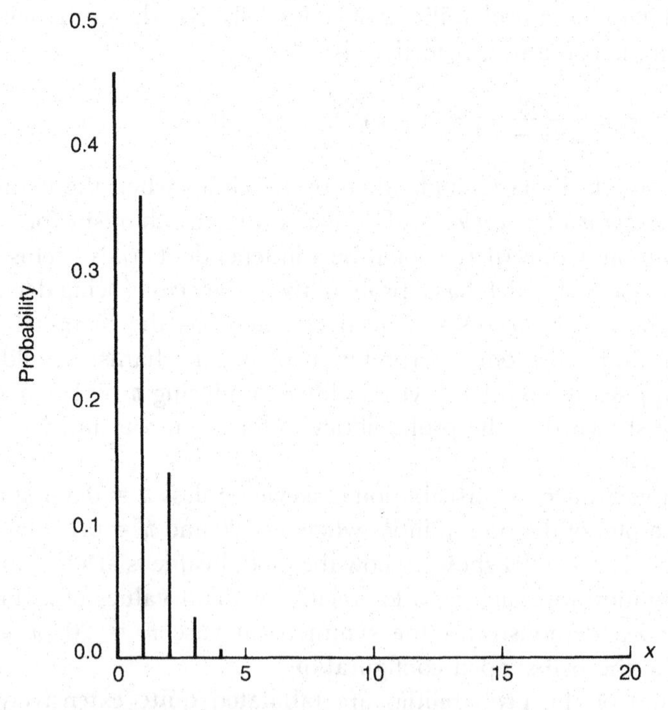

Fig. 2.4 Binomial probabilities $n = 20$, $\pi = 0.04$

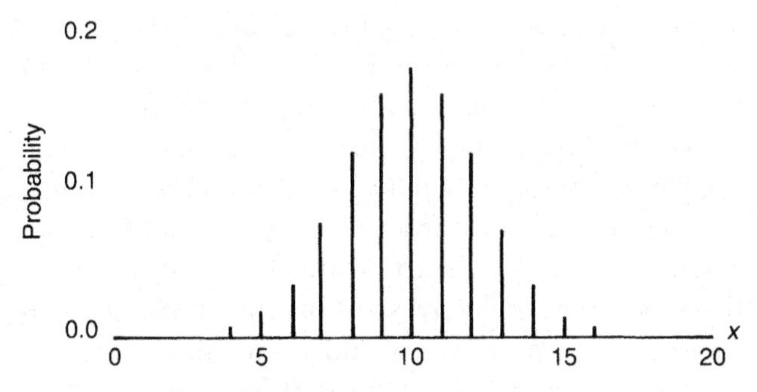

Fig. 2.5 Binomial probabilities $n = 20$, $\pi = 0.5$

example *Statistical Tables* by H. R. Neave (George Allen and Unwin, London, 1978) covers values of n up to 20 and *Handbook of Statistical Tables* by D. B. Owen (Addison-Wesley, Reading, Mass., 1962) goes to 25. For values greater than these we can usually apply an approximation, avoiding some heavy computation.

2.3.2 Populations, random variables and probability distributions

In deriving the form of the binomial distribution we have used a number of concepts which can be applied generally to our treatment of variability in measurement. Firstly, we introduce some terminology. The variable X which takes values governed by the distribution is known as a *random variable*. It is characterized by the fact that although its value cannot be guaranteed on any occasion when its experiment is performed, we can make statements about the probabilities that particular values are taken. In the above case the probabilities were defined by the binomial distribution formulae. There are many other alternatives (of which we shall discuss but three in this section), and we use the term *probability distribution* to describe the property generally.

Perhaps the most important concept stems from the initial assumption that all amplifiers have the probability π of being defective. It is equivalent to saying that there is a very large collection or *population* of amplifiers with a proportion π defective, and we select our sample of 20 from this population. In turn it is like a raffle with a very large number of tickets of which a proportion π are winners. How many winners do you get when you buy 20 tickets? The population is an abstraction. Clearly there will not be an infinite number of amplifiers made under the current manufacturing conditions, but those that are could be considered themselves as a sample from such a population, and we are simply taking a sample from a sample. Almost invariably the population has properties which are unknown to us. In our example the probability π is such a property, and one of our aims will be to make inferences about π from the evidence of our sample. This type of statement is at the heart of much statistical analysis. For example, if we observe 2 defectives out of our sample of 20 and if this is all the evidence we have, we would be justified in saying that a good estimate of π is $\frac{2}{20} = 0.1$. In doing this we are conscious of the fact that if by chance we had had one more or one less defective in our sample we would have arrived at a different estimate. Moreover, by this procedure we are restricting estimates of π to multiples of 0.05 which may not cover the true value. The general situation is summarized in Figure 2.6. The constant π is called a parameter of the population. In this case no other parameter is involved, but often populations have several unknowns.

It is also possible to think of equivalent populations. Taking the analogy of the lottery further, if we are always to buy groups of 20 tickets, it would make no difference if the lottery consisted of pieces of paper on which a number between 0 and 20 was written, the number corresponding to the number of winners. The equivalence would be exact if the proportion of tickets for each number were the probability derived from the binomial distribution. This is a compression of the

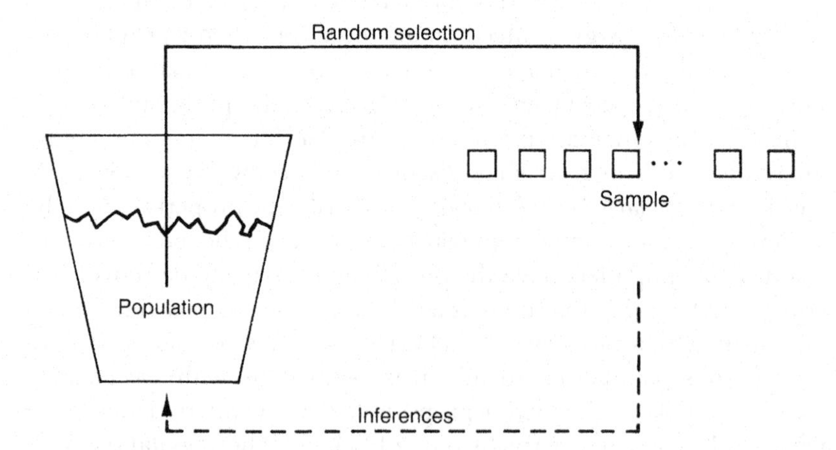

Fig. 2.6 Populations, sampling and inferences

population in which other information such as the order of the defectives within the sample has been suppressed.

A general population can be described broadly by measures of its central location and of its variability about the central location. The usual measure of centrality is the mean which, for the population, is simply the average of all of the values in the infinite collection or population. Equivalently if, as in the case of the binomial distribution, the population can only take a limited number of values (i.e. 0, 1, 2, . . ., n) and we know the probability of each, we can define the mean, μ, of the population by

$$\mu = \Sigma x \Pr(x). \tag{2.6}$$

This is the weighted average of the population values, the weights being the corresponding probabilities. Substituting the form of $\Pr(x)$ from equation (2.5) into equation (2.6) we have

$$\begin{aligned}
\mu &= \sum_{x=0}^{n} x \frac{n!}{x!(n-x)!} \pi^x (1-\pi)^{n-x} \\
&= n\pi \sum_{x=1}^{n} \frac{(n-1)!}{(x-1)!\,(n-x)!} \pi^{x-1} (1-\pi)^{n-x} \\
&= n\pi \sum_{y=0}^{m} \frac{m!}{y!(m-y)!} \pi^y (1-\pi)^{m-y}, \\
&\qquad\qquad \text{where } y = x-1, \quad m = n-1 \\
&= n\pi,
\end{aligned} \tag{2.7}$$

since the sum is the sum of binomial probabilities.

If we took several observations on a binomial random variable, say, $x_1, x_2, \ldots, x_k$, then the *average*

$$\bar{x} = \frac{1}{k} \sum_{i=1}^{k} x_i$$

is an estimate of $n\pi$. Since n, the sample size, is known, we would estimate π by $\bar{x}/n$, the average proportion defective in the samples. The question remaining is one of how good this estimate is. How likely is the true value of π to be within a given distance from the estimate? The answer to this question is beyond the scope of this book, but reference may be made, for example, to *Quantitative Approaches in Business Studies* by Clare Morris in the M & E Higher Business Education Series.

Other measures of centrality are the mode, which is the most frequently occurring value in the population, and the median, this being the value for which half the population is less, half greater. Each has circumstances where they are a more appropriate description of centrality than the mean, and they will be used accordingly.

The usual measure of variability is called the *standard deviation*, and denoted by the symbol σ. This is defined as the square root of the mean squared deviation from the population mean. This may sound complicated but since it has to be independent of the usual measure of centrality and since it has to be directly related to the generally accepted concept of variability, some simplicity has to be sacrificed. It also has some desirable properties when it comes to estimation from samples. In equation form the definition is

$$\sigma^2 = \Sigma(x - \mu)^2 \Pr(x), \tag{2.8}$$

the sum being taken over all values of x. The standard deviation is the positive square root of this expression. The parameter σ^2 is known as the variance. For the binomial distribution, using an argument similar to that used to derive the mean (equation (2.7)) but omitted here,

$$\sigma = [n\pi(1 - \pi)]^{\frac{1}{2}}.$$

The standard deviation is usually estimated from data by the sample standard deviation

$$s = \left[\sum_{i=1}^{n} (x_i - \bar{x})^2/(n - 1)\right]^{\frac{1}{2}}$$

A divisor of n is sometimes used instead of $n - 1$ in the above formula. With n of any reasonable size the difference is small.

2.3.3 The Poisson distribution

The next distribution can be derived as an extension of the binomial distribution where the sample size n becomes large and the probability of success π becomes correspondingly small. The two constants must increase and decrease respectively so that their product $n\pi$ (the mean of the distribution) remains finite and non-zero. If we denote the product $n\pi$ by μ, it can be shown that the limiting form of the probabilities is

$$\Pr(x) = e^{-\mu} \mu^x/x!, \qquad x = 0, 1, 2, \ldots.$$

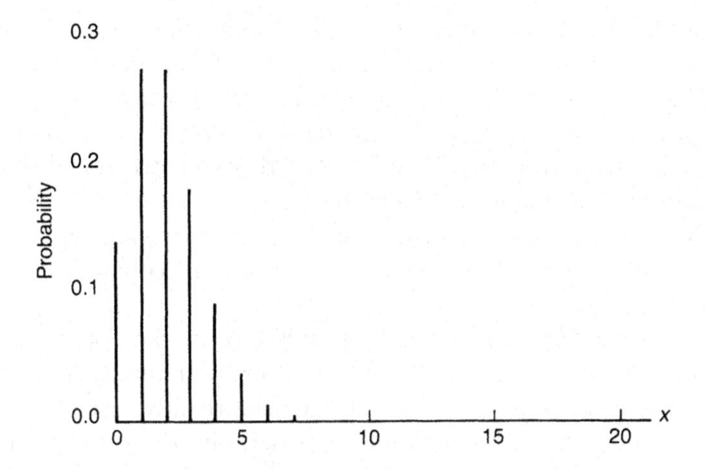

Fig. 2.7 Poisson distribution with $\mu = 2$

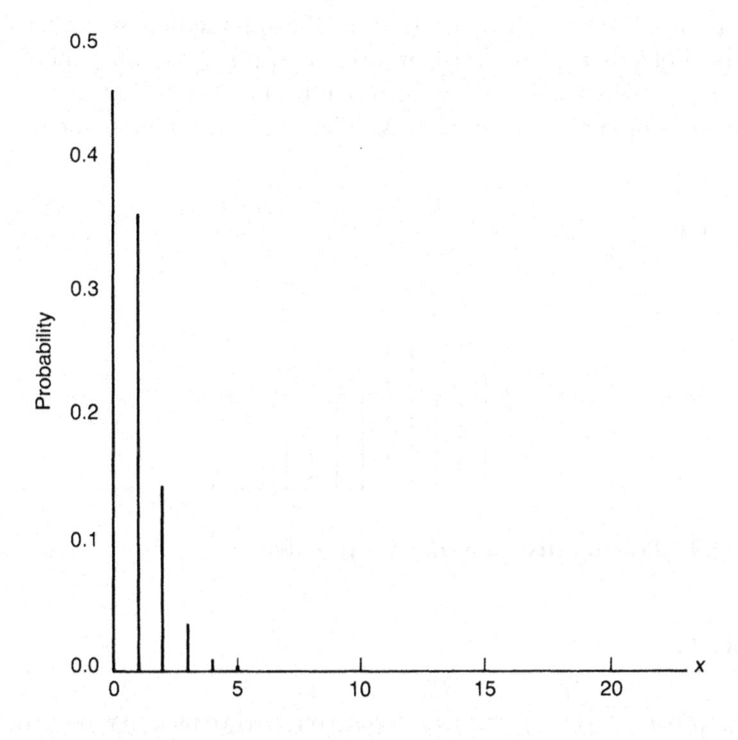

Fig. 2.8 Poisson distribution with $\mu = 0.8$

The random variable still can only take non-negative integer values, but there is now no upper limit. In shape, the distribution can appear as in Figure 2.7 when $\mu = 2.0$, or as in Figure 2.8 when $\mu = 0.8$. The model value is zero when $\mu < 1$, greater than zero when $\mu > 1$. When μ becomes somewhat larger we see a more symmetrical bell-shaped distribution as in Figure 2.9 when $\mu = 10$.

Values for the Poisson distribution have also been widely tabulated, the books of Neave and Owen cited in Section 2.3.1 containing typical tables.

As suggested in the introduction to this section, the Poisson distribution can be used as an approximation to the binomial distribution for suitable values of n and π. This is particularly advantageous since it can simplify the arithmetic quite considerably. For example, if we compare probabilities for a binomial with a sample size n of 20, and a probability π of 0.1 (hence $n\pi = 2$) with a Poisson with mean μ equal to 2, we obtain the figures shown in Table 2.3. Pairwise comparisons between Figures 2.3, 2.7; 2.4, 2.8; 2.5, 2.9, where the corresponding means have been matched, demonstrate the approximation graphically.

The Poisson distribution can also be applied to data where the limiting concept of the binomial distribution is used, but where the trials are not explicitly defined. Accident statistics tend to be of this

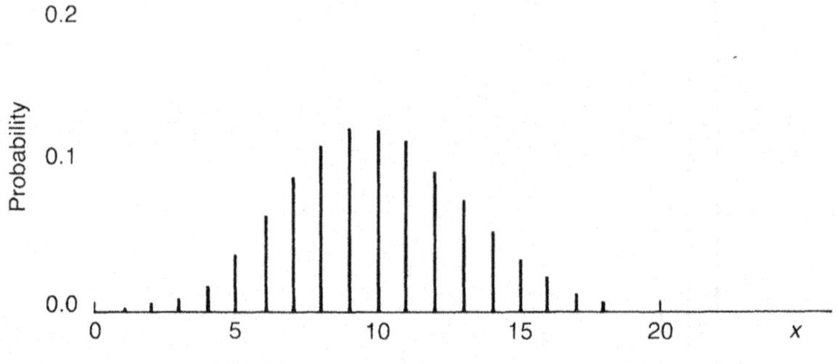

Fig. 2.9 Poisson distribution with μ = 10

Table 2.3

x	0	1	2	3	4	5	6	7	8	≥9
Binomial Pr(x)	0.122	0.270	0.285	0.190	0.090	0.032	0.009	0.002	0.000	0.000
Poisson Pr(x)	0.136	0.271	0.271	0.180	0.090	0.036	0.012	0.003	0.001	0.000

nature: over a period of time there are very many opportunities for accidents to occur, but happily few are taken. Moreover, we only recognize a trial when an accident occurs, the non-occurrence being indistinguishable from no trial. Similarly faults in a product such as a bobbin of yarn or a complete piece of equipment tend to exhibit a similar pattern of variability. A bobbin contains a long length of yarn which may be divided into small intervals each of which does or does not have a fault. Under normal production conditions faults may occur at a steady rate of, say, 2.5 faults per bobbin. If, in addition, the faults occur randomly, counts of the number of faults in any bobbin will have a Poisson distribution. In particular, the probability that a bobbin will be free of faults is then $e^{-2.5}$ or 0.0821.

The Poisson distribution is defined by a single parameter μ. The mean of the distribution (as would be expected from the corresponding result for the binomial distribution) is this value μ, and the standard deviation is $\sqrt{\mu}$. Both of these results can be found by simple algebraic manipulations and should be tried as exercises. Two hints are given. Firstly, the fact that the Poisson probabilities sum to unity, albeit in this case in an infinite series, comes into the usual proof. Also to derive the standard deviation, find the variance and take its square root.

2.3.4 The normal or Gaussian distribution

With both the binomial and Poisson distributions we have been able to describe their probability distributions by specifying the probabilities taken by all feasible values of the variables. In the case of the binomial the variable could only take values between 0 and n, and for the Poisson only non-negative integers were possible. Such distributions, in which only a finite set of values or a set of values which can be listed in a sequence are possible, are said to be *discrete*.

With some variables this is not possible. Measurement of the time for some manufacturing operation, of the length of some component part or of the weight of some delivery batch all consist of taking a reading on a continuous scale. Although our measuring devices may limit us to rounding the measurements to a discrete set of values, it is often useful to retain the concept of a reading which can take any value over some given range. Variables of this type are said to be *continuous*. Because of this definition, the probability that any particular value is taken must be zero (since there is a countless infinity of them, summing them would give an infinite total) and we must therefore adopt some alternative device to the actual probabilities to describe the distribution of a continuous random variable. Thus, instead of particular points, we specify the probability that the variable lies in any range, and we do this by equating it to the area under a curve, known as the *probability density function.* A typical example of such a curve is shown in Figure 2.10, where the probability that the random variable X lies between two values a and b is the shaded area. The curve may cover from minus to plus infinity or some restricted range. Since the probability that some value within its total range occurs must be unity, the total area under the curve must also be unity. Also the curve cannot take negative values.

If we can define the form of the curve by some function $f(x)$, then, for the benefit of readers familiar with integral calculus,

$$\begin{aligned}\Pr(a \leqslant X \leqslant b) &= \text{area under } f(x) \text{ between } a \text{ and } b \\ &= \int_a^b f(x)\, dx.\end{aligned}$$

The equality signs have been included in the probability statement, although since specific values (in this case, a and b) have zero probability, they could have been omitted without changing the remainder of the equation.

The concepts of mean, μ, and standard deviation, σ, carry over to continuous distributions. They are related to the probability density function by

$$\begin{aligned}\mu &= \int x\, f(x)\, dx \\ \sigma^2 &= \int (x - \mu)^2 f(x)\, dx.\end{aligned}$$

The integral signs have replaced the summation signs used in the definitions (equations (2.6) and (2.8)) for discrete random variables,

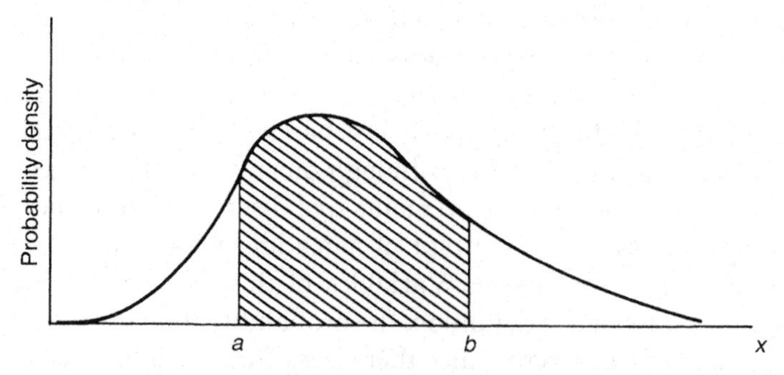

Fig. 2.10 Typical continuous probability density function

but the meaning of weighted averages taken over the whole population remains.

One particular continuous distribution is used regularly to describe the distribution of random variables. It is known as the normal or Gaussian distribution. It is defined by a particular form of the probability density function $f(x)$, but since whenever we need to make use of the distribution we have to resort to tables, the actual statement of the equation for $f(x)$ is unnecessary. Suffice it to say that it is defined by two constants μ and σ, which, as their letters indicate, are the mean and standard deviation respectively. In form it is symmetrical and bell-shaped. A typical example is shown in Figure 2.11.

As an illustration of the use of this distribution suppose that Marshall Supplies plc makes regular deliveries to a customer in Coalville some 20 miles distant. The delivery van is loaded and ready to leave promptly each day at 9.30 a.m. The time taken for the journey is of course variable, and 50 observations of the time in minutes for typical runs (i.e. excluding days of weather extremes or vehicle breakdown) are as shown in Table 2.4.

A histogram of the data classified into cells of width 5 minutes is shown in Figure 2.12. The histogram suggests a bell-shaped distribution with a reservation about the 4 observations in the 61–65 minutes cell. (Perhaps these journeys were atypical, but not realized as such when the data were recorded. They will be retained in the analysis.)

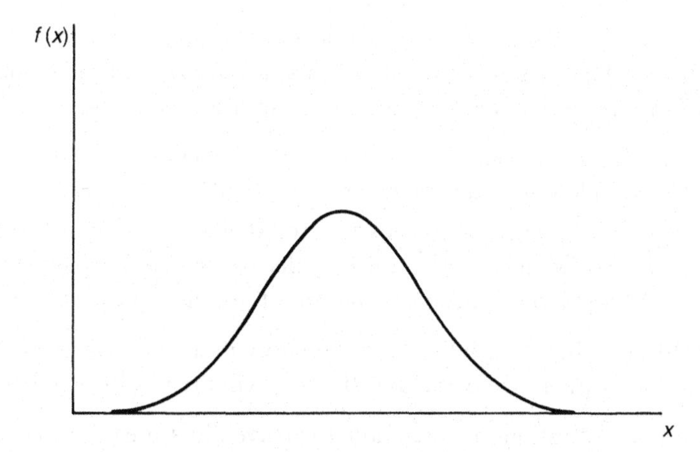

Fig. 2.11 Normal probability density function

Table 2.4

50	45	41	48	62	43	47	33	40	31
39	62	30	47	49	44	39	43	62	34
41	35	39	47	64	41	38	35	52	38
40	54	40	51	41	60	46	46	49	46
53	45	54	45	50	25	41	54	33	39

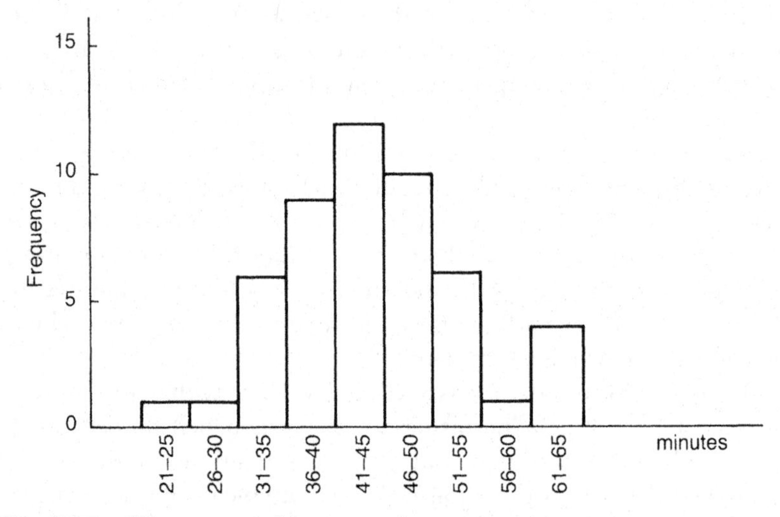

Fig. 2.12 Histogram of journey times

Suppose that we now fit a normal distribution to the data. The distribution has two unknown constants, the mean and the standard deviation, and we therefore estimate these from the data.

The sample average $\bar{x} = 2231/50 = 44.62$
The sample standard deviation $s = [3779.78/49]^{\frac{1}{2}} = 8.78.$

The customer would therefore expect the delivery van on average to arrive at approximately 10.15 a.m. In order to have his goods reception and unloading facilities prepared he might also ask questions such as

(a) What is the probability that the van will arrive before 10 a.m.?
(b) What is the probability that the van will arrive after 10.30 a.m.?

To answer question (a) we denote the travelling time by the random variable X, and therefore need $\Pr(X \leqslant 30)$. If now we assume that X

has a normal distribution with the mean and standard deviation fitted from the data, to determine this probability we convert to the standardized normal (mean 0, standard deviation 1) by subtracting the mean and dividing by the standard deviation within the inequality of the probability

$$\begin{aligned}\Pr(X \leqslant 30) &= \Pr\left(\frac{X - 44.62}{8.78} \leqslant \frac{30 - 44.62}{8.78}\right)\\ &= \Pr\left(\frac{X - 44.62}{8.78} \leqslant -1.665\right)\\ &= 0.0480, \qquad \text{using the table in Appendix 2.}\end{aligned}$$

The tables only give probabilities for positive values of the right-hand side of the inequality. If Y has a standardized normal distribution it is symmetrical about zero and therefore

$$\Pr\,(Y \leqslant -a) = \Pr\,(Y \geqslant a) \qquad \text{for any value } a.$$

Hence in this case we find the value by looking up the value 1.665 in the table. Note also that to use three decimal places we have to interpolate between two tabulated values.

Similarly for question (b),

$$\Pr(X \geqslant 60) = \Pr\left(\frac{X - 44.62}{8.78} \geqslant 1.752\right)$$

and using the same tables, this is equal to 0.0399. Thus, if we schedule the reception facilities for between 10 a.m. and 10.30 a.m., there is approximately a probability of 0.05 that the van will be waiting at the beginning of the period and also a probability of 0.04 that it will not have arrived by the end of the period.

The normal distribution does not have the recognizable generating mechanism that the binomial and Poisson distributions possess. Nevertheless, it is a useful distribution to assume for continuous random variables which tend to be spread evenly about some mean value and become sparser as they are remote from the mean.

There is one further point to make about continuous and discrete random variables. In the example used times were quoted to the nearest minute. Presumably this was because of the way in which the journeys were recorded, and this in turn may have been for other purposes such as driver bonus payments. By rounding in this way we turned what was originally a continuous variable into a discrete one, but for the purposes of the analysis we still treated it as a continuous variable. The normal distribution can, in fact, be used as an approximation to the binomial and the Poisson distributions. For the binomial the conditions are that n should be relatively large and for the Poisson that μ should be large. Since we are approximating a discrete variable by a continuous, better accuracy is obtained by correcting the values by adding or subtracting 0.5 as follows. For example, if we take a sample of 100 from a large batch which is 20 per cent defective, the number of defectives in the sample has a binomial distribution with $n = 100$, $\pi = 0.2$. This can be approximated by a normal distribution with the same mean and standard deviation, which are 20 and 4 respectively. Thus, for example, if we require the probability of 15 or fewer defectives in the sample, we find

$$\begin{aligned}\Pr(X \leqslant 15.5) &= \Pr\left(\frac{X-20}{4} \leqslant \frac{15.5-20}{4}\right)\\ &= \Pr\left(\frac{X-20}{4} \leqslant -1.125\right)\\ &= 0.130.\end{aligned}$$

2.3.5 The beta distribution

A further distribution, which we shall use in Chapter 7 in the discussion of the use of market research data, is called the *beta distribution*. A random variable X which is beta-distributed varies over a limited range, which without loss of generality we shall take from zero to one. The equation of its density function is

$$f(x) = \frac{(\alpha + \beta + 1)!}{\alpha!\beta!} x^{\alpha} (1 - x)^{\beta}, \qquad 0 \leqslant x \leqslant 1,$$

where α and β are constants greater than -1. The form of the probability density function for the case $\alpha = 2$, $\beta = 3$ is shown in Figure 2.13.

The use of this distribution will be demonstrated in Chapter 7, and the only further comment to be made is that there is a special case of the beta distribution when $\alpha = 0$ and $\beta = 0$ where the distribution reduces to what is known as the continuous rectangular distribution. This is appropriate to a random variable where all values in the range are equally likely. For example, you wish to catch a bus and the only

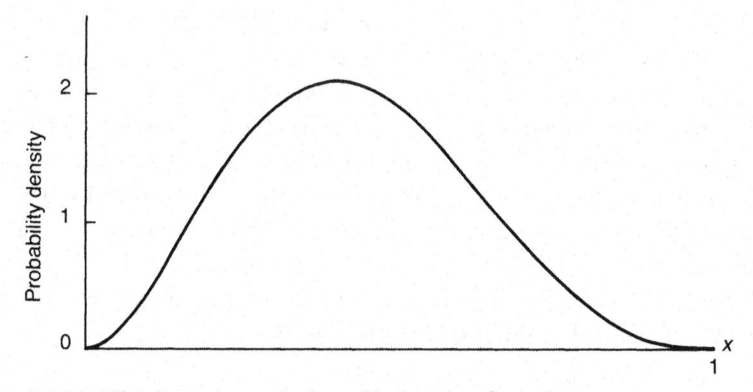

Fig. 2.13 The beta ($\alpha = 2$, $\beta = 3$) density function

information you have is that they run (exactly on time!) every 30 minutes. Your waiting time at the bus stop will then have a continuous rectangular distribution with limits 0 and 30 minutes.

There is a discrete equivalent of the continuous rectangular distribution where a random variable takes a set of values all with the same probabilities. For example, the set could be the integers 0, 1, 2, . . ., 9 all of which would be taken with probability 0.1. These would then be what are known as *random numbers*, which have a vital role in statistical techniques such as sampling, the design of experiments and simulation.

2.4 Expected values

Actions taken by a decision-maker are usually followed by outcomes over which he has no control, that is by chance outcomes. The profit or loss which the decision-maker incurs are therefore subject to these chance outcomes. He may therefore have to accept that on some occasions he will fare rather better than on others. One way of looking at the situation is to consider the long-term average, even though the particular decision only happens once. Other similar decisions will be taken, and he argues that in the long run, given his attitude to decision-making, his overall return will match the overall sum of what would be a long-term return for repeated runs of each individual decision. His criterion is therefore to consider each action from the point of view of a weighted average of the corresponding outcomes, the weights being their probabilities. A weighted average defined in this way is called an *expected value*.

Consider the following example. Rhodilan Electronics makes underwater photographic equipment and also sells special waterproof relays in batches of 20. The price of a batch is £20, but they undertake to replace any defective relay with a good one. The cost of each replacement is £5 so, for example, the return on a batch which contains two defective relays is £(20 − 10) = £10. Given that the relays on average have a proportion π defective, what is the expected value of the return on a batch?

If the random variable X denotes the number of defective relays in a batch, X has a binomial distribution with

$$\Pr(x) = \frac{20!}{x!(20 - x)!} \pi^x (1 - \pi)^{20 - x}.$$

Moreover, if there are x defectives, the return on the batch is £$(20 - 5x)$. To find the expected value of the return, we need the sum

$$20 \Pr(0) + 15 \Pr(1) + 10 \Pr(2) + \ldots + (-80) \Pr(20)$$

$$= \sum_{x=0}^{20} (20 - 5x) \frac{20!}{x!(20 - x)!} \pi^x (1 - \pi)^{20-x}.$$

Since $\Sigma \Pr(x) = 1$ and $\Sigma x \Pr(x) = \mu = 20\pi$, this expression simplifies to $20 - 100\pi$.

In general, for any function $h(x)$, the expected value of $h(x)$ is written as $Eh(x)$ and defined by

$$Eh(x) = \Sigma h(x) \Pr(x) \text{ for a discrete variable}$$
$$\text{or } \int h(x) f(x)\, dx \text{ for a continuous variable.}$$

Note that the mean of any distribution is the expected value of x and the variance is the expected value of $(x - \mu)^2$.

2.5 Bayes' formula again

Bayes' formula was introduced in Section 2.2 for the case when the outcomes were discrete entities. In Chapter 7 we shall need the equivalent formula when one of the outcomes is a probability density. Suppose that X is a random variable with the (prior) probability density function $f(x)$, and that A is any outcome. Considering any value x taken by X, $\Pr(A|x)$ is the conditional probability of A given that x has occurred and conversely $f(x|A)$ is the conditional (posterior) probability density function of X given that A has occurred. The equivalent Bayes' formula is

$$f(x|A) = \Pr(A|x)\, f(x)/\textstyle\int \Pr(A|x)\, f(x)\, dx.$$

As an illustration suppose that a car hire firm disposes of its vehicles solely on the basis of their appearance. The result of this policy is that if X is the thousands of miles driven at the time of disposal, X is claimed to have a continuous rectangular distribution between 15 and 25. The firm's cars have major accidents at an average rate of one every 5000 miles. For any car the distribution of the number, Y, of accidents follows a Poisson distribution, so that for a car disposed of after x thousand miles,

$$\Pr(y|x) = \frac{e^{-x/5}\,(x/5)^y}{y!}.$$

You are now offered a car on disposal which has had three major accidents. What conclusion can you draw about the miles driven by this car? Expressed statistically, what is the conditional density of x, given $Y = 3$?

Using Bayes' formula

$$f(x|3) = \Pr(3|x).f(x) \,/ \textstyle\int \Pr(3|x)\, f(x)\, dx$$

$$\text{where } \Pr(3|x) = \frac{e^{-x/5}\,(x/5)^3}{6}$$

$$\text{and } f(x) = \tfrac{1}{10},\ 15 \leqslant x \leqslant 25.$$

The denominator in the expression for $f(x|3)$ is a constant found by evaluating

$$\int_{15}^{25} \frac{e^{-x/5}\,x^3}{7500}\, dx$$

This is an exercise in basic integral calculus which is left to the reader. The answer should be 0.1911. The posterior probability density function is therefore

$$f(x|3) = \frac{x^3\, e^{-x/5}}{(6)\,(125)\,(10)\,(0.1911)}$$

$$= 0.0006977x^3\, e^{-x/5}, \qquad 15 \leqslant x \leqslant 25.$$

Figure 2.14 is a plot of the prior probability density function $f(x)$ and the posterior probability density function $f(x|3)$. It shows that

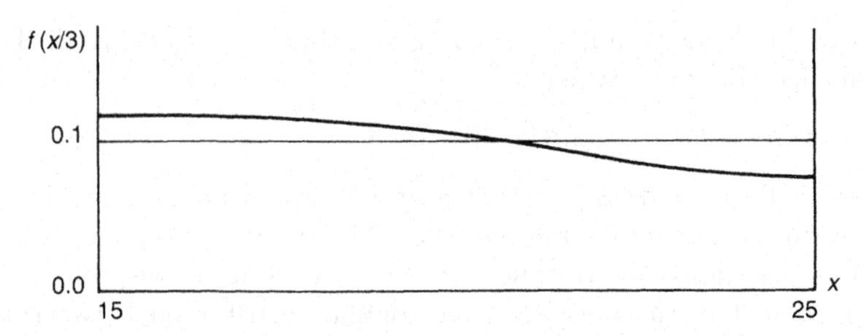

Fig. 2.14 Posterior distribution given three major accidents

with cars having between 15 000 and 25 000 running miles and an accident rate of one every 5000 miles on average, a car which has had only 3 accidents should be toward the lower end of the mileage range. This information certainly does not preclude a higher mileage but there is a detectable shift in the density function toward the lower values. The marginal extent of this shift can be measured by the fact that the prior distribution has, of course, a mean value of 20, and evaluation of the mean value of the posterior distribution

$$0.0006977 \int_{15}^{25} x^4 \, e^{-x/5} \, dx \text{ results in a mean of 19.61.}$$

A similar analysis for a car which has had no major accidents produces the posterior probability density function $f(x|0) = 4.645 \, e^{-x/5}$, $15 \leqslant x \leqslant 25$ (plotted in Figure 2.15). The mean of the distribution is 18.43.

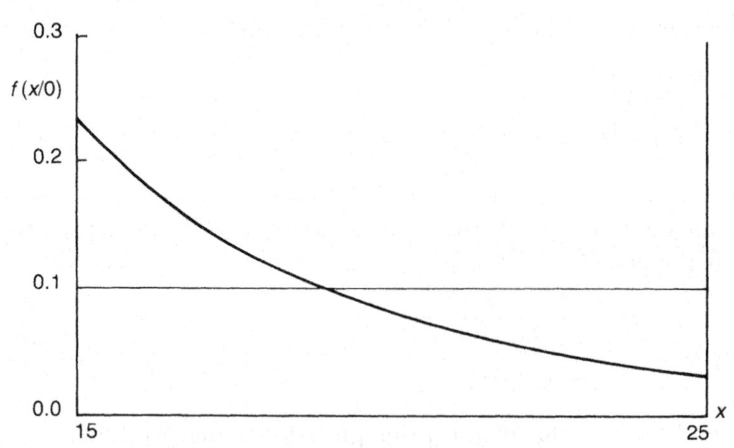

Fig. 2.15 Posterior distribution given no major accidents

This last section has demonstrated the way in which the decision-maker's information is modified in the light of some further observation. In the example the prior distribution is presumably based on observations. We shall use the methods of this section in Chapter 7, where prior distributions are more subjective, and also they are applied to elements such as proportions of a population who will buy a product. These are parameters of the population and therefore not directly observable. Nevertheless, the application of Bayes' formula provides a fruitful way of modifying subjective views when the more objective evidence of actual observations is available.

2.6 Further reading

There are very many excellent books on the subject of statistics. If the reader requires amplification of the material in this section, he should make sure that he finds a book written for his level of mathematics. A selection which he may find useful is as follows:

Bierman, H., Boneni, C. P. and Hausman, W. H., *Quantitative Analysis for Business Decisions*, Richard D. Irwin, Homewood, Illinois (1977).

Edwards, P. J., Ford, J. S. and Lin, C-Y., *Probability for Statistical Decision Making*, Prentice-Hall, Englewood Cliffs, New Jersey (1974).

Mood, A. M., Graybill, F. A. and Boes, D. C., *Introduction to the Theory of Statistics* (3rd edn), McGraw-Hill, New York (1974).

Moroney, M. J., *Facts from Figures*, Penguin, Harmondsworth (1951).

Stafford, L. W. T., *Business Mathematics*, M & E Handbooks, Pitman, London (1979).

Probabilities

Hamburg, M., *Statistical Analysis for Decision Making*, Harcourt Brace, New York (1970) (Chapters 1, 2).

Lindley, D. V., *Making Decisions*, John Wiley, London (1971) (Chapters 2, 3).

Moskowitz, H. and Wright, G. P., *Operational Research Techniques*, Prentice-Hall, Englewood Cliffs, New Jersey (1979) (Chapters 2, 3).

Tversky, A. and Kahneman, D., 'Judgement Under Uncertainty: Heuristics and Biases', *Science*, **185**, 1124–1131 (1974); reproduced in Kaufman, G. M. and Thomas, H. (eds), *Modern Decision Analysis*, Penguin, Harmondsworth (1977).

2.7 Problems and practical exercises

1 Bag A contains 4 white and 2 red balls. Bag B contains 3 white and 5 red balls. Given that one ball is drawn randomly from each bag, find the probability that:

(a) both are white,
(b) both are red,
(c) one is white and one is red.

Suppose now that one of the bags is selected randomly and a ball is then drawn randomly from it. If the ball is white, what is the probability that A was selected?

2 A system consists of two legs in parallel, each leg consisting of three switches in series. To be complete, all three switches in a leg must function. The full system will function provided that at least one of the legs functions.

The designers require a system which has a probability 0.99 of working when initially assembled. If all switches have the same probability of functioning, what must this probability be to achieve the design target?

3 A firm operates two production lines in the manufacture of quaternions. Line A runs twice as fast as line B, but produces 6 per cent defective quaternions, whereas line B produces only 4 per cent defectives. A quaternion is selected randomly from the mixed output of the two production lines.

(a) What is the probability that it was made on line A?
(b) The quaternion is tested and found to be defective. How would you modify your answer to (a)?

4 A batch of 20 items is submitted for inspection. Items are drawn sequentially from the batch, up to a maximum of five. As soon as the first defective is found the batch is rejected, but if no defectives are found in the five items drawn, the batch is accepted. What is the probability of accepting a batch containing:

(a) one defective,
(b) four defectives?

5 A manufacturer receives supplies of components from two sources, A and B. He knows from experience that the components from source A are on average 3 per cent defective, but has no direct information about the quality of the components from source B. Over a trial period he has received components from the two sources in the ratio of 60 per cent from source A, 40 per cent from source B.

(a) If the overall proportion defective of components is 5 per cent, what is the proportion defective from source B?
(b) A component is selected randomly from amongst the defectives. What is the probability that it came from source B?

6 Faults in a synthetic yarn occur randomly at an average rate of 1 in 100 000 m. The manufacturer packages his yarn in bobbins containing 100 000 m.

(a) What proportion of his output will be fault-free?
(b) What would be the effect on the proportion of fault-free bobbins if he reduces the size to 50 000 m of yarn?

7 An aircraft has 300 seats available for booking. Experience indicates that on average 1 per cent of all seats booked are not claimed when the aircraft leaves.

(a) What is the probability of 3 or more vacant seats at take-off when the aircraft is fully booked?
(b) What are the implications of this argument in terms of the number of bookings which should be accepted for any journey?

8 Stereo Graphic Equalizers are subject at the time of final inspection to surface faults and to functional faults. A surface fault costs £2 and a functional fault £5 to repair. Surface faults and functional faults have independent Poisson distributions with means 2 and 0.8 respectively.

(a) What proportion of the Equalizers will have a total repair cost of £10 or more?

(b) If Equalizers with a repair cost of £10 or more are scrapped, what proportion of those scrapped has no functional fault?

9 The lengths of telephone calls from a certain office were noted and the results are shown in Table 2.5, giving the times in seconds.

Table 2.5

141	43	203	104	82	63	24	84	41	86	47	43
100	53	139	147	137	186	214	106	150	109	170	172
194	124	175	177	162	129	128	219	40	105	48	65
105	154	154	35	149	54	104	109	119	74	140	104
168	127	191	30	109	88	104	207	38	164	182	120
166	53	145	29	112	143	49	199	130	52	109	77
142	75	146	105	125	112	40	126	67	49	90	140
132	118	134	133	159	123	161	112	157	104	92	112
151	98	156	117	156	190	122	135	116	96	163	116
186	155	106	153	69	105	136	106	131	118	94	121

Plot a histogram of these data. Calculate the sample average and sample standard deviation. What is the median value of these data?

10 The time to manufacture an item can be assumed to be normally distributed with mean 20 minutes and standard deviation 2 minutes. What is the probability that any item

(a) takes longer than 24 minutes,
(b) takes less than 17 minutes,
(c) takes between 18 and 25 minutes?

11 The filling process of 1000 g bags of flour must be set so that no more than 5 per cent of the bags have contents weighing less than 1000 g. The process is variable. It is possible to set the mean level at any weight desired, without affecting the standard deviation, and previous experience shows that the weights are always normally distributed about the mean level set. Currently a mean setting of 1008 g is producing 12 per cent bags below 1000 g.

(a) At what level should the process be set to achieve no more than 5 per cent bags below this level?

(b) At this level of setting, what is the probability that a bag will contain more than 1020 g?
(*Hint:* Use the current situation data to find the standard deviation.)

12 Using the data from the example in Section 2.5, determine the posterior probability density function of disposal mileage for a car which has had one major accident. What is the mean value of this distribution?

3
Theory of games

OBJECTIVES

The general aim of this chapter is to show how a decision analyst formulates a problem in terms of his strategies, the strategies of the opposition and the outcomes of using any of his strategies against any of the opposition's. The reader should then be able to 'solve' a simple two-person zero-sum game either in the form of an optimum pure strategy or, more usually, in the form of a probability distribution over a range of strategies. Simple extensions to non-zero-sum games are then discussed.

3.1 Introduction

Most of us understand what is meant by a game, and the mathematical formulation does not depart substantially from the popular conception. The game may be a simple one like noughts and crosses (or tic-tac-toe as it is sometimes known). Two players each have a sequence of actions to take, at the end of which either one player is agreed to be the winner or the game is drawn. It is possible to specify sequences of actions for each player which will guarantee that they do not lose, and consequently, the game should always be a draw. It is a game of little theoretical interest because each player has full information on the state of the game and can therefore frustrate the other's intentions to win. Both draughts (checkers) and chess suffer from similar limitations and could therefore be similarly dismissed. The reason why chess is such a good game is that there are so many different actions available to each player at each move of the game that we cannot explore each to its win or lose outcome. In noughts and crosses it is quite feasible to specify the non-losing actions, so that a substitute player (or a computer) could play the game for you. A similar specification is not yet possible for chess, for which all aficionados of the game must be thankful.

It is possible, of course, to play a game like chess where groups of people resolve the different moves made. In effect the group behaves as a single player and the model treats them accordingly. A distinction has to be made for games like monopoly or poker where – at least in the initial stages of the game – there are several players competing against each other. It is a situation which can give rise to coalition between players, who then have a further decision to take on when a coalition no longer serves their interests. Such games are known as *N*-person games ($N>2$) and because of the likelihood of coalitions, of which there will be very many indeed, the results derived from these games have not reached anything like the precision that we find in two-person games. Although it is accepted that in practical commercial situations (the context aimed for with this book), there are many competing interests, we are really only concerned with an individual member of the competing group. It is often possible therefore to take the situation as our member against the rest and approximate the model by a two-person game.

In introducing the games of monopoly and poker a further element has been brought into the model. No longer does each player have full information on the game as he did in chess. Note that full information does not mean that each player knows what the other player will do. It means that he knows precisely the state of the game, and can therefore formulate his actions to counter any action taken by his opponent. In both monopoly and poker there is an additional element for which the players cannot make full allowance. This is the chance factor in either the roll of the dice or the draw of the cards. Each player not only has the opposition to contend with but also has to cope with the whims of Nature. Some actions will be better than others even in the presence of this uncertainty, but it has to be recognized that the same action taken on different plays of the game can now produce different outcomes.

Most games have a strong chance element present and, in fact, in a game like bridge much of the skill lies in the way that the players reduce the risks resulting from the random patterns in which the cards have been dealt. The same is true for the commercial situation. For example, the outcome of a marketing decision concerning, say, the timing and the promotion of a new product launch depends not only on the actions of competitors in the same market but also possibly on the effects of weather conditions on raw material prices, possibly on government legislation or possibly on the general buoyancy of the economy. In all such cases we might think of Nature as being the main opponent. The distinction between this situation and the straight two-person competitive situation is not in the structure of the game but rather in the way it is played. In particular we would speak of a *zero-sum* game as one in which each player's gains are the other's losses. In such a situation it is reasonable to assume that the players will each attempt to maximize their gains. They do this in a way which allows for an identical motivation in their opponent and hence there is a strong tendency to think in terms of countering the worst that he can do. Against Nature, although there is sometimes a feeling that events actually do conspire against one, it is more realistic to 'play the odds'; that is, the player attempts to assess the probabilities of the various outcomes of Nature, and then bases his action on the resultant risks. In neither case – in a game against a conscious opponent or in a decision situation against Nature – is there an unequivocally best action to take. Much depends on the proclivity of the decision-maker towards risks, as will be demonstrated in the next chapter.

The remaining sections in this chapter are concerned with two-person games. The zero-sum version is important and relatively straightforward, but following this the non-zero-sum model is introduced to show how, paradoxically, an action which is apparently inferior for an individual can, with cooperation from the other player, be mutually beneficial.

3.2 Two-person, zero-sum games

The basic model for the two-person, zero-sum game denotes the players as I and II. Each player has a finite number of *actions* from which to select. Corresponding to each action for each player there is a *payoff*, which is the amount Player I gains from Player II. This may be an actual payment from II to I or it may be the amount Player I takes from a fixed total prize, the remainder going to Player II. Stated in this way, the game may be described in matrix form, as the following example demonstrates.

Suppose that two companies (Players I and II) are competing for business from two customers A and B. The profits from the customers are 4 and 8 units respectively. The customers are to be visited on the same two consecutive days, one on each day. Each customer will give all his business to the first company to visit him. If both companies visit on the same day, the business is shared.

Each player has two possible actions. He can visit A first and then B, or vice versa. Denote the actions by AB and BA respectively. The total profit is 12 units, and whatever Player I does not gain is gained by Player II. In this sense it is a *zero-sum* game. The model can therefore be summarized in matrix form as in Table 3.1, with the payoffs as the profits gained by Player I.

Table 3.1

		Player II actions	
		AB	BA
Player I actions	AB	6	4
	BA	8	6

Player I now argues as follows. If Player II plays AB, my best action is to play BA. Also, if Player II plays BA, my best action again is to play BA. I will therefore play BA.

In this case Player I would be foolish to take any other action than BA since no matter what happens his payoff will be at least as large as that for action AB. The action AB is said to be *dominated* or inferior.

Applying the same arguments to the actions of Player II leads to a similar conclusion. For Player II the action AB is dominated, since in this instance it would result in a greater payoff to Player I than action BA, irrespective of the action taken by Player I. The result is that both players would visit customer B first, resulting in a profit of 6 units to Player I, and by subtraction a profit of 6 units to Player II. Any departure from this action would give an immediate advantage to the opposing player.

Situations of this simple form are rather special cases. They are characterized by what is known as a *saddle point* in the payoff matrix, where one element of the matrix is simultaneously the minimum of a row and the maximum of a column. Whenever a saddle point exists, it will always follow that each player will have a single action which he should always take if he wishes to maximize his payoff. This single action is known as a *pure strategy*, where the term strategy is introduced to describe the instructions which a player needs in order to play the game. We shall see shortly that there are occasions when his choice of action can depend on some chance mechanism over his range of actions. This will be called a *mixed strategy*.

Returning to the example there is a further concept to be introduced. This is known as the *value* of the game, and is the minimum amount which Player I can guarantee to gain on average from repeated plays of the game. It is also the maximum amount which Player II need lose on average from repeated plays of the game. In the case of a payoff matrix with a saddle point, the value will be the payoff at the saddle point, and it should always be the actual payoff achieved if both players play optimally. In general terms it is the limit which Player I should be prepared to pay to be allowed to play this game. In the example, the value of the game is, of course, 6.

Now suppose that the customers in our example behave in a different manner. The company denoted by Player I is well-established and has experienced sales representatives, whereas Player II is relatively new in the market. The result of this is that whenever they both arrive at a customer on the same day, the entire business goes to Player I. Adopting the same interpretation of payoff, the game is now described by Table 3.2.

Table 3.2

		Player II actions	
		AB	*BA*
Player I actions	*AB*	12	4
	BA	8	12

Neither player now has a dominated action, there is no saddle point and therefore no pure strategies. Player I might favour *BA* since the worst he will receive is 8 units, whereas with *AB* he could receive 4. If he does this and the same game is repeated, Player II would react by choosing action *AB*. This in turn could persuade Player I to switch to *AB*, which then would lead Player II to take action *BA*. The argument has now come full circle. It is clear that no player can afford to stick entirely to one action. He has to mix the actions in a way which will not allow the opposition to take advantage of any tendency towards an unfavourable emphasis on any specific actions. The way in which this is done now follows. Firstly, though, a new dimension has been brought into the problem and it requires comment. The arguments, both preceding and following, imply that the same game is to be played many times. This is often not the case. The alternative then is either to follow the argument at the beginning of this paragraph, making a judgement of the point at which the 'he thinks I'll do this, so he will do that' argument terminates, or to accept that although there is no optimal pure strategy, a single action has to be taken and to let some optimal chance mechanism select the action. We shall see that if this latter approach is adopted, the opposing player can gain no advantage from knowing that it is being used. In broad terms it can also be argued that if a sequence of similar decisions is being taken, in the long run the outcome will be optimal.

Suppose, therefore, that Player I decides to take action *AB* with probability p, and therefore action *BA* with probability $1 - p$. If Player II plays *AB*, the expected payoff is $12p + 8(1 - p) = 8 + 4p$. Alternatively, if he plays *BA*, the expected payoff is $4p + 12(1 - p) = 12 - 8p$.

We are now arguing in terms of expected values, the concept introduced in Section 2.4. Justification for this lies in consideration of the payoff received on average in a long run of plays of the game, following the above argument.

The expected payoffs are now plotted in Figure 3.1. Player I's decision now is to select a value of the probability p. His expected payoff will be the corresponding value on the payoff line of whichever action Player II takes. Clearly, Player II will try to obtain the lower payoff for his opponent and, once he has fathomed which value of p Player I is using, will play *AB* if $p < \frac{1}{3}$ and *BA* if $p > \frac{1}{3}$. (The value $p = \frac{1}{3}$ is the p value of the intersection of the two expected payoff lines.) To counter this, it is clear that Player I should play the mixed strategy corresponding to $p = \frac{1}{3}$, that is *AB* with probability $\frac{1}{3}$, *BA* with probability $\frac{2}{3}$. Any departure from this will allow Player II to reduce his (Player I's) expected payoff. The actual expected payoff received by Player I is the common ordinate when $p = \frac{1}{3}$, that is $\frac{28}{3}$.

Player II is similarly placed with no pure strategy. Suppose that he decides to play *AB* with probability q and *BA* with probability $1 - q$. Note that in this second version of the problem there is no longer the symmetry between the two players' priorities, and although the arguments for the two players are similar, the results will not be identical.

If Player I plays *AB*, Player II's payoff will be

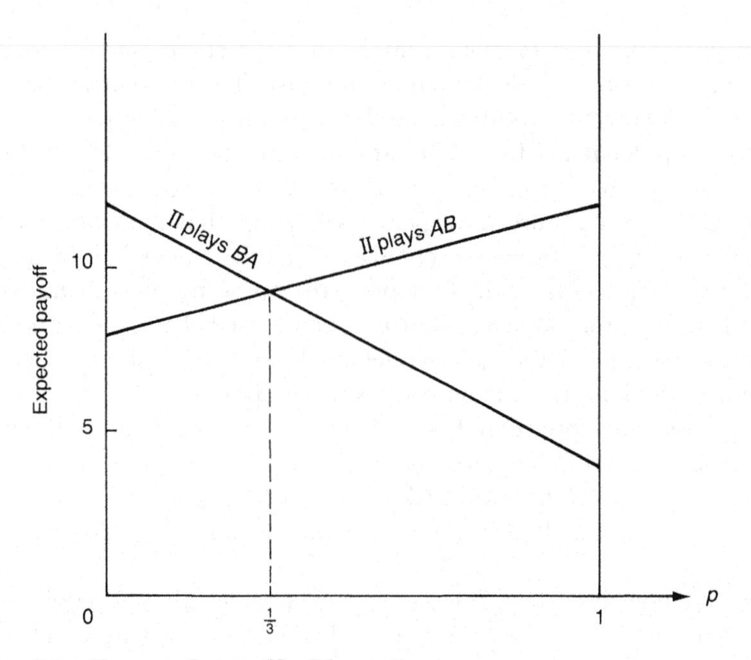

Fig. 3.1 Expected payoffs, Player I

$$12q + 4(1 - q) = 4 + 8q.$$

If Player I plays *BA*, his payoff will be $8q + 12(1 - q) = 12 - 4q$. Plots of these lines are shown in Figure 3.2.

The argument now is that it is in the interests of Player II to keep this expected payoff as low as possible, whereas Player I will be able to ensure that the payoff lies on the upper boundary segments. Player II therefore goes for the q value at the intersection of the lines, which in this case is $q = \frac{2}{3}$. The corresponding value of the payoff is again $\frac{28}{3}$.

Thus the 'solution' to the game is that Player I takes action *AB* with probability $\frac{1}{3}$, *BA* with probability $\frac{2}{3}$; Player II takes action *AB* with probability $\frac{2}{3}$, *BA* with probability $\frac{1}{3}$. With these mixed strategies the average payoff to Player I will be $\frac{28}{3}$.

This common value $\frac{28}{3}$ is now the value of the game. Using his optimal mixed strategy, Player I can guarantee this return on average, and similarly Player II can guarantee that on average Player I will never receive more than this amount.

The existence has been demonstrated of optimal mixed strategies and of a value for the game in the example. It can be proved that

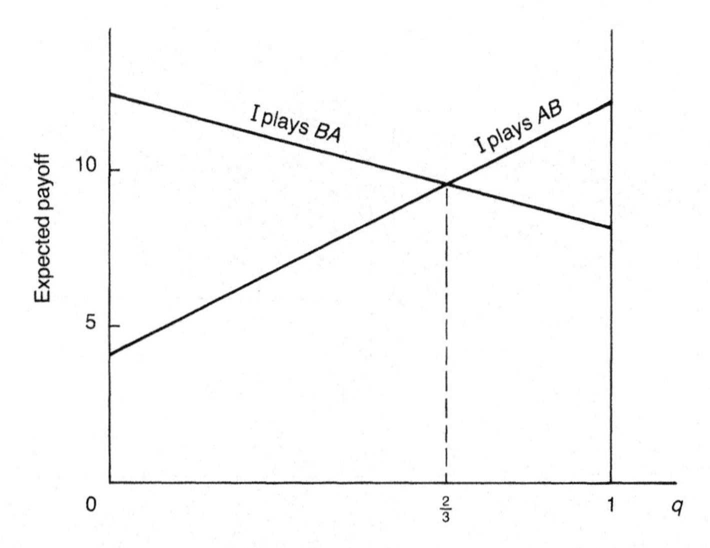

Fig. 3.2 Expected payoffs, Player II

these properties exist for any two-person zero-sum game where each player has a finite number of actions. The proof is beyond the scope of this book, but the more mathematically sophisticated reader can refer to the books by Blackwell and Girshick (1954) or by Luce and Raiffa (1958) in the references at the end of this chapter.

An alternative way of viewing the problem in the example is that each player is really selecting the probability distribution over his actions, which in this case is simply the selection of a p or q value. The probabilities of the various payoffs being realized are then as shown in Table 3.3. Hence the expected payoff is

$$12pq + 4p(1 - q) + 8(1 - p)q + 12(1 - p)(1 - q)$$
$$= 4(3 - 2p - q + 3pq)$$

Table 3.3

Player I action	*AB*	*AB*	*BA*	*BA*
Player II action	*AB*	*BA*	*AB*	*BA*
Probability	pq	$p(1 - q)$	$(1 - p)q$	$(1 - p)(1 - q)$
Payoff	12	4	8	12

A plot of this surface is shown in Figure 3.3. The surface has a saddle point at $p = \frac{1}{3}$, $q = \frac{2}{3}$ (verifiable either graphically or by the standard method of differential calculus), showing that any departure by

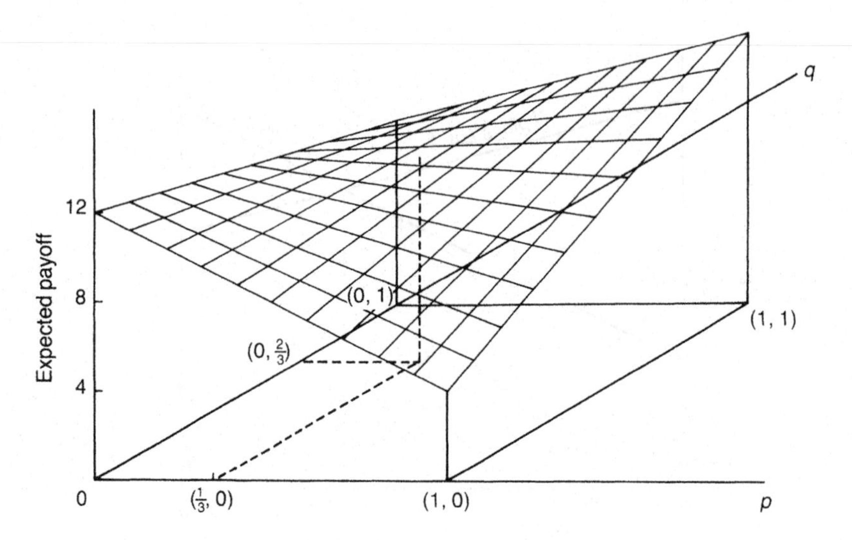

Fig. 3.3 Expected payoff surface

either player from the saddle point values can be exploited by the opponent. The situation is therefore similar to that of the simple game where the actual payoff matrix had a saddle point. The reasoning is the same, although the context requires more thought.

Any two-person zero-sum game with a number of actions for each player can be 'solved' in the sense of finding an optimal mixed strategy and the value of the game. Note that a pure strategy is a mixed strategy with all of the probability concentrated on one action. We have demonstrated how this solution can be found when each player has only two strategies. The extension to games where one player has two strategies but the opponent has more than two is straightforward.

Suppose now that our two companies (Players I and II) are competing for business from three customers *X*, *Y* and *Z*. We return to our assumption that if visits are made on the same day, the profit is shared equally. The profits from *X*, *Y* and *Z* are 6, 8 and 4 respectively. Player I feels obliged to leave customer *Z* to the last day, so as not to offend customers *X* and *Y*. He has therefore only two actions, denoted by *XYZ* and *YXZ*. Player II has no such inhibitions and can visit the customers in any order, resulting in the six possible actions *XYZ*, *XZY*, *YXZ*, *YZX*, *ZXY* and *ZYX*. The payoff matrix is now as shown in Table 3.4.

We should first note that Player II has three actions (*XYZ*, *ZXY* and *ZYX*) which are dominated and which he will therefore never play. This reduces the game to the matrix shown in Table 3.5.

Table 3.4

		Player II					
		XYZ	*XZY*	*YXZ*	*YZX*	*ZXY*	*ZYX*
Player I	*XYZ*	9	11	8	6	14	14
	YXZ	10	8	9	10	11	14

Table 3.5

		Player II		
		XZY	*YXZ*	*YZX*
Player I	*XYZ*	11	8	6
	YXZ	8	9	10

No further reduction using dominance may be achieved. Player I now has to decide on a mixed strategy where he will play *XYZ* with probability p, and *YXZ* with probability $1-p$. His payoffs when Player II uses each of his three remaining actions are

Player II	action	*XZY*	*YXZ*	*YZX*
	payoff	$8 + 3p$	$9 - p$	$10 - 4p$

When plotted these appear as in Figure 3.4.

Player II can keep his opponent to the lower boundary *ABCD*. Player I therefore chooses the maximum value of this boundary, which occurs at *B*, the intersection of the payoff lines for the actions *XZY* and *YXZ* of Player II. The value of p at this point is $\frac{1}{4}$ and the corresponding expected payoff is $\frac{35}{4}$. Thus the optimal strategy for Player I is to take action *XYZ* with probability $\frac{1}{4}$, and action *YXZ* with probability $\frac{3}{4}$. The value of the game is $\frac{35}{4}$. To find the optimal strategy for Player II we note from Figure 3.4 that to hold his opponent to this value he should only play actions *XZY* or *YXZ*, and never *YZX*. The probability distribution over his actions $(q, 1 - q)$ can be found as before, since we have now reduced the game to the 2×2 action format. The reader should verify that this results in the value of $\frac{1}{4}$ for q.

In cases where one player has only two possible actions, it is always possible to reduce the number of actions considered by his opponent to no more than two. The exception to this rule arises when the expected payoff lines for more than two actions meet in a single point

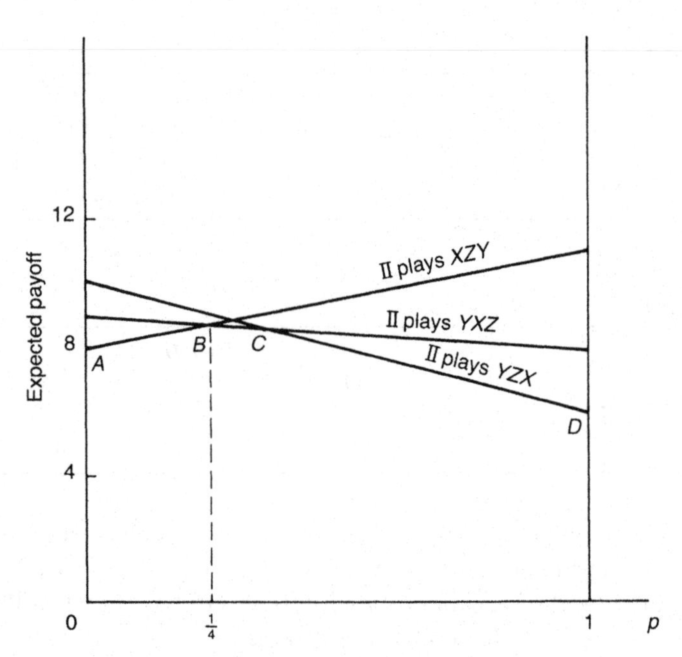

Fig. 3.4 Expected payoffs, Player I

which is also the highest point on the lower boundary. In such cases the player's optimal strategy will not be unique.

The general two-person, zero-sum game is more difficult to solve. If we are lucky we can use the dominance principle successively to reduce the number of actions to two for at least one of the players. For example, suppose that two computer manufacturers (Players I and II) are trying to sell their systems to two different banks. Manufacturer I has four salesmen available but Manufacturer II has only three. Each manufacturer must decide independently how many salesmen to assign to each bank, sending any number from zero to their individual limit. Each bank will buy one computer system and the manufacturers believe that a bank will buy from the manufacturer sending the larger number of salesmen. If the numbers are equal (including zero) because Manufacturer I has a better reputation, the bank will buy from him. Taking the number of systems sold by Manufacturer I as the payoff, the problem may be described as a two-person, zero-sum game as shown in Table 3.6, where each action consists of values (i, j), i being the number of salesmen assigned to the first bank, and j the number assigned to the second.

Table 3.6

		Player II			
		(0,3)	(1,2)	(2,1)	(3,0)
	(0,4)	2	1	1	1
	(1,3)	2	2	1	1
Player I	(2,2)	1	2	2	1
	(3,1)	1	1	2	2
	(4,0)	1	1	1	2

Player I, however, would never play (0,4) since (1,3) is always at least as good. Also (3,1) is always at least as good as (4,0). The game is thus reduced to Table 3.7.

Table 3.7

		Player II			
		(0,3)	(1,2)	(2,1)	(3,0)
	(1,3)	2	2	1	1
Player I	(2,2)	1	2	2	1
	(3,1)	1	1	2	2

Player II now has two dominated actions, namely (1,2) and (2,1), reducing the game still further to Table 3.8.

Table 3.8

		Player II	
		(0,3)	(3,0)
	(1,3)	2	1
Player I	(2,2)	1	1
	(3,1)	1	2

In this reduced form, Player I has his action (2,2) dominated by both of the other actions, and hence the game finally becomes as shown in Table 3.9.

Table 3.9

		Player II	
		(0,3)	(3,0)
Player I	(1,3)	2	1
	(3,1)	1	2

It can now be shown that the optimal strategy is for each player to play the two actions with probability $\frac{1}{2}$, and that the value of the game is $\frac{3}{2}$.

To solve the more general model use is made of the equivalence between this problem and that of the linear programming model. Linear programming is to be covered in more detail in Chapter 8, and therefore at this juncture the emphasis will be on the model formulation rather than the solution.

The approach depends on the existence of the value of any finite game and the relationship between the value and the players' optimal strategies. Suppose therefore that Player I has m actions, Player II has n actions and that when actions i and j respectively are played, the payoff is r_{ij}. Then if Player I has a mixed strategy $(p_1, p_2, \ldots, p_m)$ it follows from the definition of the value of the game that

$$p_1 r_{1j} + p_2 r_{2j} + \ldots + p_m r_{mj} \geqslant v \tag{3.1}$$

for all j between 1 and n, i.e. for all of the actions taken by Player II. The optimal strategy for Player I is given by the corresponding values of $p_1, p_2 \ldots, p_m$ when v is maximized. There are the additional constraints on the p_i values, namely

$$p_1 + p_2 + \ldots + p_m = 1 \tag{3.2}$$

and

$$p_i \geqslant 0 \tag{3.3}$$

Equations (3.1)–(3.3) are of the standard form of the linear programming model, where a linear function of the variables – in this case the very simple function of v alone – is to be optimized, subject to linear constraints on the variables ((3.1 and (3.2)) and the non-negativity constraints (3.3). The value v need not be non-negative, and to cover this possibility we can write $v = v_1 - v_2$, where $v_1 \geqslant 0$, $v_2 \geqslant 0$. It is often the case that all the payoff quantities are positive, and therefore the value of the game must also be positive. Even where this is not the case, we can, for the purpose of solving this game, add some positive quantity to all of the payoffs, thereby ensuring a positive value. The optimal strategies for the players are unaffected by such a transformation. It is as if Player II adds some fixed bonus to each payoff, and then receives back this bonus after the game has been played. The assumption that v is positive allows us to divide by v in the equations (3.1) without changing the direction of the inequalities. Also writing $u_i = p_i/v$, equations (3.1) become

$$u_1 r_{1j} + u_2 r_{2j} + \ldots + u_m r_{mj} \geqslant 1, \qquad j = 1, \ldots, n,$$
where $v = p_1/u_1 = p_2/u_2 = p_m/u_m = 1/(u_1 + u_2 + \ldots + u_m)$

(by *componendo et dividendo*), is to be maximized. Equivalently $u_1 + u_2 + \ldots + u_m$ is to be minimized. Again $u_1 \geqslant 0, u_2 \geqslant 0, \ldots, u_m \geqslant 0$.

This formulation is also that of a linear programme, and has the advantage that it concerns one less variable (v). The optimal mixed strategy for Player II can be found in the same way. There is in fact no need to solve a further problem, as it will be shown that this is what is known as the *dual* problem to the original and its solution can be read directly from that of the original problem.

Suppose, as an illustration, we take what might appear to be a simplified version of the computer manufacturers' problem, where each manufacturer has only two salesmen. Under the same assumptions of buying from Player (manufacturer) I when the same numbers are assigned to the same bank, the payoff matrix is now as shown in Table 3.10.

Table 3.10

		Player II		
		(0,2)	(1,1)	(2,0)
	(0,2)	2	1	1
Player I	(1,1)	1	2	1
	(2,0)	1	1	2

There is now no dominance of actions to reduce the problem, and it turns out to be more difficult to solve than the previous problem. Using the terminology defined in the previous paragraph, we have to minimize

$$u_1 + u_2 + u_3 \tag{3.4}$$
$$\text{subject to } 2u_1 + u_2 + u_3 \geqslant 1 \tag{3.5}$$
$$u_1 + 2u_2 + u_3 \geqslant 1 \tag{3.6}$$
$$u_1 + u_2 + 2u_3 \geqslant 1 \tag{3.7}$$
$$\text{and} \quad u_1 \geqslant 0, u_2 \geqslant 0, u_3 \geqslant 0 \tag{3.8}$$

The value of the game, v, is the reciprocal of the minimized $u_1 + u_2 + u_3$, and the optimal strategy for Player I will be $p_1 = u_1 v$, $p_2 = u_2 v$, $p_3 = u_3 v$.

The problem defined by relationships (3.4)–(3.8) can now be solved by linear programming. In this particular case we can argue to the solution from 'first principles'. If we add the left and right sides of the inequalities (3.5)–(3.7) we find that

$$4u_1 + 4u_2 + 4u_3 \geqslant 3,$$
$$\text{or } u_1 + u_2 + u_3 \geqslant \tfrac{3}{4}.$$

This is the expression which is to be minimized. Hence, if we can find values of u_1, u_2 and u_3 which satisfy the individual inequalities (3.5)–(3.8) and which also sum to $\frac{3}{4}$, no further reduction in this sum can be achieved and the minimum has been found. Clearly the values $u_1 = u_2 = u_3 = \frac{1}{4}$ satisfy the requirements, leading to a value $v = \frac{4}{3}$ and an optimal strategy for Player I of $(\frac{1}{3}, \frac{1}{3}, \frac{1}{3})$. The optimal strategy for Player II is also $(\frac{1}{3}, \frac{1}{3}, \frac{1}{3})$, as can be seen from the symmetry of the payoff matrix.

In this example Player (manufacturer) I had the advantage of winning the business when the same numbers of salesmen were assigned. Had the competition been fair, each manufacturer would expect on average to win the business from one bank, which would result in a value of 1 for the game. The value of $\frac{4}{3}$ found above is a measure of the advantage which is assumed to accrue from the reputation of Player I. It is a matter of a $33\frac{1}{3}$ per cent increase in his business. Note that both players would use the same strategy.

This has been an introduction to the basic model of the two-person zero-sum game and the way in which such games should be played. Most business decisions do not have the purely competitive aspect which, as we have seen, makes it necessary to think in terms of countering the opponent's actions when a strategy or action is chosen. Usually there is a strong chance element which virtually specifies the outcome or payoff, and, because of this, other ways of playing the game are more appropriate. This will be the general theme of the next chapter, but before proceeding to this, some attention will be given to non-zero-sum games.

3.3 Two-person, non-zero-sum games

Perhaps the most famous illustration of a two-person, non-zero-sum game is the so-called 'prisoner's dilemma'. There are a number of versions of this problem, but the situation is broadly as follows. You and an associate have been apprehended for a robbery. The police have enough evidence to convict both of you for this crime, but they have a strong suspicion that this is just one of a series of robberies which you have both committed. You and your associate are being kept in separate cells and you are both made the following offer. If you turn Queen's evidence and confess to the series of robberies, thereby convicting your accomplice, you will be allowed to go free immediately. If you do not turn Queen's evidence then you will certainly be convicted for the last robbery and receive a one-year sentence. As far as your accomplice is concerned, your confession will lead to him being given a five-year sentence. Otherwise he will receive the same one-year sentence for the last robbery. You are, of course, not being permitted to confer, nor to know what he decides before you have made up your mind. Naturally you will be concerned with your own fate following your accomplice's decision. If he turns Queen's evidence and you do not, then you will receive the five-year sentence. If, on the other hand, both of you confess, then a trial and conviction will be unavoidable, although a slightly reduced sentence of four years for both of you seems highly likely. What, then, do you do?

A little thought produces the following conclusion. If your accomplice confesses, then if you do not confess also you can expect to receive a five-year gaol sentence. Alternatively, if you do confess, the sentence will be four years, and therefore in this circumstance you would be better off confessing. Now suppose that your accomplice does not confess. If you do not confess as well, your gaol sentence will be one year. Alternatively again if you confess, you will be allowed to go free. Once more you would be better off confessing. The solution is clear – confess!

We can draw up the problem as a game with a payoff matrix whose values are the expected gaol sentence *for you*, as shown in Table 3.11. Your second strategy is dominated by the first and this has led you to your decision to confess.

Your accomplice is also well-versed in game theory and has drawn up an identical payoff matrix and has come to the same conclusion. You both, therefore, confess and go to gaol for four years, which is

Table 3.11

		Accomplice	
		confesses	*does not confess*
You	*confess*	4	0
	do not confess	5	1

ample time to ponder on the differences between zero-sum and non-zero-sum games. If only both of you had decided not to confess, you would both have received but a year's sentence. Such actions required cooperation, which was explicitly denied you, and trust, which was absent. The degree of trust needed was quite significant, since either of you could have changed actions beneficially and walked away free.

A similar situation can arise in advertising decisions. Using a rather simplified description, suppose that two firms operate in a particular sales territory. The territory has a total value of £10 000 profit which will be divided between the two firms. If neither firm conducts an advertising campaign the profit is split equally, the same split occurring if both conduct advertising campaigns. Should only one firm advertise, the profit will be split in a 4:1 ratio. The cost of an advertising campaign to either firm is £1000. Thus, if one firm advertises but the other does not, the profit to the firm advertising will be £8000 − £1000 = £7000. The other firm will make a profit of £2000. Taking either firm, the payoff matrix (in £000) will be as shown in Table 3.12.

Table 3.12

		Competitor	
		advertises	*does not advertise*
Firm	*advertises*	4	7
	does not advertise	2	5

Again we have the curious feature of dominance leading both firms to conduct advertising campaigns, thereby reducing their profits to £4000 from the £5000 level which they would attain if they colluded over their decisions.

Collusion over decisions in non-zero-sum games may not be practicable, ethical or even legal. The real lesson from such situations is that the players develop a form of cooperation where it is appreciated that a totally selfish strategy is not in their long-term interests. This is apparent in many commercial situations where understandings emerge between competing bodies. It has been suggested that evolution itself operates in this way, where organizations in a common environment do not exercise fully their peculiar advantages at the expense of others. Axelrod at the University of Michigan has carried out a number of experiments with human beings which demonstrated the benefit of cooperation with repeated plays of a prisoner's dilemma type problem. These have been reported in an article by D. R. Hofstadter 'Metamagical Themas' in the May 1983 issue of *Scientific American.*

3.4 Further reading

The following books treat the theory of games and its allied topics in more depth. They range widely in the mathematics required for understanding, the first listed book in particular being quite demanding.

Blackwell, D. and Girshick, M. A., *Theory of Games and Statistical Decisions*, John Wiley, New York (1954).

Chernoff, H. and Moses, L. E., *Elementary Decision Theory*, John Wiley, New York (1959).

Gass, S. I., *Linear Programming*, McGraw-Hill, New York (1958).

Hadley, G., *Linear Programming*, Addison-Wesley, Reading, Mass. (1962).

Luce, R. D. and Raiffa, H., *Games and Decisions*, John Wiley, New York (1958).

Vajda, S., *The Theory of Games and Linear Programming*, Methuen, London (1956).

Williams, H. P., *Model Building in Mathematical Programming*, John Wiley, Chichester (1978).

Williams, J. D., *The Compleat Strategist*, McGraw-Hill, New York (1954).

3.5 Problems and practical exercises

1 Two companies (Players I and II) are competing for business from three customers X, Y, Z. Using the context of one of the examples in Section 3.2, each company is to visit the customers on the same three days and both must plan their visits in ignorance of the other's strategy. Once more as a matter of policy, Player I must leave customer Z to the third day, a fact which is well known to Player II. The profits to be made from customers X, Y and Z are 6, 8 and 4 respectively.

In this situation it is assumed that Player I is a well-established reputable company, so that if both companies visit a customer on the same day, all of the profit goes to Player I.

Draw up a payoff matrix and show how the game can be reduced to one where each player has two strategies.

Find the best mixed strategy for each player and determine the value of the game.

2 Refer to the situation in Exercise 1. Clearly after the first visit Player II would know Player I's strategy. Suppose now that he could if necessary amend his own strategy. Would he in fact do so?

3 Two companies each have two salesmen who are to compete for two contracts. Each company must therefore allocate one salesman for each contract. For company *A*, salesman Robinson will win the contract, no matter which salesman company *B* allocates in opposition. Their second salesman, Lloyd, will win the contract if company *B* allocates salesman Hargreaves in opposition, but not if salesman Foster is allocated. (It follows therefore that for company *B* Hargreaves is doomed to failure no matter whom he is up against. The company is in fact bidding for the contract simply to show continuing interest.) The profits from the two contracts are £10 000 and £5000.

Express the situation as a two-person zero-sum game, and evaluate the best strategies for the two companies.

4 Show that the intuitive strategy of always sending the best salesman to the most lucrative contract is optimum if the order of ability is Robinson, Foster, Hargreaves, Lloyd.

5 In formulating its advertising policy for a particular region for the next year a firm selects between three levels of activity – high (H), medium (M) and low (L). The firm can assess the effects of each level

Table 3.13

		Competitor	
		N	C
	H	120	50
Company	M	110	60
	L	80	70

dependent on the policy adopted by its only competitor for this market. It is assumed that the competitor will either adopt his normal advertising policy (N) or concentrate extra resources (C) on this region. The returns to the company (in £000 net of advertising expenses) are as shown in Table 3.13. Why is this not a zero-sum game?

If the odds are 2:1 that the competitor will concentrate on this region, which strategy should the company adopt?

Discuss the appropriateness of the model if there are several competitors in the region.

4
Criteria for decision-making

OBJECTIVES

From this chapter the reader should have an understanding of the different criteria which can be used by the decision-maker in deciding between actions in the light of resulting consequences. He should understand the importance of taking into account information on the probabilities of actions taken by competitors; or on the probabilities of events occurring by chance (Nature) when he is trying to anticipate the future.

4.1 Properties of a good decision criterion

In the sections following a number of criteria for decision-making will be suggested. These offer ways in which a decision problem, viewed as a game against Nature, may be solved. Although the ways differ in their underlying philosophies and therefore in the actions they imply, they must satisfy certain desiderata in accordance with an acceptable level of plausibility. Three such properties will be proposed, which, although almost trivially reasonable, do on occasions require some remedial step to avoid the decision-maker being put in an untenable position.

The first desideratum is that any action for the decision-maker which is dominated should be removed from consideration. Dominance is used in the sense (defined in Section 3.2) that he could do at least as well with a single alternative action, no matter which action Nature takes. We shall speak of Nature taking actions with no necessary implication of malevolence toward the decision-maker. Two points should be made about this property. Firstly, it is no longer appropriate to apply the concept of dominance sequentially, as was done in Chapter 3 with the illustration involving computer manufacturers and banks. Nature does not necessarily behave in such a self-interested way. Secondly, there are criteria which depend only on extreme payoffs.

With such a criterion where, for example, the same action by Nature produces the smallest payoff for two different actions by the player, it might be felt that the actions were equally pessimistic, whereas when the other payoffs were examined one action dominated the other. It would therefore be absurd to play the dominated action.

For example, a firm has to decide on its advertising campaign for the following quarter. It has a choice between television, newspaper and poster advertising, there being only sufficient funds for one medium. The actual return on the advertising outlay is measured by the estimated total exposure of the campaign, that is the number of potential customers who have the opportunity to see each medium. This will depend on the weather, poor weather favouring television viewing and newspaper reading, whereas better weather gets people outdoors. Estimates of the actual exposures (in thousands of potential customers) are given for four broad but explicit descriptions of the weather in Table 4.1.

Table 4.1

		Weather			
		Poor	*Moderate*	*Good*	*Excellent*
	Television	200	190	170	130
Medium	*Newspapers*	180	160	150	130
	Bill posters	110	140	140	190

Based on these figures the medium of newspaper advertising is dominated by television advertising, and therefore newspaper advertising should be eliminated from further consideration. Nature also has a dominated strategy – moderate weather. Clearly we do not deduce that the weather will never be moderate since the firm's payoff will always be no greater if the weather is good. The second point made above concerns a strategy which might plausibly argue that the firm wishes to choose the action which maximizes the smallest return. With TV or newspaper advertising the smallest return would be 130 (excellent weather) and with bill poster advertising it would be 110 (poor weather). The criterion therefore argues that either TV or newspaper advertising would be optimal, and it does not seem to matter which of these two actions is taken. The concept of dominance shows that, notwithstanding the decision criterion, it would be unreasonable to use newspaper advertising.

The second desideratum is that if under any decision criterion action *a* is preferred to action *b*, and if also action *b* is preferred to action *c*, then action *a* must be preferred to action *c*. Again this seems to have an irreproachable logic, but consider the following logic. Three candidates, Cooper, Foster and Swan are being interviewed for a job. The selection committee has three members, Messrs Worthington (Chairman), Davenport and Marston. After the interviews the committee finds itself deadlocked as each of its members favours a different candidate. The conversation then proceeds as follows:

WORTHINGTON The only way to resolve this problem is to take the candidates in pairs and to vote. For example, taking Cooper and Foster, I prefer Cooper to Foster.

DAVENPORT Chairman, I agree. Cooper was better than Foster.

MARSTON I felt the opposite to be the case, but I see that I am out-voted.

WORTHINGTON Good. The committee then, by a majority verdict prefers Cooper to Foster. Now let us take Cooper and Swan together. I actually felt that Swan was better than Cooper.

MARSTON Here, chairman, I agree with you.

WORTHINGTON Well, that just about settles it. Swan is preferred to Cooper no matter how you vote Davenport, and since Cooper was preferred to Foster, Swan must get the job by majority verdicts of the committee.

DAVENPORT Agreed.

MARSTON Agreed.

After the committee meeting had ended, Davenport happened to mention that he would have voted against Swan in comparison with Cooper, but that it made no difference. He then happened to mention that curiously enough he also preferred the 'last' candidate, Foster, to Swan. To their surprise, Marston also had preferred Foster to Swan, indicating that a majority verdict would have thereby restored the deadlock. It was not difficult to deduce that each member of the committee had behaved rationally and that their chairman had astutely manipulated the procedure to match his own preferences. The committee had ranked the candidates

WORTHINGTON 1 Swan 2 Cooper 3 Foster

DAVENPORT	1 Cooper	2 Foster	3 Swan
MARSTON	1 Foster	2 Swan	3 Cooper

Apart from using the chairman's prerogative, the deadlock was in fact complete. The error made by the committee was to use a preference criterion which did not take into account all of the information available. It did not use the ranks of the candidates within each assessor. Alternatively, the procedure itself was incomplete. In coming to a decision when only two of the three possible comparisons had been made, the committee did not treat each possible action in a balanced fashion. Had it done so the committee would have realized that there could be no majority verdict for any candidate. A decision procedure which produces results where A is preferred to B, B is preferred to C and C is preferred to A is said to be incoherent. For a fuller treatment see Lindley (1971).

Finally, a third desideratum is that the preference between any two actions should not depend on the presence or absence of a third. The classic example of this is the case of the man who went into a restaurant and was told that the only dishes left on the menu were steak and fish. 'Very well,' he said, 'I will have fish.' A few moments later, before the man had been served, a delighted waiter returned to inform him that the chef was also able to offer chicken. 'In that case,' said the man, 'I will have the steak.'

One possible explanation of the man's behaviour is that he believed that serving fish is a specialist's business and, much as he preferred fish to steak he would not eat it in a restaurant where the menu was too diverse, that is where more than one alternative dish was offered. The two decision situations were therefore quite different, operating under different rules and producing different preferences. The presence of a further action for the decision-maker changed the probability of at least one of Nature's actions, namely that he would not enjoy a fish dish. Much more care has to be attached to the definition of Nature's actions, incorporating in this particular case the number of dishes offered.

4.2 The maximin criterion

The first of the decision criteria to be considered is one that has been mentioned in the previous section and also used in the two-person zero-sum game situation. Suppose that the decision-maker has to select from m actions $a_1, a_2, \ldots, a_m$; Nature has n actions $b_1, b_2, \ldots b_n$; the payoff when actions a_i and b_j respectively are taken is r_{ij}. When action a_i is taken by the decision-maker, his payoff will be one of the values $r_{i1}, r_{i2}, \ldots, r_{in}$. If r_{i0} is the smallest of these, then clearly he will always obtain a return which is no less than this value. Taking this argument a stage further, he then argues that if he chooses the action which has the largest minimum or guaranteed payoff, he will have the satisfaction of knowing that whatever happens, that is whatever Nature does, his return will be at least this maximum of the minima. Expressed mathematically, this requires that the decision-maker takes action a_z, where

$$r_{z0} = \max_i \min_j r_{ij}.$$

Suppose, for example, that the firm with the decision on the selection of its advertising medium now assessed the return as in Table 4.2. The minimum payoff values for the three actions are 140, 150 and 130, showing that the maximin strategy is action 2, newspaper advertising.

Table 4.2

		Weather			
		Poor	*Moderate*	*Good*	*Excellent*
	Television	220	210	190	140
Medium	*Newspapers*	180	170	160	150
	Bill posters	130	160	160	160

The maximim criterion arises from a very pessimistic or conservative attitude on the part of the decision-maker. It assumes that Nature is likely to take a most uncooperative action and that the decision-maker therefore has to counter this approach as well as he can. Against a competitive opponent in a two-person zero-sum game this is a reasonable way to play the game, and in Chapter 3 we saw how a strategy which can now be identified as maximin led to pure strategies for both players when the payoff matrix had a saddle-point, and to the concept of an optimum mixed strategy when it had not. In such a situation, using the same argument, Player II would be reversing the order of the minimization and maximization, so that, although his criterion is the same, the appropriate label would be minimax.

The conservatism of the maximin criterion is its main drawback. Each action is judged solely on its worst payoff, other payoffs being ignored. In the example, it may be noted that the television medium comfortably betters newspaper advertising in all situations other than the occurrence of excellent weather. The chance that a spell of fine weather will occur, drawing people away from the television sets, may or may not be remote. This chance aspect of the problem is not considered. Any action by Nature, no matter how remote, is included in the payoff matrix and can have a crucial influence. Indeed, since the maximin criterion by definition is dealing with extreme outcomes, it is inevitable that the criterion will lead to a decision based on extremes. Thus in the example we could have rejected the action of television advertising simply because, when the model of the problem was formulated, a strategy for Nature was included which had little chance of actually happening. Although this is a fair criticism of the maximin criterion there is a danger in the argument, arising from the temptation to remove a strategy of Nature so that a particular strategy of the decision-maker is excluded from consideration. The cure would then be worse than the complaint, and it seems that if the maximin strategy is to be used, the drawback described here simply has to be accepted.

A further criticism, which follows on from the above argument, arises when the states of Nature are to be defined. For the model the states merely have to be mutually exclusive and exhaustive. They need not be 'equal' in any sense such as covering the same range of some continuous variable selected by Nature, nor as occurring with approximately equal probability. The states must cover the entire range of Nature's actions, but each state must be sufficiently homogeneous so that a single outcome or payoff is appropriate to the whole state for each of the decision-maker's actions. The modelling problem is present for any decision criterion, but it is made particularly acute for the maximin strategy with its emphasis on extremes. In the example an attempt was made to describe the way in which people's leisure behaviour depended on the weather in a certain quarter of the year. Weather conditions were described in terms of hours of sunshine and maximum temperature, and these in turn were classified under the four headings of poor, moderate, good and excellent. A five-cell classification would have given scope for more specific consideration of extreme conditions where the homogeneity requirement could become questionable. Such considerations are often present – winter heat waves or hailstones in the spring – but to give the conditions a status comparable with the more regularly met central conditions leads in the case of the minimax criterion to actions which are intuitively unreasonable.

4.3 The minimax regret criterion

This second criterion is similar to the first, but it argues that the 'correctness' of a decision should be measured not by the actual payoff received, but by the amount by which the payoff could have been increased, had the decision-maker known what the state of Nature was. It is a criterion which has in mind judgement by hindsight.

For each action of Nature and of the decision-maker we can therefore calculate what is known as the *regret*. This is the difference between the maximum payoff for Nature's action and the particular payoff for the decision-maker's action and this same action by Nature. Denoting the regret for actions i and j taken by the decision-maker and Nature respectively by ρ_{ij}, we have

$$\rho_{ij} = \max_k r_{kj} - r_{ij}.$$

Thus there is a matrix of these regret terms, where each column contains at least one zero value. In the current version of the advertising budget problem, the regret matrix is as shown in Table 4.3.

Table 4.3

		Weather			
		Poor	*Moderate*	*Good*	*Excellent*
	Television	0	0	0	20
Medium	*Newspapers*	40	40	30	10
	Bill posters	90	50	30	0

Applying an approach which, like maximin, guards against the worst action by Nature, we now determine for each of the decision-maker's actions the largest regret. Having done this we select the action with the smallest value, and term this the minimax regret action. It is now defined fully as action a_y, where

$$\max_j \rho_{yj} = \min_i \max_j \rho_{ij}.$$

In the example the maximum regrets are 20, 40 and 90 for the media television, newspapers and bill posters respectively, and therefore the television medium ($i = 1$) would be recommended under the minimax regret criterion. Note that this differs from the maximin action (newspapers), demonstrating a fact which can be proved, namely that the maximin and minimax regret criteria do not necessarily result in the same action being taken.

The minimax regret criterion has some of the disadvantages of the maximin criterion, but often to a lesser extent. If, for example, an extreme action by Nature is fairly uniformly bad for all the decision-maker's actions, this will not necessarily be the main influencing factor in the decision. For both criteria only the relative value of each payoff has to be considered, but for the minimax regret criterion the relativity is only *within the column.* How much worse these values generally are when compared with other more moderate actions by Nature then becomes irrelevant.

One disadvantage peculiar to this criterion is illustrated by a further adaptation of the advertising budget example. Suppose now that the payoff had been assessed as in Table 4.4. Suppose also that initially only the television and newspaper media were considered. The regret matrix is then as shown in Table 4.5, and the action which minimizes the maximum regret is therefore action 1 (television).

Table 4.4

		Weather			
		Poor	*Moderate*	*Good*	*Excellent*
	Television	220	210	190	140
Medium	*Newspapers*	180	170	160	150
	Bill posters	130	160	160	190

Table 4.5

		Weather			
		Poor	*Moderate*	*Good*	*Excellent*
Medium	*Television*	0	0	0	10
	Newspapers	40	40	30	0

Now, as an afterthought, the bill posters action is introduced. The complete matrix of payoffs now produces the regrets shown in Table 4.6. The minimax regret action is now action 2 (newspapers). The

Table 4.6

		Weather			
		Poor	*Moderate*	*Good*	*Excellent*
	Television	0	0	0	50
Medium	*Newspapers*	40	40	30	40
	Bill posters	90	50	30	0

situation is similar to that used in Section 4.1 to illustrate the third desideratum for decision criteria, namely that of preferences between two actions not depending on the presence or absence of a third. Unfortunately, the concept of regret is bound to suffer from this drawback, since it is measuring the loss against the known optimum which can easily change as more alternative actions are introduced. In this case the dilemma could be resolved by taking all paired comparisons, but frustratingly this way out of the difficulty cannot be guaranteed. Simply changing the payoff for the television and excellent weather actions from 140 to 105 produces a self-contradictory set of paired comparisons which now violates the second desideratum. Verification of this result is left to the reader.

4.4 The maximax criterion

The maximax criterion presents a means of selecting an action which is in direct contrast with minimax. The procedure in full argues that for each of the decision-maker's actions, the largest payoff is noted. The action is then selected which has the largest of these values. The criterion therefore simply amounts to scanning the payoff matrix to find its largest value and then taking the corresponding action. In the example this would result in action 1 (television) being taken.

It is a totally optimistic criterion, ignoring all other payoffs and taking no account of the probability that Nature will take this particular favourable action. It is very much the approach of the desperate gambler and, considered in isolation, merits no further mention.

4.5 The Hurwicz criterion

In an attempt to achieve a compromise between the pessimism of the maximin criterion and the optimism of the maximax criterion, Hurwicz proposed a criterion which is a weighted average of the two extremes. For each of the decision-maker's actions i, the maximum payoff R_{i0} and the minimum payoff r_{i0} are weighted to form

$$h_i = \alpha r_{i0} + (1 - \alpha) R_{i0},$$

where α is a constant, $0 \leqslant \alpha \leqslant 1$. The optimal Hurwicz-$\alpha$ action is then a_x, where

$$h_x = \max_i h_i.$$

The criterion depends critically on the chosen value of α. The index is a measure of the balance between the pessimism and the optimism of the decision-maker. A small value of α indicates an optimistic attitude, where in the limit a value of zero results in the maximax action. Conversely a value close to one would produce a pessimistic action, where in the limit a value of one gives the maximin action precisely. Note that although this strategy requires the weighting of two of Nature's actions by weights which could be interpreted as probabilities, it follows that since the criterion compares different actions by the decision-maker these weights generally apply to different actions by Nature throughout the comparisons. Derivation of α is not therefore simply a matter of assessing the relative likelihoods of two of Nature's actions. It is a genuine measure of the decision-maker's pessimism (or optimism), and as such it is of rather questionable value. For a general value of α, the example would require comparisons between

Television	$140\alpha + 220\,(1 - \alpha) = 220 - 80\alpha$
Newspapers	$150\alpha + 180\,(1 - \alpha) = 180 - 30\alpha$
Bill posters	$130\alpha + 160\,(1 - \alpha) = 160 - 30\alpha$

A plot of the weighted payoffs against α is given in Figure 4.1.

Note that the optimum Hurwicz-α action depends on the selected value for α in the following way:

$0 \leqslant \alpha \leqslant 0.80$	action 1 (television)
$0.80 \leqslant \alpha \leqslant 1$	action 2 (newspapers)

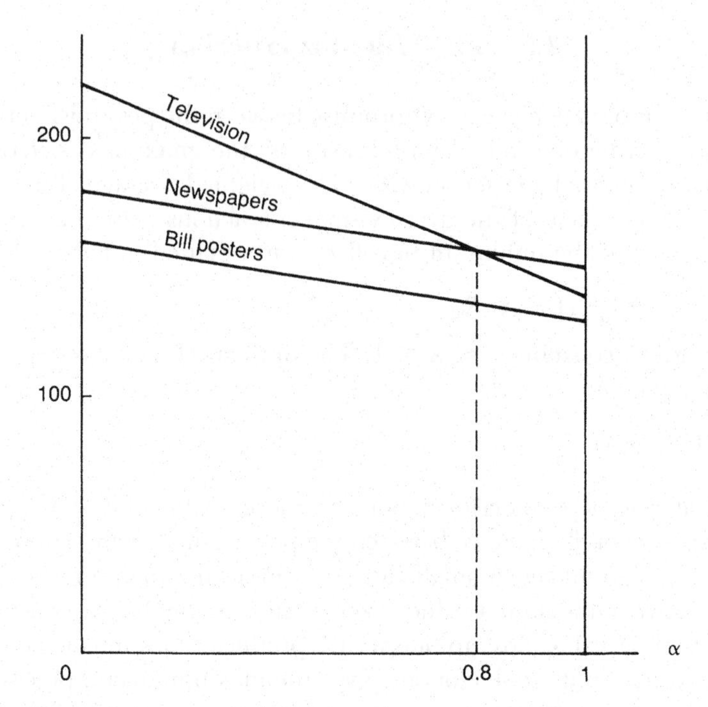

Fig. 4.1 Weighted payoffs for the Hurwicz criterion

At $\alpha = 0.8$, the criterion is indifferent between actions 1 and 2. Also action 3 (bill posters) is 'dominated' in the Hurwicz sense, as there is no value of α which would lead to the use of this action.

In attempting to obtain the best of the maximin and maximax criteria, the Hurwicz strategy has unfortunately not succeeded in shedding the serious disadvantage of their criteria, namely their dependence on extreme payoffs. In introducing the parameter α the criterion involved the decision-maker in a further decision which could be of crucial importance. It is difficult to imagine a practical use of the Hurwicz criterion, other than as a reinforcement of some alternative criterion.

4.6 The expected monetary value (EMV) criterion

The criteria considered so far have been expressed in terms of the payoffs, with no explicit account taken of the probabilities of Nature's actions. Some decisions therefore will have been based on outcomes which are highly unlikely and clearly this is undesirable. Four criteria will now be proposed which depend to varying extents on the probabilities of Nature's actions. The first and also the most popular is usually called the Expected Monetary Value or EMV criterion.

Suppose, therefore, that we can assign probabilities $q_1, q_2, \ldots, q_n$ to Nature's n actions. These probabilities may be subjective and based on expert opinion, or they may result from the analysis of past data (for example from market research) and therefore have a strong objective component. We shall return to this issue in Chapter 7 when market research and the value of information are discussed. For our present purposes the probabilities represent the total amount of information available to the decision-maker about Nature's actions.

With the payoff r_{ij} when the decision-maker and Nature take actions i and j respectively, we can calculate a weighted average payoff for the decision-maker's action i:

$$A_i = q_1 r_{i1} + q_2 r_{i2} + \ldots + q_n r_{in}.$$

The EMV criterion requires that the decision-maker takes action w, where $A_w = \max_i A_i$.

Clearly, if the same decision problem is met on a large number of occasions, the EMV criterion will find the action which, when repeatedly used, gives the greatest total payoff. Most decision problems are of a 'one-off' nature or have at most a very limited number of repetitions. In such cases the applicability of the EMV criterion may be questioned. Two points may be made in defence of this criterion. Firstly, although a decision may be 'one off' it is likely that decisions of similar importance in terms of the size of their payoffs will be taken regularly by people in the same organization. Thus, although on a single occasion an action may have been taken which with hindsight appears to be a poor one, the experience has to be balanced against more favourable occasions. A good decision-maker knows that he cannot always be correct; he knows that if he is analysing his options and their consequences thoroughly his efforts will, in the long run, show up positively in the organization's balance sheet. He will also need a measure of understanding when his own action interacts badly with one of Nature's more extreme actions. The second point is that even when the decision has to be viewed as strictly 'one off' the EMV criterion is a reasonable way out of the inherent dilemma of taking the risk, as it gives an appropriate weight to each possible payoff outcome. Some action has to be taken, and nothing will remove the chance of 'being wrong'. Perhaps the main concern which the criterion obscures is the actual size of the payoffs. In a simple illustration, if Nature has two equally probable actions and if for the decision-maker's action 1, the payoffs are £1 and £(−1), i.e. a gain and a loss of £1, and for action 2 they are £100 000 and £(−99 999), then the corresponding EMVs are £0 and £0.5. Using the EMV criterion, the decision-maker would take action 2. Although the term *expected* monetary value is used, the adjective is unfortunate (it is borrowed from the theory of statistics) as, of course, the true payoff will be either £100 000 or £(−99 999). Most decision-makers, acting whether for themselves or for their organizations, would think hard before taking an action where there was a fifty–fifty chance of losing £99 999, despite the same chance of a compensating gain. To some extent this problem will be dealt with in the next chapter when utility is discussed, but whatever measure of payoff is used, the property of 'averaging out' remains.

Returning now to the original advertising medium decision, suppose that the probabilities of the classifications of weather had been assessed:

Weather:	Poor	Moderate	Good	Excellent
Probability:	0.2	0.4	0.3	0.1

The EMVs for the three actions are:

Television:	(0.2)220 + (0.4)210 + (0.3)190 + (0.1)140 = 199
Newspapers:	(0.2)180 + (0.4)170 + (0.3)160 + (0.1)150 = 167
Bill posters:	(0.2)130 + (0.4)160 + (0.3)160 + (0.1)160 = 154

Action 1 (television) is the maximum EMV action. It is comfortably ahead of action 2, showing that it would take a substantial change in the probabilities for this latter action to be considered. It is possible to carry out a more systematic study of the range of the probabilities for which the same action (1) remains the optimum. The main benefit of this would be to see if we need to reconsider any of the probabilities for a more accurate assessment, particularly if they are largely subjective. A study of this type is known as a *sensitivity analysis*.

4.7 The expected opportunity loss (EOL) criterion

The next criterion is related to the EMV in a way which parallels the relationship between the minimax regret and the maximin criteria. It assumes once more that we have probabilities for the states of Nature, but now operates on the regret matrix instead of the payoff matrix. If the regret for actions (i,j) is p_{ij}, the expected regret or expected opportunity loss when the decision-maker takes action i is

$$L_i = q_1 p_{i1} + q_2 p_{i2} + \ldots + q_n p_{in}.$$

The EOL criterion then takes action a_v where

$$L_v = \min_i L_i.$$

There is little point in discussing this criterion further, since it can be proved that it always leads to the same action as the EMV criterion. In the advertising example, the three expected opportunity losses are 2 (action 1), 34 (action 2) and 47 (action 3). Note that when these are added to the corresponding EMVs, the sum is always the same, 201. The minimization of the expected opportunity loss is, therefore, a property of the maximum EMV criterion.

4.8 The Laplace criterion

The Laplace criterion which is based on what is termed 'the principle of insufficient reason' is an attempt to use the EMV approach when the decision-maker is in a state of ignorance about the probabilities of Nature's actions. The proposal is to assume then a uniform distribution over Nature's actions, and to adopt the action which has the largest weighted average payoff. Equivalently it is the decision-maker's action which has the largest total payoff, a rather bizarre measure which some may feel reflects the true merits of this criterion. For the example used in recent sections of this chapter, the criterion amounts to the selection of the largest value amongst 760 (action 1), 660 (action 2) and 610 (action 3). The television medium once more is chosen.

The criterion has two points of contention. The first is the impression created that ignorance can be expressed by an assumption that all outcomes are equally likely. This assumption is, of course, a very specific statement about the likelihood of the outcomes and it should not obscure the fact that if the decision-maker is in this unfortunate position he cannot overcome it by a mathematical sleight of hand. He has either to resort to one of the criteria which are not based on the probabilities of Nature's actions, or else to carry out some investigation to improve his knowledge. As far as the latter is concerned, we shall see in Chapter 7 how a prior position, possibly of ignorance as defined here, can be modified to a posterior distribution in the light of experimental evidence.

In fairness to James Bernouille (1654–1705) who first stated the principle of insufficient reason, the claim for this principle was merely that if there is no evidence to believe that one outcome is more likely than any other, then all outcomes should be judged to have the same probability. This is certainly not the same thing as complete ignorance.

The second point concerns the way in which Nature's actions are defined. The discussion in the final paragraph of Section 4.2 mentioned that it was often possible to define Nature's actions in more than one way. Suppose, therefore, that in one formulation of the problems it was decided to partition one of Nature's actions into two virtually equivalent actions, where the payoffs for the decision-maker's actions were effectively identical. Under the Laplace criterion this would double the influence that the original action of Nature had on the selection, and clearly it is undesirable that the criterion should be so strongly affected by an issue at the problem formulation stage. Ideally we might specify Nature's actions by the uniqueness of their set of payoffs against the decision-maker's list of actions. Any of Nature's actions which produced the same payoff would be merged into a single action. In practice this could be difficult to apply and could, for example, in a decision dependent on weather conditions, lead to a very large set of actions. There is also the possibility that two quite distinct (in the sense of their description) actions have the same payoffs and reason would then dictate against their combination.

4.9 The maximum modal payoff criterion

Intuitively this is a very reasonable criterion to adopt. Again a probability distribution over the actions of Nature is required, and the decision-maker can then determine which is the most likely (modal) action to be taken. He determines his action as that which gives the largest payoff against the modal action. Using the probabilities (0.2, 0.4, 0.3 and 0.1) assessed for the EMV criterion in the example in Section 4.6, the modal action by Nature is to produce 'moderate' weather. Against this the best payoff (210) comes from the television medium.

The criterion suffers from the same difficulty with the definition of Nature's actions that the Laplace criterion experiences. In the example, if 'moderate' weather had been classified as either 'average' or 'temperate', then the probability of 0.4 allocated originally to 'moderate' would have to be split between the new classifications, possibly in equal amounts. This now makes the 'good' classification into Nature's modal action, presenting the possibility of a different action by the decision-maker under the same criterion. In this the separation or amalgamation of two actions of Nature can make the criterion incoherent, a property which may be abhorrent to the mathematician but which can be avoided if the criterion is used sensibly. Suppose, for example, that a property developer is interested in building an hotel to meet the needs of a new city airport. He can only make the project pay if he buys the land now, before the decision on the location is made. There are three sites being considered and there is also the possibility of postponing the airport development for the time being. The developer has four sites for his hotel, each of which has varying accessibility for the proposed airport sites. He can assess the return on his investment for each of his sites for each action of the airport development body. Without going into detail on these returns, it must surely be reasonable to argue that if his researches indicate that one particular site is more favoured than the other actions, he should seriously consider building his new hotel where the greatest return will be accrued for this modal action. This may not be the overriding criterion to be used, but it should add weight to the argument. In this example the possibility of amalgamation or separation of actions does not arise. (It is assumed that no two airport sites can be close enough together to give the same payoff for each hotel site.) It would therefore be a pity to reject the criterion because of its incoherence when dealing with other applications.

4.10 Overview of decision criteria

Eight decision criteria have been described in this chapter which may leave the reader with the feeling that not only does he or she have a decision to make but he also has a decision on how he is going to make it. It is unlikely that each decision criterion will result in the same action being taken, and it is certainly not suggested that the choice should go to the most popular one. What should happen from consideration of the various criteria is that the decision-maker will clarify his objectives and his attitude towards risk. This insight into the mechanism of decision-making is where the benefit lies.

It is not difficult to find fault with each criterion individually and indeed the treatment presented here may appear to be negatively biased. Clearly, some criteria merit more serious consideration than others, and of the ones for which probability statements on the actions of Nature can be made, the EMV criterion appears to be the most popular. Failing this ability, it is likely that recourse will be made to the maximin criterion. In supporting roles we have modal payoff (where appropriate) and minimax regret criteria.

In the following chapters on decision trees and on their applications, the EMV criterion will, in the main, be used to compare the merits of actions. It must always be borne in mind that these are averages and we should not lose sight of the actual gains or losses which can occur when a particular sequence of decisions is taken. Thus, although the criterion leads to a specific recommendation, it is worth while examining the runners-up to see if their spreads of gains or losses may not be more attractive, even though they are inferior on average.

4.11 Further reading

Bunn, D. W., *Analysis for Optimal Decisions*, John Wiley, New York (1982), Chapter 4.

Ewart, P. J., Ford, J. S. and Lin, C-Y., *Probability for Statistical Decision Making*, Prentice-Hall, New Jersey (1974), Chapter 12.

Lindley, D. V., *Making Decisions*, John Wiley, London (1971).

Luce, R. D. and Raiffa, H., *Games and Decisions*, John Wiley, New York (1958), Chapter 13.

Moore, P. G., *Risk in Business Decision*, Longman, London (1972), Chapter 5.

Moore, P. G. and Thomas, H., *The Anatomy of Decisions*, Penguin, Harmondsworth, Middlesex (1976), Chapter 3.

Radford, K. J., *Modern Management Decision Making*, Prentice-Hall, Reston, Va. (1981).

4.12 Problems and practical exercises

1 Refer to Exercise 1 in Section 3.5. Suppose that Player I can now visit customer Z, so that he now has six possible actions, identical to these of Player II. Again if the two players visit the company on the same day, all the profit goes to Player I.

Draw up the payoff matrix and evaluate the maximin, minimax regret and Laplace strategies for Player I. Which action would you recommend?

Carry out a similar analysis for Player II. Explain why the maximin and minimax regret concepts do not produce useful results. Find Player II's Laplace strategy. What would happen if both players used their Laplace strategies?

2 A food canning company has the option to agree a price at the beginning of the season for the peas that it will need for the year. If the season turns out well there will be a glut and the price will drop. Conversely a poor season will mean that the price will be higher than that offered at the beginning. There will always be sufficient peas for the company's needs and the company will always buy. Expressing the season as good, average or poor the problem can be summarized in terms of price paid (pence per kilo) as in Table 4.7.

Table 4.7

		Season	
	Good	*Average*	*Poor*
Agree price now	20	20	20
Wait	12	22	28

A further alternative being considered by the company is to agree to a price of 20p now for half of their needs and to buy the remaining half at the season's price.

Show that the maximin philosophy would cause the company to agree to the 20p price. Show also that the minimax regret argument would lead the company to the strategy of agreeing the price for half now and the other half at the season's price.

The company now wishes to consider an EMV strategy and hence estimates the probabilities π_1, π_2 of good and average seasons respectively. The probability of a poor season is therefore $1 - \pi_1 - \pi_2$.

What is the minimum value of π_1 which would persuade the company to wait?

Show that there can be no combination of π_1 and π_2 for which the half now, half later strategy can be optimum under the EMV criterion.

3 Melchester City Football Club orders programmes for its home matches in batches of 10 000, 20 000 or 30 000 at costs of £1000, £1600 or £2000 respectively. Programmes sell at 20p each. The order for each match has to be placed a week in advance and any programme not sold on the day has no value. Sales of programmes depend on the match attendance which, in turn, depends on the attractiveness of the visiting team, the weather and the date.

For simplicity assume that programme sales will be 10 000, 20 000 or 30 000. Find the maximin and minimax regret strategies for programme orders.

Next Saturday Melchester will be playing arch rival Casterbridge Athletic. The board at Melchester estimate that there is a 0.6 probability of 30 000 programme sales, 0.3 of 20 000 and 0.1 of 10 000. What is the optimum EMV strategy under these conditions?

4 A pharmaceutical firm has developed a food additive which it believes could inhibit tooth decay. Clinical trials are about to begin. If the firm goes ahead with the trials the cost will be £1 million. This will be lost completely if the trials are negative, but will result in a discounted profit (i.e. not including the cost of the clinical trial) of £4 million if they are positive.

The firm now receives an offer from an overseas company to share equally in the expense of the clinical trial on the understanding that they will have the right to market the product if it is successful. This will reduce the discounted profit for the firm to £2.5 million.

If π is the firm's assessment of the probability of success, over what range of π should the firm accept the offer of the overseas company?

The firm actually believes that π is approximately 0.5, which in the light of the answer to the above, should mean that the offer will be refused. However, they are prepared to accept a straight cash offer for the privilege of meeting half of the clinical trial costs and enjoying the marketing rights. What is the smallest offer that they should entertain?

5 The government is planning a major motorway development in an area where there is land available to build a hotel/restaurant. Local protesters are raising objections to the government's plans and it is by no means certain that they will go ahead. If the development is confirmed, the hotel will yield a profit equivalent to a present value of £20 million. On the other hand if the land is bought but the development does not go ahead the present value of the resultant loss is £5 million. There is an alternative site for this investment which will yield a return of £7 million irrespective of the government's decision on the motorway development.

What is the smallest value of the probability that the development will go ahead which will cause you to buy the land? How sensitive is your answer to the assumed return of £20 million?

6 Imagine that you are a college tutor who decides to entertain his class of 20 students on a particular summer evening. You plan an outdoor barbecue and would like to advertise it as such. You could if necessary accommodate the group inside your home and provide a cold buffet and you also have a large garage which, although rather drab, would serve as a shelter for outdoor entertainment. The barbecue itself could be protected from rain.

Use the ideas of this chapter to model the decision problem you face, and assess what you would actually do in terms of your utility and your attitude to risk.

5
Utility theory

OBJECTIVES

From this chapter the reader should obtain an understanding of the need for the concept of utility. He should appreciate that it is a concept in common use, although explicit definition of the utility function for any person or organization presents problems. The use of the concept in insurance is explained, and the meanings of risk averse and risk seeking are illustrated.

5.1 The need for the utility measure

In Section 4.6 a case was made for the use of the Expected Monetary Value (EMV) criterion to select an action. To apply this criterion we needed not only the probabilities of the various outcomes, but also the corresponding payoffs. The impression was given that these payoffs were expressed in terms of actual money. If this were the case exclusively, the criterion would be of very limited use, since there are many instances where the consequences of outcomes may not be measured unequivocally in cash flows. A firm may, for example, decide to invest in better social facilities for its employees. The motivation for this decision may be a desire on the part of the firm's executives to work with contented people, but there will presumably also be a financial return in that people work more effectively when consideration is given to their working environments. No exact measure of the actual return will be possible. In some circumstances it may be possible to obtain an estimate of the increase in profitability, but clearly this does not tell the full story. It may, in fact, be the case that the expected return is less than the actual investment in the facilities. This does not necessarily mean that the investment should not be made. There are other benefits such as the possibility of working in a pleasant environment, the improved health of the staff, increased loyalty, etc. Similar considerations arise when organizations give money to charities, sponsor sporting events or send their employees to conferences. Clearly there is a 'publicity factor' where the prominent use of the organization's name will be reflected in maintained or increased sales of its wares, but this is rarely the single return sought from such a decision. The true benefit is not so easily quantifiable.

The same argument can be put forward with even more conviction when personal decisions are contemplated. We ourselves take many actions not because of a financial benefit, but more for the enjoyment or satisfaction we derive from their outcomes. We go to cinemas, theatres, football matches, and we take holidays in places where it will cost more than living at home. We do such things in the belief that the satisfaction we derive will justify the cost, and in the full knowledge that the actual financial return will be zero. Our payoff, which we balance against a financial outlay, is measured in terms of what we perceive to be the experience resulting from the chosen action. We are making some subjective comparison between the precise amount we pay out and the enjoyment of the result, and in this sense implicitly putting a lower bound on our assessment of the value of this enjoyment. For example in attending a football match we are faced with the problem of deciding how much we shall pay to enter the ground. Different sections of the ground will have charges which range from £1 to £10, and the choice depends on a number of factors such as our enjoyment of witnessing the match, our value of comfort during the match and our alternative uses for money. At the cheaper end of the range we shall have to stand up for the duration of the game, if it rains we shall get wet, we have the risk of our view being obscured by taller spectators and we shall be in one of the less advantageous viewing positions. The more expensive entrance charge will guarantee a seat under cover close to the half-way line. The proprietors of the football ground will have done an analysis to anticipate the public's valuation of the relative comforts of the different parts of the ground, and our balance of costs against comforts is likely to be very marginal. The one aspect of the analysis which remains individual concerns the alternative uses for money. If it rains we all get wet largely to the same extent and we all enjoy the comforts of an unrestricted view, but a person with what for the time being we shall loosely describe as 'a lot of money' sees less point in enduring discomfort in order to save money than one of his less wealthy brethren. Although in practice the opportunity is not presented, each of us could derive an entrance fee which is the upper limit which we would be prepared to pay to witness the football match or theatre performance from a particular part of the arena. This is our evaluation of the monetary value of the experience.

There are two implications from the above examples. Firstly, it is possible to put a value on what may be termed a subjective experience, and secondly such values are likely to depend on the cash resources of the individual. The concept of utility is a formal expression of these phenomena. In the initial stages of the presentation of the argument for utility, we shall concentrate on the point that each individual's valuation of a payoff depends on his or her current cash resources.

5.2 The utility of money

Although the implications of this section are important when high profit or loss actions are taken, and in particular for decisions of whether or not to take out insurance against some calamitous loss, the arguments are best made in terms of some simple gambles. For example, if you, the reader, were to ask an acquaintance to play a game where one of you tossed a standard coin, and if the coin fell heads you won five pence (or cents or any low denomination of your currency) and if it fell tails your acquaintance won the same amount. He might wish to satisfy himself that the coin was fair and, if he was of a prudent nature, he might request his share of the tossings, but usually he would be prepared to indulge your whim. A little careful calculation would show that if you persisted in the pursuit for a 100 tosses of the coin, your loss or gain is very unlikely to exceed one hundred and fifty pence (or cents, etc.). (Try this as an exercise.) Thus, although it would be pleasant to emerge from this experience with a modest gain, there is no possibility of a loss beyond what most people would count as their loose change.

Now try suggesting the same game, but changing the stakes to wins or losses of one hundred pounds (dollars). The game is still fair, but unless you have chosen a particularly wealthy acquaintance for this game, he will think hard first and will probably refuse to take part. You, yourself, would probably have similar misgivings. The reason is really very simple. Either player would enjoy the prospect of winning around one thousand pounds, which would be quite feasible with random variations using a fair coin. Assuming that the enjoyment is not marred by the opponent's discomfort, such a sum could present possibilities of a better holiday, a new car, a debt repaid earlier than expected, etc. It would permit some benefit which had not otherwise been anticipated. On the other hand, an equivalent loss could be the cause of considerable embarrassment resulting in a reduced standard of living for some considerable period. Clearly then, although the gains and losses are equally balanced in terms of the chance that they will occur, the same cannot be said for the effects that they will have. If we simply followed the EMV approach, we would say that if X_n were the gain (negative values corresponding to losses) from either version of this game where the coin was tossed n times, then X_n has expected value zero and we should be indifferent as to whether or not we played the game, no matter what the stakes were. Clearly this is unsatisfactory, and although the concept of using an expected return to discriminate between actions is worth consideration, something better than the simple monetary value is required. This argument has led to the introduction of what is known as the *utility* of an outcome. In this section we are discussing only the utility of money, so that utility, if it can be defined explicitly, will be some function of x, where x represents the capital assets of the decision-maker. The variable x is therefore the amount of money which the decision-maker could realize if he sold everything he has, or if we are talking about an organization, the actual value of that organization to its owners or shareholders. Fortunately, we shall need only a rough estimate of x for any practical use of utility. The important point is that x is not simply the individual's 'spare cash' or the amount that the organization wishes to invest in some project. The utility function is written $u(x)$.

This function has two basic properties. Firstly it is not defined for negative values of x. This amounts to saying that it is not possible to have a total amount of assets which is negative. Certainly organizations and people go into debt, but such debts will be covered by some security, so that the person to whom the debt is owed can, if necessary, recover the cash value of the debt. The individual's cash assets therefore will be non-negative. No society can operate rationally if it allows individual elements to take risks where the losses would exceed their cash assets, and the utility concept is based on the assumption that its owner operates within a rational society.

The second property is that utility is a non-decreasing function of x. Plausibly we use utility as a measure of the satisfaction we derive from our capital assets. Anything other than a non-decreasing utility function would mean that we would have a positive wish to divest ourselves of assets beyond a certain limit. Few people seem to behave in this way. Certainly, people and organizations give money to charities and other worthy causes where there will be no direct return. In so doing they suffer a corresponding loss in utility, but their behaviour can be rationalized since this loss is more than balanced by the satisfaction they derive from their gesture. It does not follow that every action we take has to be designed to maximize our expected utility. Of necessity many are, but there is also room for altruism.

5.3 Derivation of a utility function

It is argued in this chapter that each individual has his own utility function. Similarly an organization also has a utility function, which in practice is defined by the attitudes of its chief executives acting on its behalf. The utility function operates in the way that the EMV operated, inasmuch as action a_1 will be preferred to action a_2 if the expected utility of a_1 is greater than that of a_2.

As a first illustration of this property of the utility function, consider a gamble where a freely spinning pointer is to end up in one of two areas in the circumference of its spin (Figure 5.1). If it ends up in the 'win' arc, the player wins £100, but he loses the same amount if it happens to come to rest in the 'lose' arc. Just one spin of the pointer is to be played.

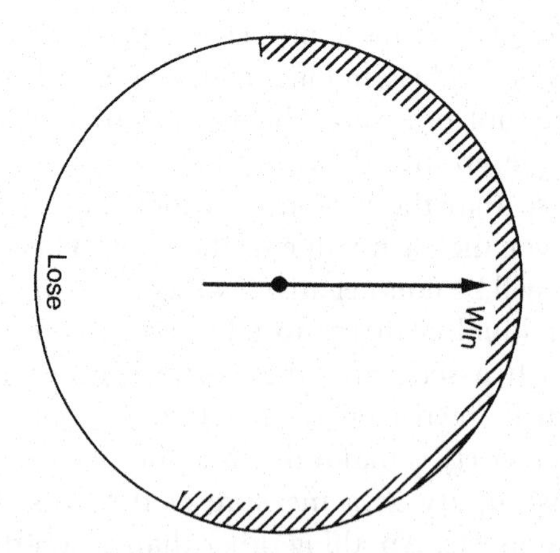

Fig. 5.1 Probability simulating spinner

In a fair game, the total circumference would be divided equally between the two arcs. Most people would, however, only play the game if the 'win' arc were longer than the 'lose' arc. Suppose, therefore, that an individual has the option of refusing the gamble. Clearly if the 'win' arc is the total circumference and if he wishes to maximize his utility, he will accept the 'gamble'. He may also be prepared to take some risk of losing. There will be a limit to this risk, expressed by the shortest length of the 'win' arc (or equivalently the longest 'lose' arc length) that he will accept. Expressing this as a proportion p of the total arc length, he is effectively saying that he is then indifferent between a gamble with a probability p of winning £100 and therefore a probability $(1-p)$ of losing £100, and keeping his assets at their current level. If this current level of assets has a monetary value of A, the expected utilities of the two options should be the same, that is

$$u(A) = p\,u(A + 100) + (1 - p)\,u(A - 100). \tag{5.1}$$

In theory, therefore, the decision-maker simply has to refer to his utility function to decide whether or not a gamble or risk should be accepted. If

$$u(A) > p\,u(A + 100) + (1 - p)\,u(A - 100)$$

the gamble would be refused; with the inequality in the opposite direction it would be accepted. In practice we have to use comparisons of preferences in the form of the above illustration in order to find out what the utility function looks like. Note also that we can choose any scale of measurement for utility in the sense that the function can be multiplied by a constant and have a further constant added to it without changing the basic property of equation (5.1). It is usual to take

$$u(0) = 0.$$

At the other end of the scale it would be convenient in making comparisons between utilities to have $u(M) = 1$, where M is the maximum attainable assets for the individual concerned. This latter concept of a maximum for the assets is difficult to accept as a specific figure. If this were not the case, the utility for any value X of the assets could be determined by asking the question 'For what probability p of gamble whose outcomes are M with probability p and 0 with probability $1 - p$ would you be indifferent to having assets X with probability 1?' Equivalently

$$\begin{aligned} u(X) &= p\,u(M) + (1 - p)\,u(0) \\ &= p. \end{aligned}$$

Mathematically this would be a neat approach, particularly as it expresses utility as a probability and thereby demonstrates its non-decreasing property. Unfortunately, the idea of a gamble whose outcomes are either untold wealth or complete financial disaster is one that we as individuals find difficult to imagine, particularly when it is to be used as a benchmark for other levels of assets. Commercial organizations would have the same difficulty. It is rather like measuring the height of a building as a fraction of the circumference of the Earth, without the benefit of more practical intermediate measure standards.

What, then, should we do in practice? The utility function $u(X)$ can conveniently be constrained to have non-negative values, and a starting point of $u(0) = 0$ is feasible. In shape it tends to flatten off as X becomes large, reflecting the common view that the satisfaction from additional wealth decreases as the level of current assets increases. In appearance, therefore, a typical utility function has the form of Figure 5.2.

A curve of this shape is said to be concave. Viewed from above 'it would not hold water.' (The opposite feature is convex or bowl-

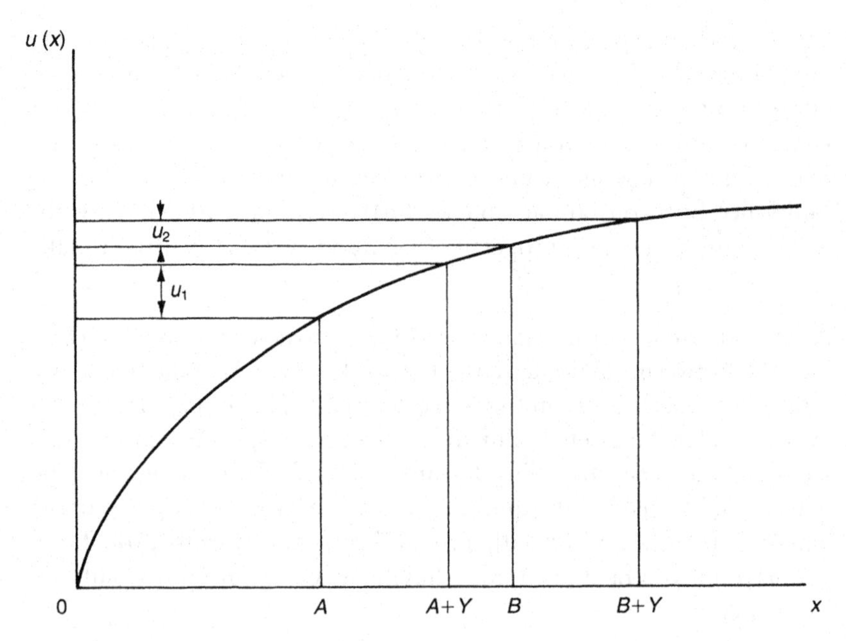

Fig. 5.2 Typical utility function

shaped.) Suppose that two individuals, Cox and Rodgers, have this same utility curve. Cox has assets at level A, whereas Rodgers' assets are at B. Both receive an additional asset of monetary value Y. The increase in utility for Cox is shown by the length U_1 on the vertical axis, whereas that for Rodgers is U_2. From the shape of the utility function it is clear that U_1 is greater than U_2, the difference reflecting the amount by which the extra money Y is more valuable to the person with less assets.

Furthermore, if Cox and Rodgers were each offered gambles where they either won and lost the amount Y with probability p and $1 - p$ respectively, Cox would require

$$u(A) \geqslant p\,u(A + Y) + (1 - p)\,u(A - Y) \tag{5.2}$$

The smallest value of p is found by solving (5.2) as an equation:

$$\begin{aligned} p &= \frac{u(A) - u(A - Y)}{u(A + Y) - u(A - Y)} \\ &= EF/GH \text{ in Figure 5.3.} \end{aligned}$$

By similar triangles EF/GH is equal to DF/DH, and this, in terms of the intercepts on the X-axis is the same as $(Y + Z)/2Y$.

$$\text{Thus } p = \tfrac{1}{2} + \frac{Z}{2Y} \tag{5.3}$$

The distance Z is defined by finding where the line from $u(X) = u(A)$ intersects the chord GD. With a curve of this shape this is always to the right of the value A, showing that p found in equation (5.3) is always greater than $\frac{1}{2}$. The individual whose utility function this is therefore always requires odds better than $\frac{1}{2}$ on a bet where the stakes are to win or lose an equal amount. Such a person is said to be *risk averse*, and his utility function would have this concave shape.

Comparison of Figures 5.2 and 5.3 demonstrates a further point. The 'Z value' for Rodgers, who has a capital asset level B greater than that of Cox, is smaller because of the reduced curvature of the utility curve. Rodgers, therefore, does not need such a large p value to be enticed into this gamble. Ultimately, when the curve flattens out to what is effectively a straight line, a value of $\frac{1}{2}$ would be acceptable. Thus the risk aversion for the same bet becomes less as the person's assets increase.

The complementary shape to the concave curve is illustrated in Figure 5.4. A similar geometrical argument to that given above shows

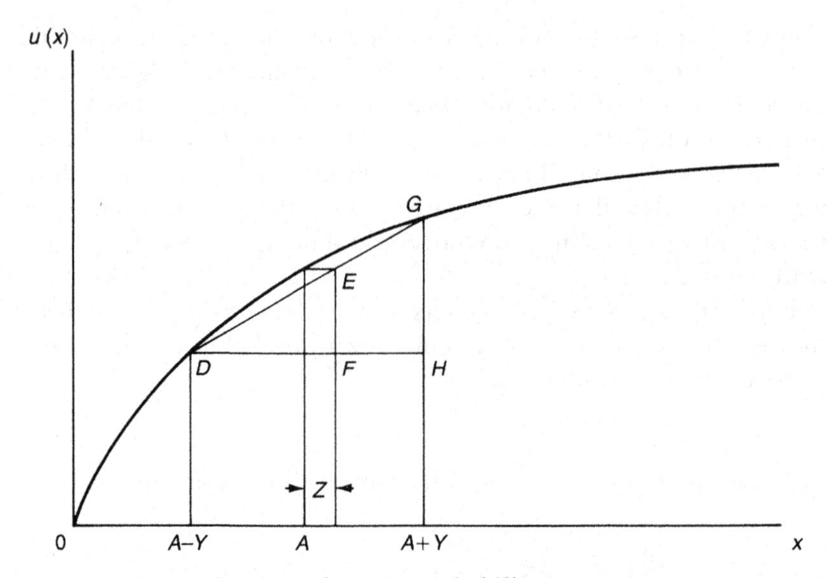

Fig. 5.3 Determination of Cox's probability

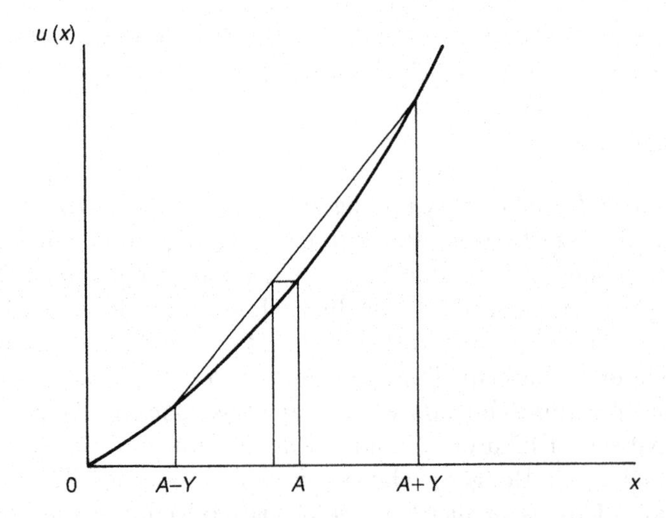

Fig. 5.4 Risk-seeking utility curve

that a person with such a utility curve would be prepared to accept a gamble in which the gain or loss would be the same, but where the probability of the gain were less than $\frac{1}{2}$. Such a person is said to be *risk seeking*. Such persons do exist. It is not uncommon for an individual at the lower end (hopefully!) of his monetary scale, to take what in terms of expected monetary value are bad risks. This is in order to extricate himself from this region of relative poverty. Often this is not successful, but once having reached a stage of reasonable wealth, the individual then becomes risk averse to maintain this position. His complete utility curve has the appearance shown in Figure 5.5. This could explain, for example, why many small firm owners who in their early days were prepared to take risks to promote their enterprise, become reluctant to take further risks once they are established.

It should therefore be possible to determine a utility curve by asking a large number of questions about the risks associated with different projects. Starting with a utility (say of 0.5) assigned to the current level of assets and a utility of zero assigned to zero assets, a unique curve can in theory be determined. In practice there are likely to be inconsistencies since the individual is only approximating to his utility when assessing the limiting probability for a particular project or gamble. It is better, therefore, to use this method in a limited way to obtain a plausible shape for the utility curve, following which the probability answers can be read off for further project questions, and checked by the individual for acceptability. We return to this approach in Section 5.5.

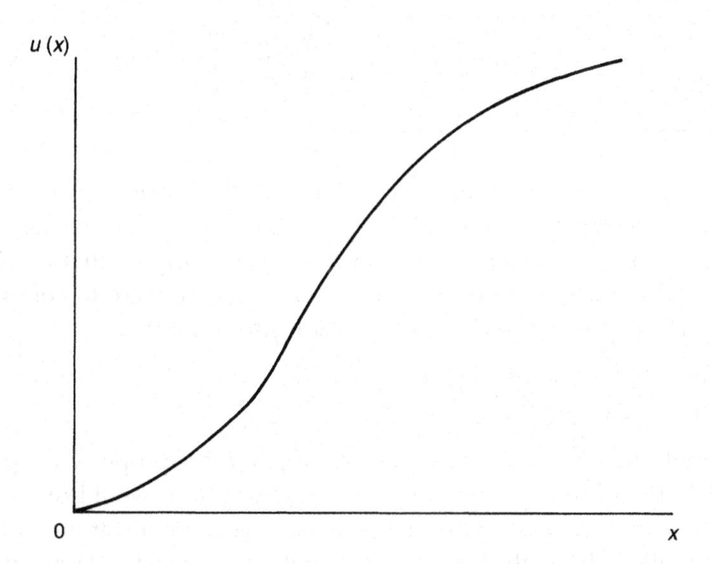

Fig. 5.5 Utility curve progressing from risk seeking to risk averse

5.4 Application to insurance

A decision common to many commercial organizations as well as to private individuals is that of taking out insurance against some major loss. For the individual it is usually on a life or on some expensive piece of property such as a house or car. The organization may similarly insure against fire or against damage caused by professional negligence.

For simplicity, take the case of the individual insuring his house against fire or damage generally. Suppose also that his current assets are of monetary value A, of which a substantial proportion consists of his house. If he decides to insure he will have to pay a premium of i for the following year. We shall assume that if there is a fire during this time, he is compensated in full. If he does not insure and there is a fire, the reduction in his assets is given by an amount B, where $B < A$.

Two actions are available: insure or not insure. We assume that if he decides not to insure he will not reconsider the problem for a further year. Nature also has two actions: fire or no fire within the next year. Expressed as a game, we have the outcomes in Table 5.1. The action

Table 5.1

	Fire	*No fire*
Insure	$A - i$	$A - i$
Do not insure	$A - B$	A

taken depends on the probability assigned to the outcome 'fire'. Note that a maximin strategy would argue in favour of insuring for any premium i up to the actual value of the house! Suppose that a probability of α is assigned to the 'fire' outcome. If we were to compare the EMV of the two actions we would insure only if

$$A - i \geqslant A - \alpha B,$$
that is if $\alpha > i/B$.

Usually this is not the case. For example, for a property of value £50 000, the insurance premium could be £100. This would require a probability of at least $\frac{1}{500}$ of a fire in any year for insurance to be worth while. Although this figure is small it is somewhat larger than the current level of risk, and with this information insurance would not be taken, even allowing for the administrative costs of the insurance company. Two factors persuade people to take out insurance. One is that this is a 'one-off' situation as far as the customer is concerned and therefore the argument of the long-term average return is inappropriate. The other is the utility function of the individual, where the effects of losing the majority of assets built up over a lifetime outweigh the monetary value of the loss. Thus, most people are prepared to pay a premium in excess of the EMV of the insurance.

In terms of utilities, we would insure if

$$u(A - i) \geqslant (1 - \alpha)\, u(A) + \alpha u(A - B),$$

or if

$$\alpha \geqslant \frac{u(A) - u(A - i)}{u(A) - u(A - B)}.$$

Figure 5.6 shows the utility curve of a risk-averse person, showing that if $\alpha \geqslant DE/DG$, he should take out insurance. If he were to rely on the EMV criterion he would require $\alpha \geqslant DF/DG$, a somewhat larger lower limit. Note, though, that these two ratios become closer when the utility curve is in its linear phase when X is large, and also when the loss insured against is small. The lesson from this is that, bearing in mind the fact that the insuring company will have to make some charge for administrative costs, insurance should not be taken

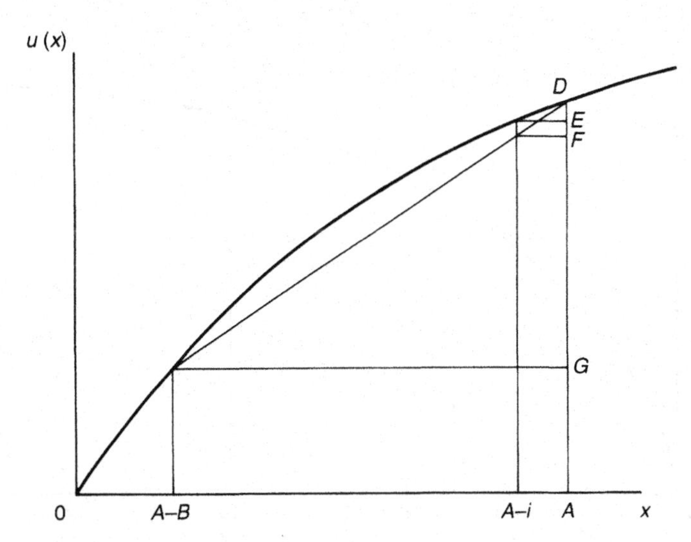

Fig. 5.6 The insurance gamble for a risk-averse person

out for anything other than a significantly disastrous loss, particularly when the prospective insurer has assets which place him on the linear part of his utility curve.

It is therefore possible for a person to take insurance which shows a positive expected utility. At the same time the insurance company can also show a positive expected utility. This is because the insurance company has very large capital assets, so that even if the company has to pay out on a claim for, say, £50 000 it will still be on the linear part of its utility curve. The insurance company can in effect be guided by the EMV criterion. For example, in the case of the £100 premium for a property valued at £50 000 the actual probability of such a fire in any year might be 0.001. The EMV of this arrangement is therefore

$$£100 - (0.001)\ £50\,000 = £50.$$

This is effectively the utility of the action of the insurance company. The contribution to overheads has to be taken from this amount, but there will still be a sufficient profit. The client is also receiving a positive expected utility, and hence it appears that both parties gain. The introduction of utilities has destroyed the 'zero-sum' aspect of the decision.

There are occasions when an insurance company is asked to insure a risk where the amount insured is a significant part of the insurance

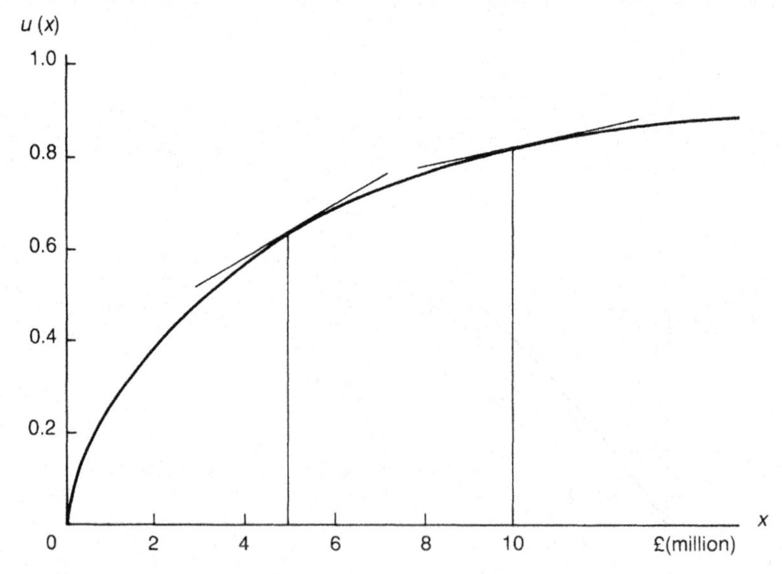

Fig. 5.7 Utility curve for insurance company

company's assets. In this case it is often prudent to share the risk with another company. Suppose, for example, an insurance company with assets equal to £10 million is asked to accept insurance on a property of value £5 million. The company's utility curve is shown in Figure 5.7. The company estimates that there is a probability 0.01 that it will be called upon for the amount insured. What premium should it charge? The EMV premium would be £50 000, but since a reduction in assets by the large amount would position the company away from the linear section of the curve, clearly this is not enough. Suppose that the premium is to be £y million. Then y must satisfy the relationship:

$$0.99\,u(10+y) + 0.01\,u(5+y) \geqslant u(10). \tag{5.4}$$

This can be solved for y by trial and error. More simply if we take linear (tangential) approximations to the curve at $X = 10$ and at $X = 5$, we have a linear equation to solve.

$$\begin{aligned} &\text{At } X = 10 \qquad u(10+y) = 0.820 + 0.026y \\ &\text{At } X = \ \ 5 \qquad u(\ 5+y) = 0.648 + 0.062y. \end{aligned}$$

The approximations are valid since the premium is a small fraction of the amount insured. Equation (5.4) now becomes

$$(0.99)\ [0.820 + 0.026y] + (0.01)\ [0.648 + 0.062y] \geqslant 0.820$$
$$y \geqslant 0.06525$$

The minimum premium should therefore be £65 250.

Now suppose that the insurance company has an arrangement to share the risks and premiums with another company. To keep matters simple, suppose also that this second company operates under the same utility function and currently has the same assets of £10 million. What premium should now be charged?

The argument follows similar lines to the above, except that we are now concerned with the possibility of a payment of £2.5 million. The equation to be solved is

$$0.99\ u(10 + z) + 0.01\ u(7.5 + z) \geqslant u(10).$$

The utility curve at a current asset level of £7.5 million can be approximated by

$$u(7.5 + z) = 0.752 + 0.038z,$$

yielding a solution $z \geqslant$ £26 030.

The minimum combined premium is double this, that is £52 060, a reduction of £13 190 on the single firm's premium. Thus, provided that the increase in administrative charges is less than this amount it will be possible to make a more attractive offer of a premium to the prospective customer if the risk is shared in this way. Note that if the second firm operates under a different utility curve or has a different level of current assets, the approach merely requires a second calculation of the same form. If the second firm happens to be operating further into the linear part of its utility curve, it should be able to offer a smaller premium than the original company.

5.5 Example of a utility function

In Section 5.3, an outline was given of a way in which the utility function of a company may be derived. This was in terms of gambles – their gains, losses and associated probabilities. Although mathematically speaking there is no difference between putting money on the spin of a roulette wheel and investing in the launch of a new product, in formulating the questions appropriate to the utility function it is advisable to use the latter context. First we should assess the current assets of the organization. To do this we would rely on the senior accountant, having explained the purpose of this need. Suppose, therefore, that we come to a figure of, say, £15 million. To this a utility value can be assigned arbitrarily. If we wish to operate in the range [0,1] a value of 0.6 could be appropriate if there is some considerable potential for growth. Otherwise a somewhat higher value of, say, 0.7 or 0.8 might be more acceptable. As mentioned previously, there is no real need to keep utilities in the [0,1] range, but intuitively it seems appropriate to have a target of 1 for the ultimate utility.

One way of proceeding is to think in terms of insurance premiums. Starting with some of the more modest losses, the chief executive (or his delegated deputy) might be asked

(a) What would be the actual cost?
(b) What is the chance that the loss would occur over the next year?
(c) Would you take out insurance against this loss, and if so, what would be the largest annual premium you would pay?

The argument now is based on the assumption that firms do not insure against what are, *relative to their* current assets, small losses. The reduction in their assets caused by such a loss is a risk they are prepared to take, virtually acting as their own insurance company. In utility terms, this means that over a reasonably well-defined range around this level of current assets, their utility curve is linear. If, therefore, this range can be established, it is likely that any insurance premium will fall within the range and the parameters of the linear relationship can be established by considering a reduction of the assets to zero.

Suppose, therefore, that we can ask the question about the level of total insurance against complete disaster within the company. (We first have to define the sort of calamity – fire, explosion, gross professional negligence – which would cause the company to lose all of its assets.) In the example an annual premium of £40 000 might be deemed appropriate, with a corresponding probability of occurrence of 0.001. The utility equation is therefore

$$0.001u(0) + 0.999u(15) = u(14.96).$$

Thus,

$$u(14.96) = (0.999)\ (0.6) = 0.5994.$$

Provided, therefore, that an asset level of £14.96 million is within the linear range of the utility function, we can now derive its parameters

$$u(X) = \alpha + \beta X,$$

$$\text{since } u(15) = 0.6 = \alpha + 15\beta$$

$$u(14.96) = 0.5994 = \alpha + 14.96\beta.$$

Solving these equations gives $\alpha = 0.375$, $\beta = 0.015$.

Payment of the annual premium of £40 000 represents a loss in assets. Similar losses will be incurred when, say, a delivery vehicle or a moderately priced piece of production equipment is lost through accident. The firm may well decide that to take out insurance against such a loss would not be worth doing, since the premium would be greater than the corresponding expected loss. The implication is that the utility function is approximately linear over this range of losses.

We now turn to losses where insurance would be considered, and obtain answers to questions (a), (b) and (c). For example, these might be as shown in Table 5.2. Note that these are not necessarily incidents against which it would be possible to take out insurance. The premium is the annual amount which the company would be prepared to pay to eliminate the consequence of the incident. Each item in this table now provides a point on the utility curve. For example, the first incident implies the equation

Table 5.2

Incident	*Loss (£m)*	*Probability*	*Max. annual premium (£)*
small building fire	0.5	0.005	3 500
factory explosion	1.0	0.001	1 300
large building fire	4.0	0.005	28 000
supplier strike	1.5	0.010	18 000
internal strike	2.0	0.010	24 000
product obsolescence	7.0	0.005	55 000

$$0.005u(15 - 0.5) + 0.995u(15) = u(15 - 0.0035),$$
$$\text{or } 0.005u(14.5) + (0.995)\ (0.6) = 0.375 + (14.9965)\ (0.015),$$

since the utility on the right-hand side of the equation falls within the linear range. Solving this we have

$$u(14.5) = 0.5895.$$

Similar equations yield utilities at other points (Table 5.3). When

Table 5.3

X	15	14.5	14	11	13.5	13	8
$u(X)$	0.6	0.5895	0.5805	0.5160	0.5730	0.5640	0.4350

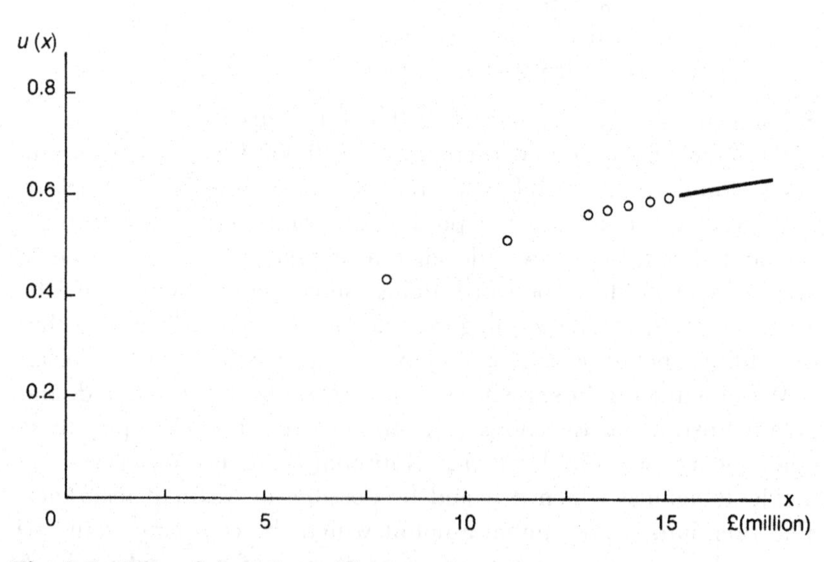

Fig. 5.8 Plot of data for utility curve

these points and the linear equation are plotted (Figure 5.8), a reasonable sketch of the utility curve can be attempted (Figure 5.9). Clearly, there will be serious gaps between the plotted points and it is unlikely that a smooth curve will go exactly through all of the points. Moreover, the method does not extrapolate beyond the level of current assets. Despite these drawbacks, this in outline would appear to be a reasonable approach to obtaining a first approximation to the utility function. It is perhaps rather sensitive to the initial determination of linear segment around the level of current assets. It is important, therefore, that the curve be checked for plausibility with the individuals concerned.

The fact that the curve can sensibly only be drawn marginally beyond the level of current assets is not too serious a drawback, since firms rarely increase their assets substantially in a short period of time. Regrettably, it is possible to lose large amounts of assets quite quickly, but it is unusual to indulge in the type of gamble which adds appreciably to the assets within, say, a year. As the assets do increase it would, of course, be possible to carry out this exercise for an increased current asset level.

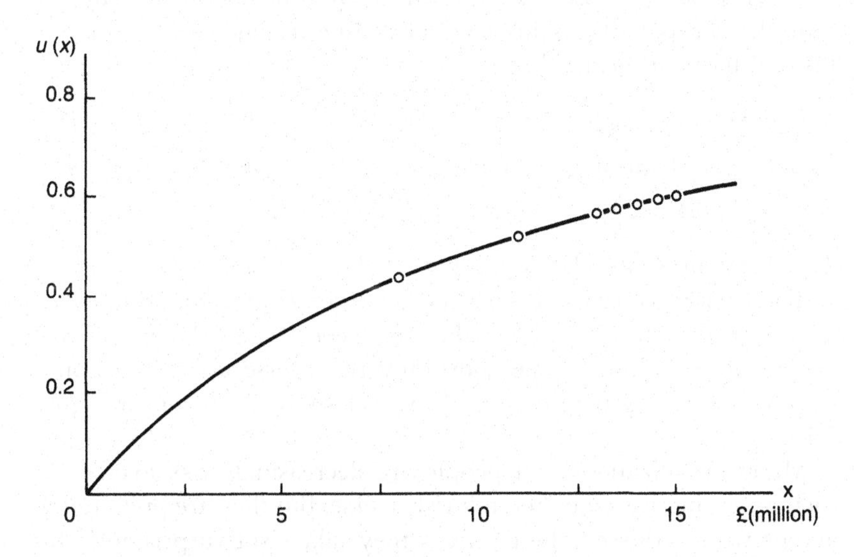

Fig. 5.9 A fitted utility curve

5.6 General comments

The approach offered in the previous section can be criticized on the grounds that it gives a spurious precision to an imprecise concept. There is no doubt that there is such a matter as utility, and that it varies from individual to individual. People's behaviour regularly exhibits risk aversion or risk seeking, and although they do not carry out the calculations to determine the risks that they will take or the price that they will pay for non-quantifiable benefits, people can be quite precise about the relationship between specific benefits and their costs. Governments and manufacturers know this when taxation or price rises on non-essential goods such as beer or tobacco are contemplated. The example of Section 5.5 is imaginary and perhaps idealized. Few analysts will deal with so receptive a client, and different circumstances will require different ploys to tease out the necessary clues to propose an actual function. An alternative approach is to postulate a pattern of behaviour which then specifies the form of the utility function. For example, the decision-maker could be 'constantly risk averse'. This means that for any gamble where there is either a gain of $+a$ or a loss of $-a$, the decision-maker would require a probability p of winning, where p depends on a but is independent of the current level of assets, X. This means that he does not change his attitude towards risk as his level of prosperity changes. In mathematical terms it means that

$$u(X) = pu(X+a) + (1-p)u(X-a) \tag{5.5}$$

for any X. Substituting the form $u(X) = 1 - e^{-\alpha X}$ in (5.5) yields

$$1 - e^{-\alpha X} = p[1 - e^{-\alpha(X+a)}] + (1-p)\,[1 - e^{-\alpha(X-a)}]$$

The terms in X cancel, leaving $p = (1 + e^{-\alpha a})^{-1}$.

The positive constant α is a scale factor for X, and is used to locate an arbitrarily chosen utility value (between 0 and 1) for, say, the current level of assets. The constantly risk-averse utility function is therefore uniquely determined. A typical shape is shown in Figure 5.10.

Many organizations and people are decreasingly risk averse. As their assets increase they remain risk averse, but they are prepared to accept worse odds on the gambles they take. Not surprisingly this rather imprecise definition of behaviour does not produce a unique utility function. It has the same general concave shape as the con-

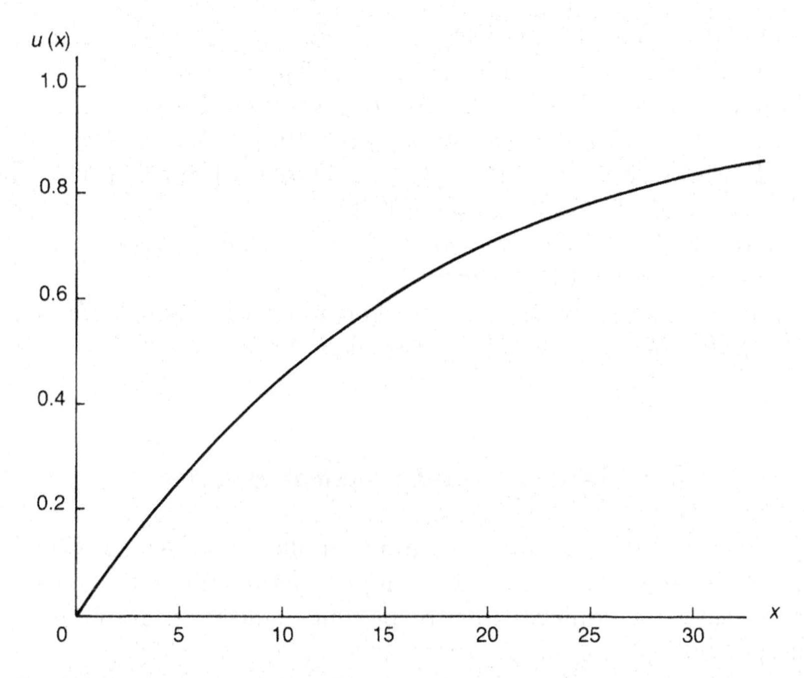

Fig. 5.10 Constant risk utility function $\alpha = 0.061$

stantly risk-averse function, but it tends to rise more steeply for low values of X, and then to flatten out. It is possible to postulate a number of algebraic forms for such curves but they tend to have rather more defining parameters (compared with the single value α for the constantly risk-averse function) and it is probably just as satisfactory to sketch such a curve by hand if the equivalent number of estimating points on the curve are available.

5.7 Further reading

Hertz, D. B. and Thomas, H., *Risk Analysis and its Applications*, John Wiley, Chichester (1983), Chapter 4.

Lindley, D. V., *Making Decisions*, John Wiley, London (1971), Chapter 5 and the Appendix.

Moore, P. G., *Risk in Business Decision*, Longman, London (1972), Chapter 12.

Moore, P. G. and Thomas, H., *The Anatomy of Decisions*, Penguin, Harmondsworth, Middlesex (1976), Chapter 9.

Mosteller, F. and Nogee, P., 'An Experimental Measurement of Utility', *Journal of Political Economy* **59** (1951). (Also reprinted in Edwards, W. and Tversky, A., *Decision Making*, Penguin, Harmondsworth, Middlesex (1967).)

Radford, K. J., *Modern Managerial Decision Making*, Prentice-Hall, Reston, Va. (1981), Chapter 2.

Swalm, R. O., 'Utility Theory – Insights into Risk Taking', *Harvard Business Review*, November–December 1966.

5.8 Problems and practical exercises

The data in Table 5.4 are taken from a utility curve which will be used in Exercises 1, 2 and 3. In using the information intermediate points can be determined either by plotting an accurate graph or by interpolation.

Table 5.4

x	$u(x)$	x	$u(x)$	x	$u(x)$	x	$u(x)$
0	0	10	0.503	20	0.798	30	0.9328
1	0.059	11	0.542	21	0.817	31	0.9404
2	0.117	12	0.578	22	0.835	32	0.9473
3	0.172	13	0.613	23	0.852	33	0.9535
4	0.226	14	0.645	24	0.867	34	0.9591
5	0.277	15	0.675	25	0.881	35	0.9640
6	0.327	16	0.704	26	0.893	36	0.9684
7	0.374	17	0.730	27	0.905	37	0.9724
8	0.420	18	0.754	28	0.915	38	0.9759
9	0.463	19	0.777	29	0.924	39	0.9790
						40	0.9817

1 An insurance firm with current assets at £30 million is asked to accept a policy on the successful launch of a space satellite. The sum insured is to be £10 million and informed opinion tells the firm that the chance of a successful launch is 0.95. What should be the minimum premium charged?

Use the given table of utilities where x is measured in £ million.

Now suppose that the risk can be shared with a similar insurance firm with the same level of current assets and the same utility curve. What would be an appropriate premium?

Given that the second firm had assets of £35 million, show that the premium required would increase. Why is this so?

2 A firm with current assets of £15 million is considering two projects. Project *A* will produce a return of £5 million with probability 0.4, and of £1 million with probability 0.6. Project *B* will produce a return of £7 million also with probability 0.4, and a zero return with probability 0.6.

Assuming the form of the utility function at the beginning of the chapter, show how the firm's decision would change if maximum expected utility were substituted as the criterion for maximum EMV.

3 Using the given utility function, what is the largest level of assets at which a gamble: win 4 with probability 0.6, lose 4 with probability 0.4, would be considered?

4 A property owner has six buildings valued at a total of £45 million. He has taken out insurance against fire on each building and estimated the probability that any building will be burned down in any year. Table 5.5 summarizes the amount insured, the annual premium and the probability of loss.

Table 5.5

Building	*Amount insured* £million	*Annual premium* £	*Probability of loss in any year*
A	11	20 000	0.001
B	7	15 000	0.001
C	2	15 000	0.005
D	1	1 500	0.001
E	4	7 500	0.001
F	20	40 000	0.001

Assuming that the utility for the total amount £45 million is 0.6 and for the residual total when building *F* is lost (i.e. £25 million) is 0.4, calculate the utilities at the values corresponding to the loss of each building. Assume that the utility function is linear in the region covered by the payments of the annual premiums.

Do these figures suggest any anomalies in the premiums paid? Also what is the effect of different choices of the base utilities for £45 million and £25 million?

5 Some countries raise funds by national lotteries. Why are such lotteries popular?

6 Why does the average person insure his house and his car but not his wrist watch?

6
Decision trees

OBJECTIVES

This chapter uses material from the previous chapters to introduce a simple graphical way of analysing a decision problem. By the end of the chapter the reader should be able to:

(a) draw up a decision tree
(b) allocate costs and benefits
(c) determine probabilities of the tree branches
(d) find the maximum EMV strategy
(e) carry out a simple sensitivity analysis on costs and probabilities
(f) carry out a simple stochastic analysis.

6.1 The sequential nature of decisions

Few decisions are taken in isolation in the sense that an action is selected in order to optimize a single immediate outcome. Even with gambling it is usually the case that the gains or losses resulting from one 'spin of the wheel' influence the stakes to be placed on the next spin, and therefore in deciding how any particular gamble is to be taken, the player is thinking ahead to the subsequent plays. Certainly in most political and industrial negotiation situations, the participants should be thinking ahead over several rounds of action selection. For example, when in the Autumn of 1939 Great Britain and France decided to present Germany with the choice of withdrawing her troops from Poland or being in a state of war, they were aware that Germany would almost certainly take the latter course of action. Similarly, when the United States took a similarly positive step against Cuba at the time of the missile crisis in 1962, they were also prepared for whatever choice Cuba would make. On the industrial scene, the decision to close a factory may lead to a variety of possible actions by the work force, ranging from full acquiescence to a worker occupation.

Dependent on the action taken, management's next round of actions could consist of taking legal action, waiting, dismissing employees, re-opening the factory, etc. The sequence of actions and counter-actions could go on for several rounds, and a good decision-maker anticipates the counter-measures which his actions are likely to provoke at each round of the confrontation.

In this chapter we shall examine decision outcomes where there is a finite number of rounds, at the end of which an objective consequence in the form of a gain or loss in utility can be assessed. The model is appropriate for more general decision situations similar to the examples described in the previous paragraph where consequences (even in the form of utilities) are difficult to measure to the satisfaction of all parties concerned. In such cases the analysis will have to be more subjective than that which now follows.

6.2 Drawing up a tree

The Moss Side Molasses Mining Company is considering the suggestion that the food additive 'plaquoff' will, when added to its sugar products, inhibit the development of dental caries in human beings, particularly among adolescents. Laboratory tests conducted so far have been encouraging and the decision has now to be taken whether or not to proceed further with the testing of plaquoff.

Before plaquoff can be incorporated into the firm's products, it must be established that it is both safe and effective. This is achieved by carrying out further and more extensive laboratory trials, following which (if it is deemed advisable) clinical trials will be held using controlled groups of human beings. The clinical trials will only be permitted if the toxicity aspects of the further laboratory trials prove to be satisfactory.

The company has three stages of action-taking. At the first stage it can decide to go ahead with the further laboratory trials or to abandon the project completely. If it decides to go ahead, it will have the results of the trials on hand when it comes to the second stage. Here it can decide to go ahead once more, only this time with the clinical trials, or again to abandon the project. Certainly if the toxicity results are unsatisfactory it will have to abandon, but there may also be a case for abandoning if the effectiveness of plaquoff now appears to be doubtful. Finally, following the clinical trials, the decision will then have to be taken to launch the product or to abandon it.

Let us now return to the first decision. Since there are two actions available to the decision-maker this can be shown as two branches coming from an initial node (see Figure 6.1). Note that a square has been used to depict the node. This is to distinguish the node from nodes where actions are to be taken which are not under the control of the decision-maker, in which case the nodes will be depicted by circles.

Although the action 'abandon plaquoff' terminates the decision process as far as this project is concerned, the alternative of carrying out further laboratory trials now produces outcomes concerning the additive's toxicity and its efficacy. These outcomes could be demonstrated in the tree as branches from a single node, but in this case it is easier to follow if the two stages are treated separately. It has been assumed that the toxicity outcome will be either 'acceptable' or 'unacceptable', the latter classification requiring termination of the project. The efficacy outcomes have been agreed as 'good', 'moderate' or 'bad', with clear implications for the commercial viability of the food additive. The decision tree is now as shown in Figure 6.2. Note that since an unacceptable toxicity means that the project ends, there is no point in extending the tree by the efficacy outcomes from the terminal node.

Further stages of actions and outcomes now follow from the interpretation of the project description. It is assumed that the clinical trials will have outcomes which can be simplified to the dichotomy 'positive' or 'neutral' (by this time it can be assumed that the additive cannot be harmful). Finally, there is the outcome describing the economic viability of plaquoff. Again for simplicity it will be assumed

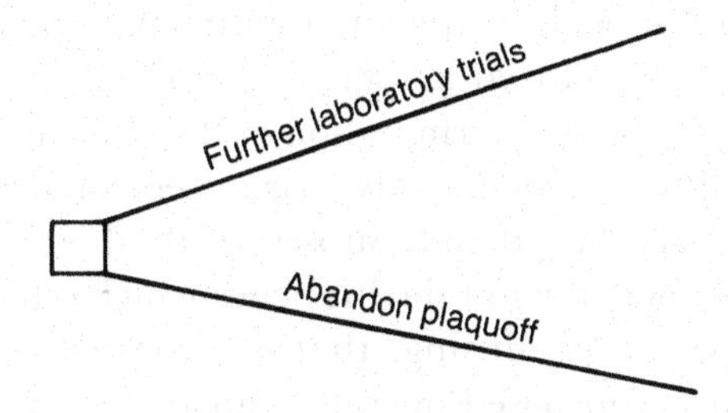

Fig. 6.1 The first decision

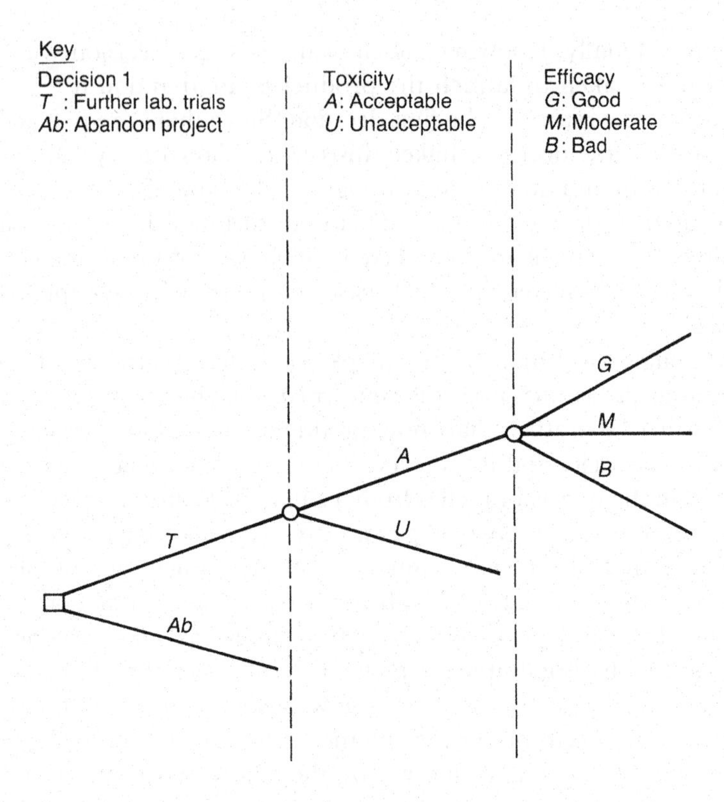

Fig. 6.2 Consequences of the first decision

that this can be represented as 'success' or 'failure'. The complete tree is shown in Figure 6.3.

The tree, like any other mathematical model, is a simplification of the actual situation. Often there are opportunities to change an action at an intermediate stage. A clinical trial may, for example, run for three years, and if it is clear at some intermediate point of time that the additive is not going to work, then obviously a decision to abandon is taken before the planned termination of the clinical trial. To incorporate all such possibilities into the basic tree could complicate the analysis with unnecessary detail. Subsequently there could be a case for more detailed modelling of the branches which our initial analysis will identify as being promising, that is of branches which are included in the initial proposed overall strategy.

There is also the problem of how much to include in the decision tree. In the discussion above it was pointed out that it was unnecessary

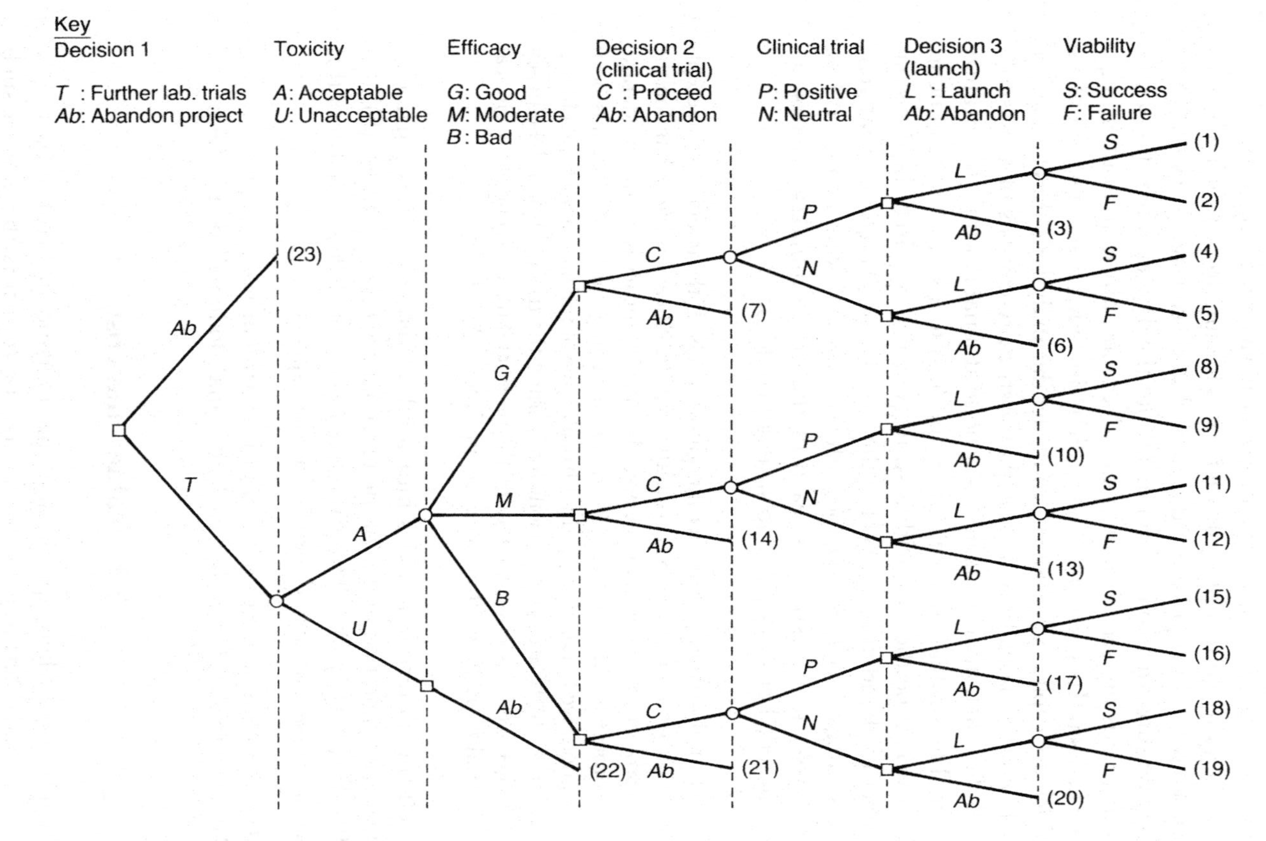

Fig. 6.3 Decision tree, plaquoff project

to consider the efficacy outcomes of the laboratory trials if the toxicity was unacceptable. A similar argument might have been applied to the 'bad' outcome when the efficacy itself was considered. Why should the company contemplate further trials leading to a launch when the further laboratory trials produced such a discouraging result? There may be substance to this argument, but the danger is that the *decision analyst* finds himself actually taking decisions, albeit in a rather negative fashion. It is generally advisable to include all actions, save those which are explicitly forbidden, and to let the analysis do the elimination.

It now remains to point out that a strategy consists of a prescription of the action to be taken at any node attainable within the strategy itself. The strategy may be simple and lead to a single terminal node. This would be the case with a strategy which required the decision-maker to abandon the project at the initial action stage. Alternatively, the strategy may have to cope with a number of eventualities. It has, of course, to be explicit at each of the decision-maker's nodes encountered. For example, one strategy could be:

> Initially, carry out further laboratory trials. If the toxicity is unacceptable or if the toxicity is acceptable but the efficacy is bad, abandon the project. If the toxicity is acceptable and the efficacy is good or moderate, proceed to the clinical trial. At the next stage, if the efficacy has been good, we launch the product irrespective of the outcome of the clinical trial (there are other reasons for holding the trial). Also, if the efficacy has been moderate, we launch the product if the outcome of the clinical trial is positive. Otherwise the project is abandoned.

Of the 23 terminal nodes indicated in Figure 6.3, this particular strategy would end at one of the terminal nodes (1), (2), (4), (5), (8), (9), (13), (21) or (22). The attractiveness to the decision-maker of this or any other strategy depends on two aspects:

(a) the consequences at each of its terminal nodes, and
(b) the probabilities that each terminal node will be reached.

6.3 Values and costs

Each branch, where appropriate, has an associated cost or return, so that for each terminal node the cost or return is the sum of such values for its component branches. It is convenient to associate these costs with the intermediate branches rather than attempt to assess each total separately.

Since the object is to find what will be in some sense the best strategy it is only necessary to include relative or marginal costs. Costs, such as general overheads, which are incurred no matter what decision is taken can be omitted from the analysis, although they will, of course, affect the financial viability of the organization. In the problem facing the Moss Side Molasses Mining Company for example, it could be asserted that the cost of abandoning the plaquoff project initially is zero. Certainly some costs will already have been incurred, but these are losses which cannot be recovered directly no matter what action is now taken. Suppose that the company estimates that the further laboratory trials will cost £50 000. For the moment we shall assume that all costs can be estimated and no explicit allowance will be made for errors in these estimates. A fuller account of the modifications which can be introduced when the costs are treated as random variables is given in Section 6.7. Usually a fairly rough estimate of a cost will be sufficient. If ultimately the choice between strategies is close, it may be thought necessary to investigate a cost further or to carry out a sensitivity analysis on the cost (again see Section 6.6). It may be of course that the choice simply is close, and refining the costs is only giving a spurious objectivity to a choice where there is no real difference. A photo finish in a horse race may be of interest to the punters, but it does not really establish that one horse is significantly faster than another over the distance.

The laboratory trials will take a year. Over such a relatively short period it is questionable whether or not to take into account the 'time effect' of money. Certainly a sum spent at the end of the year actually costs less than the same sum spent at the beginning, if only because of the argument that it could have been invested in the mean time providing the company with benefit of an interest return. Formally it is possible to discount all costs and values to the present, so that fair comparisons can be made. The procedure is known as the Net Present Value (NPV) method. In its simplest form cash flows (either gains or losses) are assumed to occur at particular points in time – usually at the end of each year – and are discounted to the present by what is termed 'an agreed minimum earnings rate'. The implication of this is that the company would expect to have a return on capital investment projects of at least this assumed rate. If, therefore, there are cash flows of C_0 initially and $C_1, C_2, \ldots, C_n$ over the n years of the life of the project, then the net present value is given by

$$\mathrm{NPV} = C_0 + \frac{C_1}{1+r} + \frac{C_2}{(1+r)^2} + \ldots + \frac{C_n}{(1+r)^n}, \qquad (6.1)$$

where $100r\%$ is the minimum earnings rate or discount factor. A fuller discussion of this and other methods of investment appraisal will be given in Chapter 13.

Suppose, therefore, that the company decides on a discount rate of 10 per cent and makes the assumption that half of the costs of the further laboratory trials will be incurred at the beginning of the year, the remainder at the end. Using equation (6.1),

$$\mathrm{NPV} = 25\,000 + 25\,000/1.1 = 47\,727.$$

Turning now to the costs of the clinical trials, these will take three years; if the decision to go ahead with the trials is taken, they will start immediately the laboratory trials finish, that is in years 2, 3 and 4. The estimated costs involved are £150 000, £175 000 and £200 000 respectively. Allowance has been made for inflation and for incremental changes in salaries. It is further assumed that these costs are incurred at the beginnings of the corresponding years, resulting in

$$\mathrm{NPV} = 150\,000/1.1 + 175\,000/(1.1)^2 + 200\,000/(1.1)^3 = 431\,255.$$

The other monetary values involved in the project are the costs of launch and the return from sales of products with the plaquoff additive. The launch will involve newspaper and television advertising as well as display material for the retail outlets. If the launch is to go ahead it will happen four years hence. The estimated cost at that time is £500 000. Discounted at the same rate this amounts to

$$500\,000/(1.1)^4 = 341\,507.$$

Resolving the problem of determining the monetary return when the additive is either a success or a failure is rather more difficult. If plaquoff is a success, profits from the company's products will be enhanced for several years, until presumably a further technological advance in this area is made. On the other hand, failure will result in reduced profits and also there is the possibility that the public may lose confidence in the company's products. To place a monetary value on either of these outcomes is bound to involve some subjectivity.

One approach is to borrow the concepts described in the previous chapter (Section 5.5) where questions were posed about the appropriate level of insurance premium for a given circumstance. In the example now being considered, we might take the abandonment of the project as the origin of gains or losses. If the project is abandoned at any time before the launch, the company anticipates sales of its products at a level governed by the company's estimates of market trends. The company now has to answer two questions.

(a) Suppose that plaquoff is successful. Somebody now comes and offers to buy the secret of plaquoff on the understanding that neither he nor Moss Side Molasses will ever manufacture the substance. It will disappear completely from the market. What is the minimum amount that Moss Side would accept?

(b) Now suppose that plaquoff is a failure. An offer now comes to 'turn the clock back' so that, as far as the public is concerned, plaquoff never existed. How much would Moss Side be prepared to pay for this piece of magic?

The answers to (a) and (b) are, of course, the gain and loss respectively corresponding to success and failure of plaquoff. They really amount to asking how much success is worth and how much failure will cost, in each case measured against the return from continuing with the existing products. The questions have been phrased to avoid any discussion of the likelihoods of failure or success and in this sense the argument differs from that of the insurance premium. We are not asking about insurance against failure. To do this would involve not only the assessment of the cost of failure but also of its probability. Probabilities will be dealt with in the following section.

Suppose that in the example the answer to (a) is £5 000 000 and to (b) is £2 000 000. Discounted to the present time these are £3 415 067 and £1 366 027 respectively. We are now in a position to place costs and values on all appropriate branches on the tree, as shown in Figure 6.4. Note that costs are shown with negative values.

Figure 6.4 has been augmented by calculating the total of all costs along each branch. One of these costs will occur when the decision strategy is taken. By a careful choice of strategy the decision-maker can ensure that it is one from a restricted and in some sense optimal set. Knowledge of the probabilities that the branches from nodes not under his control will be selected then

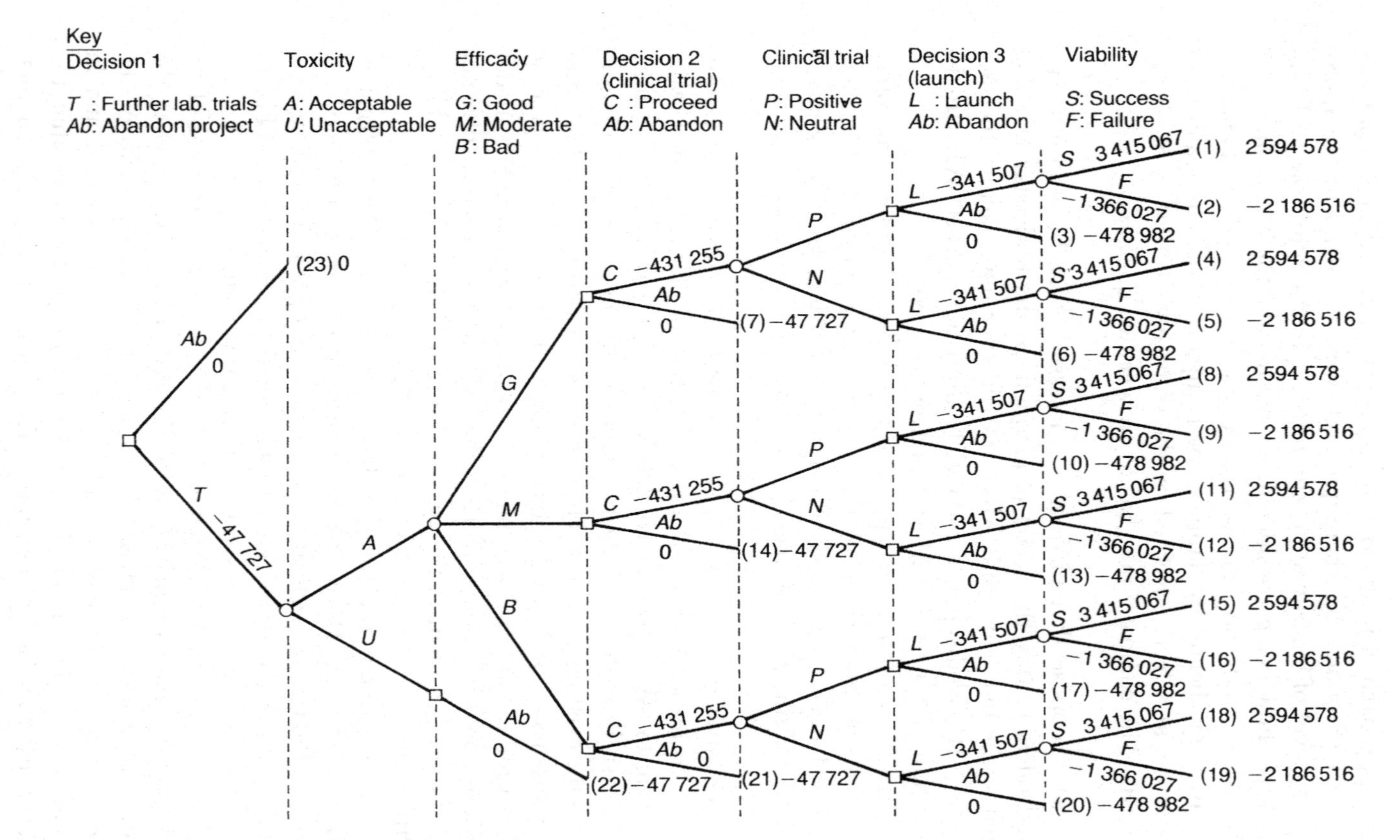

Fig. 6.4 Decision tree with costs and values

allows him to assess the probability that each total cost within his set occurs. This is the subject of the next section.

6.4 Probabilities

Most of the probabilities needed for a decision tree are of a subjective nature (see Section 2.1). It is unusual either to be able to postulate a mechanism which will define the probabilities required or to have sufficient past experience to make objective predictions. In the example there are four stages where probabilities are required. These are:

Further laboratory trials

(a) What are the probabilities that the toxicity will be acceptable or unacceptable?

(b) What are the probabilities that the efficacy will be good, moderate or bad?

Clinical trials outcome

(c) What are the probabilities that the trials will be positive or neutral?

Economic viability

(d) What are the probabilities that the additive will be a success or a failure?

The issues are complicated by the fact that the outcomes at each stage are unlikely to be independent. For example, if the further laboratory trials produce a good result as far as efficacy is concerned, the probability of a positive result from the clinical trial is surely enhanced when compared with the probability of a similar result when the efficacy is moderate or bad. Similar arguments can be put forward for the relationship between the probabilities of the outcomes from the clinical trial and the economic viability. The probabilities needed for the branches from the 'circle' nodes are likely to be *conditional* (Section 2.2) on the particular node of the decision tree.

The probabilities are therefore to depend on the opinions of the experts attached to the company. If we are fortunate, they will be prepared to give their judgements in the forms required; that is, they will offer probabilities in a left-to-right sequence across the decision tree. For example, they will give a value for the probability of a neutral outcome to the clinical trial given a moderate efficacy verdict from the laboratory trials and an acceptable toxicity level. As we progress across the tree the complexity of the questions posed to the experts increases. The final set of questions will consist typically of asking for the probability that plaquoff is economically viable given that the toxicity is acceptable, the efficacy is good and the clinical trial has a neutral outcome. Moreover, at the origin of the tree the questions, although simple to state, can pose demands on the mental agility of the expert. It may be better to look at the problem more directly. Before we go into this in more detail there is one further point to make. The conditional probabilities simplify to unconditional probabilities if the outcomes considered are independent. In the plaquoff example it could plausibly be argued that toxicity is independent of all the other outcomes being considered. The effectiveness of plaquoff in reducing tooth decay is independent of the effect that it has on a person's general health. With this assumption we can then ask for the probability that the material has an acceptable toxicity, and subsequently request probabilities for the remaining outcomes with no reference to the toxicity outcome.

Suppose now that in the example we define the outcomes as follows:

A : toxicity acceptable — U : toxicity unacceptable
G : efficacy good — M : efficacy moderate
B : efficacy bad
P : clinical trial positive — N : clinical trial neutral
S : viability success — F : viability failure

The individual in charge of the laboratory trials feels that there is a good chance that the toxicity will be acceptable and has been persuaded to put the odds on this happening as four to one.

$$\Pr(A) = 0.8 \qquad \Pr(U) = 0.2$$

We now require $\Pr(G|A)$, $\Pr(M|A)$, and $\Pr(B|A)$, but because of the independence assumption, these can be formulated as unconditional probabilities $\Pr(G)$, $\Pr(M)$ and $\Pr(B)$. Following the determination of these quantities, we next require $\Pr(P|G)$, $\Pr(N|G)$, $\Pr(P|M)$, $\Pr(N|M)$, $\Pr(P|B)$ and $\Pr(N|B)$; and for the final stage $\Pr(S|PG)$, $\Pr(F|PG)$, $\Pr(S|NG)$, $\Pr(F|NG)$, $\Pr(S|PM)$, $\Pr(F|PM)$, $\Pr(S|NM)$, $\Pr(F|NM)$, $\Pr(S|PB)$, $\Pr(F|PB)$, $\Pr(S|NB)$ and $\Pr(F|NB)$. The complexity of assessing a probability like $\Pr(S|NG)$ has already been mentioned. Furthermore, the assignment of a value to the unconditional probabilities, such as $\Pr(G)$, may involve a train of thought which argues that if plaquoff is a success then the laboratory trials will have a certain probability of yielding a 'good' result. On the other hand, if it is a failure, the laboratory trials will have a certain lesser probability of a 'good' result. We arrive at an overall probability of a 'good' result by weighting these two answers by the believed probability of success and failure respectively. Rather than go through this procedure, it is more logical to ask the component questions directly, and thence to determine the required conditional probabilities.

We might, therefore, begin by asking our research chemists, biochemists, etc. for their estimate of the probability that plaquoff will be commercially viable. Suppose that they come to a consensus that it is 0.7, i.e. $\Pr(S) = 0.7$ and $\Pr(F) = 0.3$. We then ask for the probabilities of the various combined outcomes for the further laboratory trials and for the clinical trial, given that they are working on a substance such as plaquoff whose economic viability is known to be successful. The same questions are then asked for a similar substance whose economic viability is now known to be a failure. The results of the questions could be tabulated as in Table 6.1.

Table 6.1

Given economic			*Outcomes of trials*			
viability	*PG*	*NG*	*PM*	*NM*	*PB*	*NB*
success (*S*)	0.40	0.15	0.20	0.10	0.10	0.05
failure (*F*)	0.05	0.10	0.10	0.20	0.20	0.35

Note that the probabilities must sum to unity in both rows. The values tabulated above are conditional probabilities of the form $\Pr(PG|S)$ etc., whereas for the decision tree conditional probabilities in the opposite direction are required. These can now be derived using the laws of probability. The final stage probabilities such as $\Pr(S|PG)$ can be derived directly using Bayes' formula (equation (2.4)). The initial stage probabilities such as $\Pr(G)$ can be derived if we note that, since P and N are mutually exclusive and exhaustive, G occurs if and only if either G and P occurs or if G and N occurs.

$$\Pr(G) = \Pr(PG) + \Pr(NG), \qquad \text{by the addition law (2.1)}$$

Using the law of total probabilities (twice) (2.3)

$$\begin{aligned}\Pr(G) = {} & \Pr(PG|S)\Pr(S) + \Pr(PG|F)\Pr(F) \\ & + \Pr(NG|S)\Pr(S) + \Pr(NG|F)\Pr(F).\end{aligned}$$

This is now in terms of probabilities given by our team of experts.

Finally, in the intermediate stage

$$\Pr(P|G) = \frac{\Pr(PG)}{\Pr(G)} \quad \text{by the multiplication law (2.2)}$$

$$= \frac{\Pr(PG|S)\Pr(S) + \Pr(PG|F)\Pr(F)}{\Pr(G)}.$$

Again we have an expression in probabilities either given initially or subsequently derived. It is left to the reader to verify that the values shown in Table 6.2 result.

Table 6.2

Outcome	G	M	B	$P\|G$	$N\|G$	$P\|M$	$N\|M$
Probability	0.430	0.300	0.270	0.686	0.314	0.567	0.433
Outcome	$P\|B$	$N\|B$	$S\|PG$	$F\|PG$	$S\|NG$	$F\|NG$	$S\|PM$
Probability	0.481	0.519	0.949	0.051	0.778	0.222	0.824
Outcome	$F\|PM$	$S\|NM$	$F\|NM$	$S\|PB$	$F\|PB$	$S\|NB$	$F\|NB$
Probability	0.176	0.538	0.462	0.538	0.462	0.250	0.750

Note that there is a plausibility to the relative magnitudes of these figures. For example, the outcomes *PM* for the trials should indicate a higher probability of economic viability than outcomes *NM*, or *PB*. If reversals to this logic do appear, it may be worth trying to identify the cause or causes within the original probabilities to see if the experts may wish to modify their views.

Considerable skill is required to obtain the probabilities for the decision tree. Care must be taken to decide which questions to ask, in the hope of simplifying the task of the interrogated expert. Care must be taken in the way that the questions are asked, in particular clarifying the situation when a conditional probability is required. Finally, there is the need, illustrated above, to ask questions so that the answers will lead uniquely to the probabilities required in the decision tree. Too few questions (e.g. omitting the question about the probability of economic viability) will mean that a unique solution cannot be found. Too many questions (e.g. supplementing the set used by asking for the probability of a positive outcome from the clinical trial given a good efficacy outcome from the further laboratory trials) could lead to the embarrassment of contradictory estimates of some of the decision-tree probabilities derived from different sets of the given probabilities. In mathematical terms we would have more equations than unknowns.

This part of the analysis can be criticized by the view that in asking for these probabilities we are asking for the impossible. How can anyone say, particularly in a 'one-off' situation, what the chance is that a product like plaquoff will succeed? Our discussion of subjective probabilities in Chapter 2 should by now have prepared us for this attack. Decisions have to be taken whether or not we can quantify the consequences – even a decision to do nothing is a decision. Whereas without the formal analysis of the decision tree, the decision-maker may act on hunch or intuition; what we have done here is to force on him the discipline of quantifying his hunch. Single estimates of the probabilities should not be left uncritically. When we have found our best strategy we should be asking whether or not it would have changed if some of the probability estimates were marginally changed. If not, then there will be a reassuring robustness about the strategy adopted; but if our strategy changes, we may have to think further about the particular subjective probability to which our solution is so sensitive. For this reason the subject of sensitivity analysis is introduced in Section 6.6.

For the present we now have values for all the relevant costs, profits and probabilities for our decision analysis. In decision-tree form they are summarized in Figure 6.5 where, to distinguish the different types of value, the probabilities are underlined.

6.5 Application of decision criteria

To appreciate the power of the decision-tree approach, the reader might attempt the exercise of listing all the strategies available to the decision-maker. Recall that each strategy consists of an instruction of the action to be taken at each decision node attainable from actions already taken. Taking only the simplified outcomes used in the previous sections, in the example used to illustrate this chapter there would be 6563 different strategies. Admittedly, some are logically absurd; for example, a strategy which requires that the product be launched if the efficacy has been bad and the clinical trial outcome neutral and at the same time requires abandonment of the project if the efficacy has

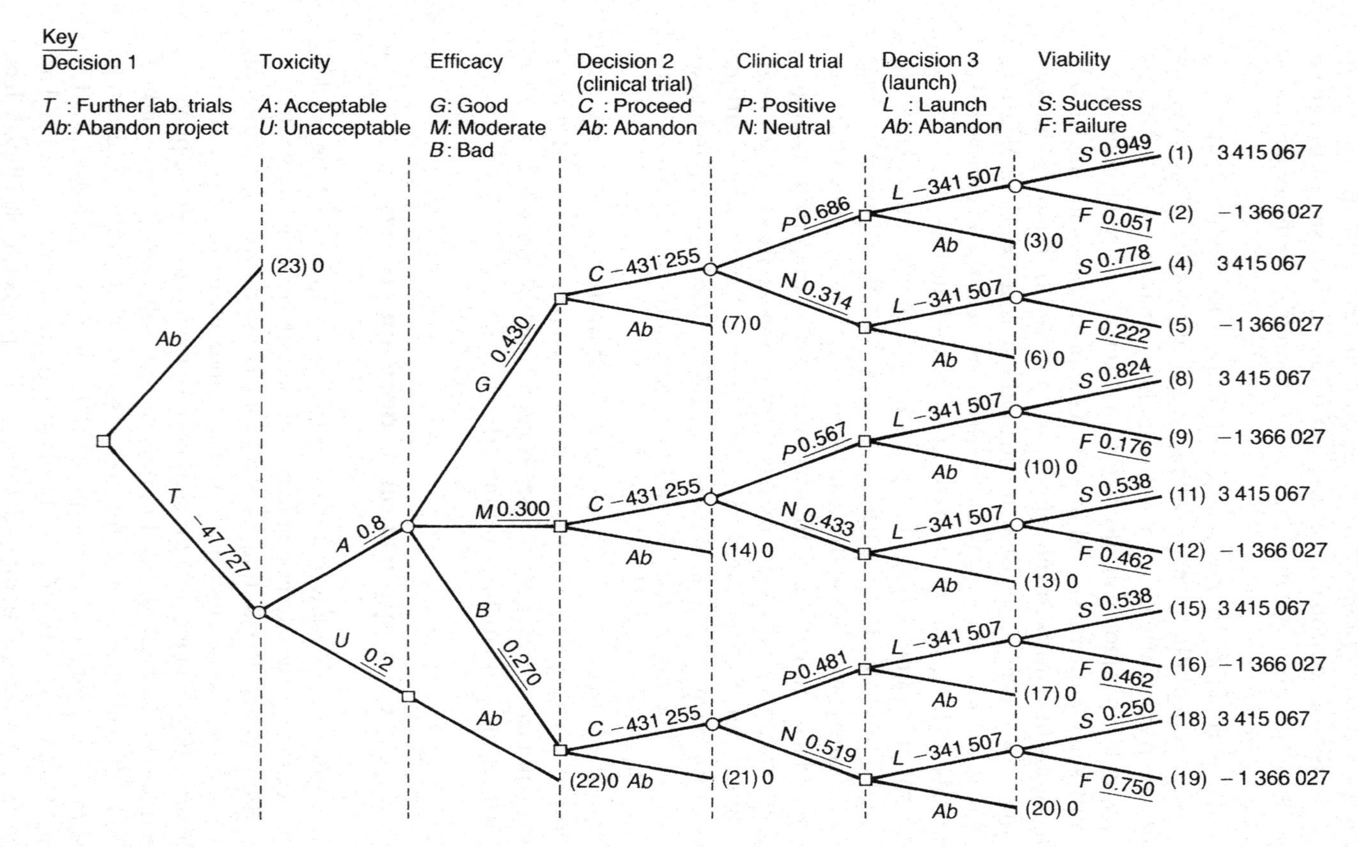

Fig. 6.5 Decision tree with costs and probabilities

been good and the clinical trial positive is clearly perverse. To sort through the strategies retaining only those which would be regarded as sensible by the firm's management is still a daunting task.

For the time being let us retain the concept of the large number strategies available to the decision-maker. In Chapter 3 we considered the equivalence of the decision situation to that of a two-person game, where we might call the opposition 'Nature'. Nature in our present illustration selects actions on toxicity, efficacy, clinical trial outcome and economic viability. There would be a total of $2 \times 3 \times 2 \times 2 = 24$ combinations of actions, each corresponding to a strategy of Nature, except for the fact that if Nature selects the unacceptable action for toxicity, the subsequent actions are irrelevant as far as our decision is concerned. There are therefore only 13 meaningful strategies for Nature, which in the notation introduced in the previous section are: *U, AGPS, AGPF, AGNS, AGNF, AMPS, AMPF, AMNS, AMNF, ABPS, ABPF, ABNS* and *ABNF*. Corresponding to each of the decision-maker's strategies played against each strategy of Nature there is a unique payoff which can be determined from the appropriate costs and (where applicable) profits. For example, if the decision-maker abandons the project initially, the payoff is zero no matter what Nature does. A more complicated strategy for the decision-maker would be:

(a) Proceed with further laboratory trials. If the toxicity is unacceptable, abandon the project.
(b) Otherwise, if the efficacy is good proceed with the clinical trial but launch the product irrespective of its outcome.
(c) If the efficacy is moderate proceed with the clinical trial, but launch the product if its outcome is positive. If the outcome is neutral abandon the project.

Table 6.3

U	*AGPS*	*AGPF*	*AGNS*	*AGNF*
−47 727	2 594 578	−2 186 516	2 594 578	−2 186 516
AMPS	*AMPF*	*AMNS*	*AMNF*	
2 594 578	−2 186 516	−478 982	−478 982	
ABPS	*ABPF*	*ABNS*	*ABNF*	
−47 727	−47 727	−47 727	−47 727	

(d) If the efficacy is bad, abandon the project without proceeding to the clinical trial.

The payoffs for this strategy can be tabulated (Table 6.3).

Now let us consider the application of the decision criteria discussed in Chapter 4. Take first the maximin criterion. Clearly this is to abandon the project without any further trials producing a payoff of zero irrespective of Nature's actions. Any other strategy by the decision-maker would incur a cost, and, if Nature so chooses, the project will have a negative return. This 'abandon' or 'do nothing' action is typical of the pessimism which the maximin criterion generates.

As far as the minimax regret criterion is concerned, it must follow that any abandonment of the project other than at the beginning or that required by an unacceptable toxicity will not be considered. The regret criterion, which, let us remind ourselves, does not take any account of probabilities and therefore of any encouragement that intermediate outcomes may have presented, would argue in this case that any abandonment after the first must be more expensive. It must therefore have a greater regret than the initial abandonment. It follows, therefore, that the only component of Nature's actions which is of any consequence for the regret criterion is whether or not the project is economically viable. If it is viable, then with hindsight the best strategy to adopt is to launch irrespective of all other outcomes. Conversely, if it is not viable, the best strategy, again with hindsight, is to abandon the project before any further laboratory trials take place. These two strategies of the decision-maker dominate all others for the minimax regret criterion. With the former the outcome will be either a gain of £2 594 578 (and therefore zero regret) or a loss of £2 186 516, which when compared to initial abandonment amounts to a regret of this same quantity. With the latter strategy of initial abandonment, there will be a regret of £2 594 578 whenever the launched project is successful, but a regret of zero when the project is a failure. For these two strategies the maximum regrets are therefore £2 186 516 and £2 594 578 respectively, all other strategies having larger maxima. The minimax regret strategy must therefore be to proceed with the trials and launch irrespective of their outcomes, the only proviso being that the project be abandoned if the toxicity is unacceptable. In summary, the minimax regret criterion is saying that, since the company stands to gain more by a successful project than it would lose by a failure, it should always launch. In these circumstances the criterion is hardly likely to be considered seriously.

The maximax criterion where the decision-maker is continually seeking the best outcome produces the same strategy as the minimax regret criterion in this particular example.

The Laplace criterion where all of Nature's actions are assumed to have the same probability can also be summarily dismissed. The example highlights one of the criterion's major drawbacks, namely the listing of Nature's strategies. If the toxicity in the further laboratory trials is unacceptable, the decision-maker has no further interest in the efficacy of the laboratory trials, the clinical trial outcome and the economic viability. Indeed, he proceeds no further with the project. Should, however, all of these separate actions by Nature be included in the Laplace list – in which case there are 24 strategies and the total probability of unacceptable toxicity is $\frac{1}{2}$ – or should they be counted simply as the single strategy, producing a total of 13 and a probability of unacceptable toxicity of $\frac{1}{13}$? Fortunately, we do not need to resolve this dilemma as the criterion itself is inappropriate for this example.

For the Hurwicz criterion any strategy, other than 'abandon initially' which did not include the possibility of a launch action, would contain only negative payoffs and would therefore be dominated by the 'abandon initially' strategy itself. The remaining strategies would therefore always have a maximum payoff of £2 594 578 and a minimum of −£2 186 516. The Hurwicz strategy would amount to abandoning the project initially for low values of the index α. For the remaining values of α it fails to discriminate between all of the 'launch' strategies, and thus fails to answer the real problem for the decision-maker.

In considering the final two criteria (we shall not concern ourselves with the expected opportunity loss criterion because of its established equivalence to the expected monetary value criterion) we shall use the probabilities derived in the previous section. These, therefore, are the first criteria which take into account the company's beliefs in the various intermediate and final outcomes, and, as such, should carry greater plausibility.

The first is the maximum modal payoff criterion, where the decision-maker acts to maximize his gain, assuming that Nature will always take the most likely action. Since Nature's probabilities are conditional we shall need to refer to Figure 6.5, which also contains the costs and profits as they occur. Comparisons of strategies can now be carried out sequentially, working across the decision tree from right to left, eliminating branches as they are proved inferior by this criterion. For example, suppose that the decision-maker finds himself at point X on the decision tree (see now Figure 6.6). If he launches, the probability of success has been estimated to be 0.949 and this clearly is the most likely outcome. He acts therefore as if the launch will be successful, producing a profit of

£3 415 067 − £341 507 = £3 073 560
(= profit from successful launch − cost of launch).

This he then compares with zero, the return from abandoning the launch, and hence decides to go ahead with the launch. It is useful to mark the branches which have thus been barred with some appropriate sign, such as —╫—. Repeated applications of this method of choice allow the decision-maker to proceed across the decision tree, thereby building up a strategy which is optimal by this criterion. Note that on occasions some of the choices found will not actually be used. For example, a bad efficacy outcome leads to abandonment of the project, making the choices already found in this part of the tree, assuming the project proceeds, superfluous. This is unavoidable. A complete analysis showing profits and costs is shown in Figure 6.6. For the relevant probabilities, refer to Figure 6.5.

The resultant optimum strategy using the MMP criterion is as follows:

Carry out further laboratory trials.
If the toxicity is unacceptable, abandon the project.
If the toxicity is acceptable and the efficacy is either good or moderate launch the project irrespective of the outcome of the clinical trial.
If the toxicity is acceptable and the efficacy is bad, the project is abandoned without the clinical trial.

The modal return from this strategy is £2 594 578 which is, as we have already seen in Figure 6.4, the NPV of the net profit if the venture is successful.

The second criterion using the assumed probabilities is the Expected Monetary Value (EMV) Criterion. To apply this criterion a right-to-left pass through the decision tree may again conveniently be used, and on this occasion instead of considering the modal action by Nature we take the weighted outcomes using the probabilities as weights. Again, using the decision tree eliminates dominated strategies by

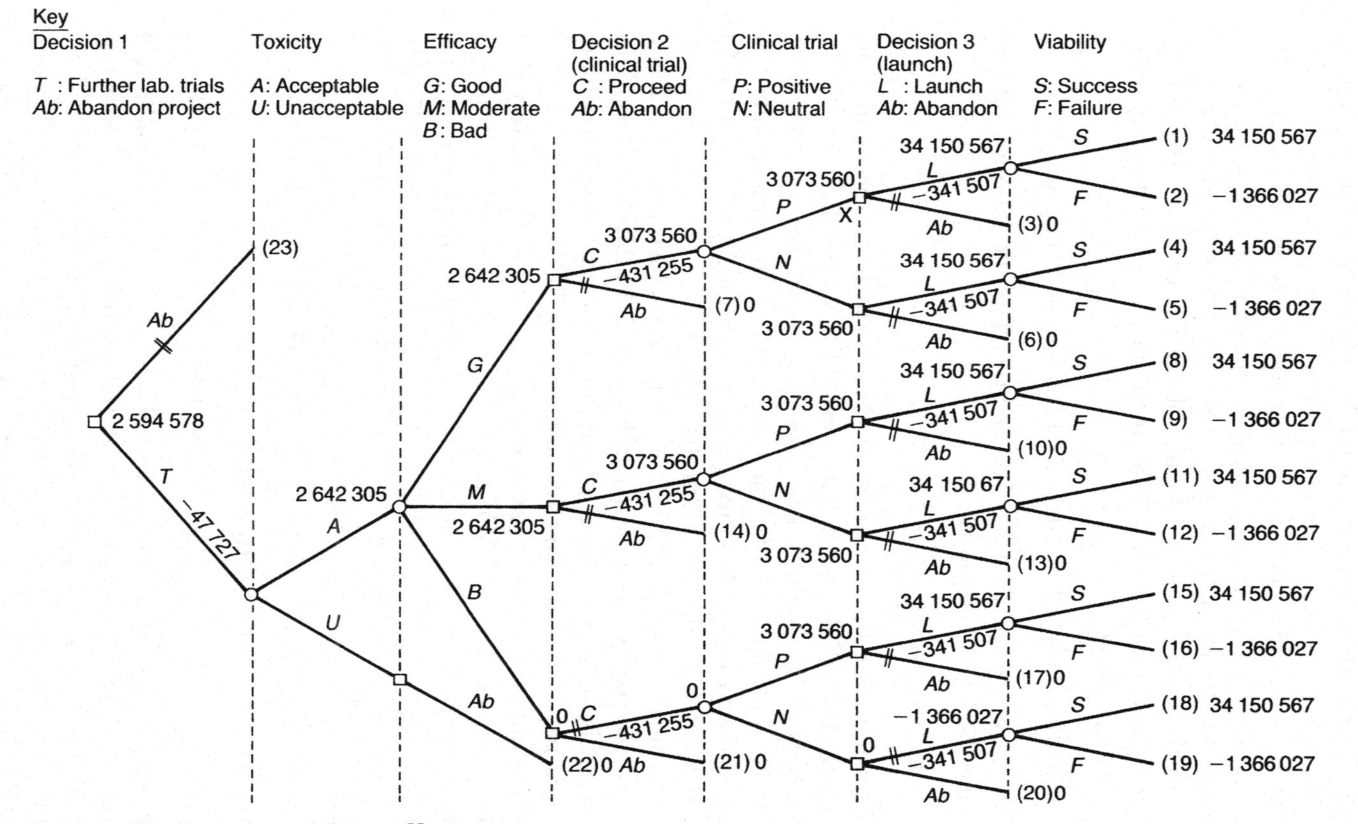

Fig. 6.6 Maximum modal payoff strategy

barring branches which are inferior by this criterion. It is permissible in this case to use either the totalled costs or profits at the end of each branch (as in Figure 6.4) or the costs or profits as they occur (as in Figure 6.5). The latter is preferred in this demonstration; the values considered are therefore the expected values subsequent to the node where the calculations are taking place. For example, referring to node X in Figure 6.7, the EMV at the node immediately following the launch decision is

$$(0.949)(3\,414\,067) + (0.051)(-1\,366\,027) = 3\,171\,231.$$

The cost of the launch is 341 507 and therefore the expected value of taking the launch action is

$$3\,171\,231 - 341\,507 = 2\,829\,724.$$

The value of the abandon action is zero, and therefore the comparison leads to taking the launch action, and the EMV at node X is 2 829 724. Progressing from right to left across the tree in this way, therefore, gives not only the EMV for each node, but the action to be taken. A complete analysis is presented on the tree in Figure 6.7. (Again refer to Figure 6.5 for the branch probabilities.)

The final EMV is £979 136. The strategy which gives this EMV can be read off the decision tree in Figure 6.7. In this case the optimal strategy is the same as the one found using the criterion of maximum modal payoff, a reassuring and not unusual phenomenon.

It is important that the analyst presents his results in a way that demonstrates how the EMV arises. For example, here the project is abandoned with a loss of £47 727 if either the toxicity is unacceptable or the efficacy verdict is bad. Otherwise the project proceeds right through to its launch, where either it will be successful with an NPV of

$$£3\,415\,067 - £341\,507 - £431\,255 - £47\,727 = £2\,594\,578,$$

or it will be a failure with a NPV of

$$-£1\,366\,027 - £341\,507 - £431\,255 - £47\,727 = -£2\,186\,516.$$

In other words the project may yield a return of over two-and-a-half million pounds, a loss of over two million pounds or a loss of just under fifty thousand pounds. The estimated probabilities for these outcomes can also be determined from the decision tree. They are 0.476, 0.108 and 0.416 respectively. The derivation of these values is

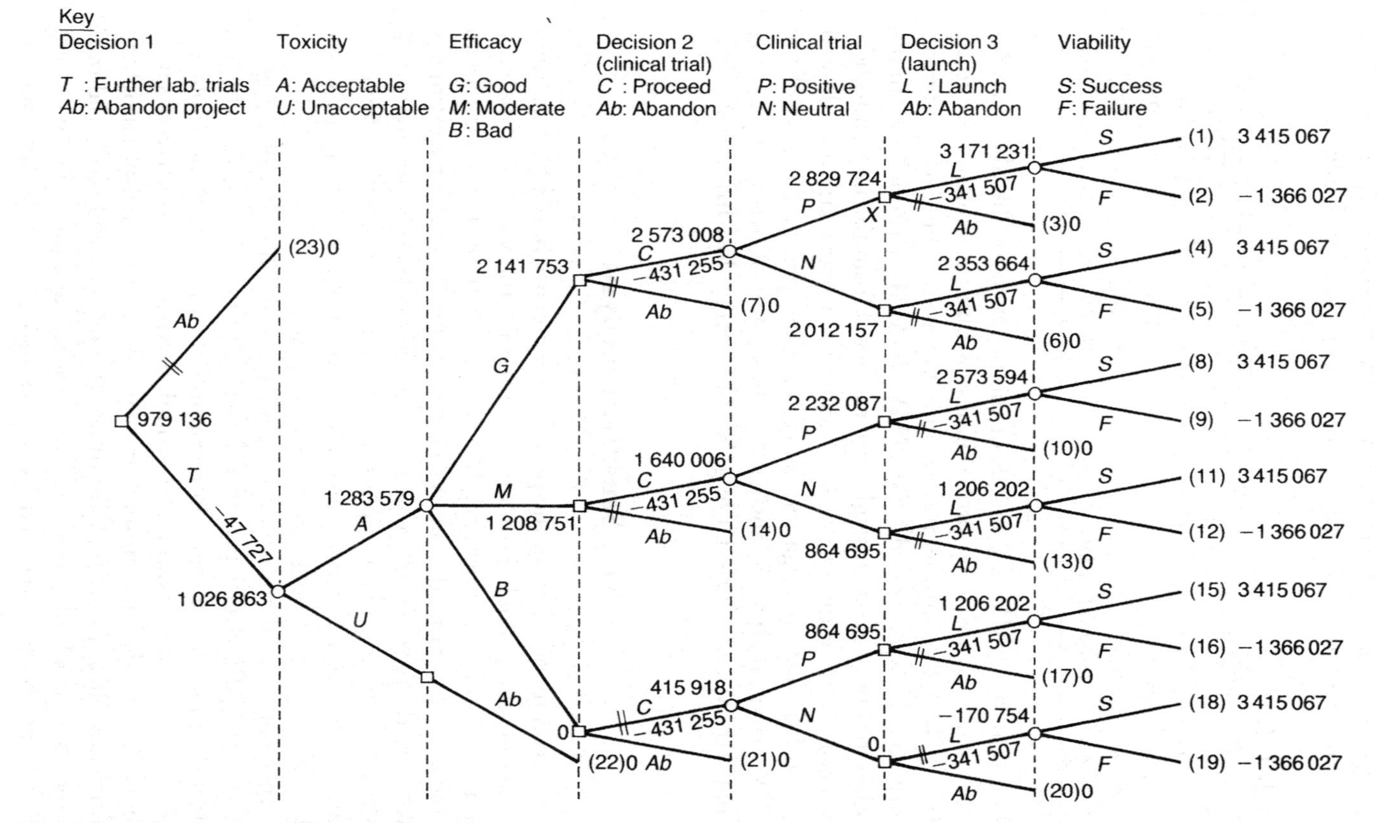

Fig. 6.7 Maximum EMV payoff strategy

left as an exercise for the reader; note that they must sum to one. The effect of the strategy can therefore be summarized (Table 6.4).

Table 6.4

Outcome	*Probability*	*NPV of profit or loss (£)*
Launch successfully	0.476	2 594 578
Launch unsuccessfully	0.108	−2 186 516
Abandon after further lab. trials	0.416	−47 727

Note that the EMV is the sum of the products of the above NPVs with their corresponding probabilities (within the limits of rounding errors). It is by consideration of this table that the decision-maker must make up his mind. The EMV or MMP criteria are simply means whereby good strategies are extracted from the large numbers available. The resultant strategies should not necessarily be adopted without further investigation, since, for example, in the case of the EMV criterion there is always the possibility of a large loss being balanced by a similarly large gain and thereby lost from view.

6.6 Sensitivity analysis

The recommendations in the previous section are based on a decision model in which single estimates are used for the various probability and cost values. It is reasonable to ask, therefore, for the extent to which the recommendations would change if different single estimates were used, particularly when the differences are only marginal. Although the argument applies both to costs (including the discount rate for the NPV calculations) and to probabilities, it is to the latter that attention will be given in this discussion.

In most cases there is only a relatively small number of plausible strategies for the decision-maker. Others may be dominated or contain absurd sequences of logic. For example, in the Moss Side Molasses Mining Company example it would need a very contrived set of circumstances for an optimal strategy which consisted of launching when the efficacy was bad and there was a neutral clinical trial outcome, but which at the same time required abandoning the project for a good efficacy followed by a positive clinical trial outcome. If, therefore, the choice of strategy is limited, it is likely that the same strategy will be optimal under a range of values of the various constants in the model. The investigation of this range is termed *sensitivity analysis*.

Returning to the example of this chapter, in the original problem description a prior value of 0.7 was placed on the probability that the additive was to be economically viable. How sensitive was our solution to this probability? Suppose for example that the chance of success had been estimated as fifty–fifty, so that now $\Pr(S) = 0.5 = \Pr(F)$. If we assume that all of the remaining costs and probabilities are unaltered, we obtain a revised table of unconditional and conditional probabilities (Table 6.5).

Table 6.5

Outcome	G	M	B	$P\|G$	$N\|G$	$P\|M$	$N\|M$
Probability	0.350	0.300	0.350	0.643	0.357	0.500	0.500
Outcome	$P\|B$	$N\|B$	$S\|PG$	$F\|PG$	$S\|NG$	$F\|NG$	$S\|PM$
Probability	0.429	0.571	0.889	0.111	0.600	0.400	0.667
Outcome	$F\|PM$	$S\|NM$	$F\|NM$	$S\|PB$	$F\|PB$	$S\|NB$	$F\|NB$
Probability	0.333	0.333	0.667	0.333	0.667	0.125	0.875

The maximum modal payoff criterion has, in this case, the disconcerting property of producing a tie on two separate occasions – between the positive and neutral clinical trial outcomes after the moderate efficacy result, and between the good and bad efficacies. Since both occasions amount to a choice between launching and abandoning, the normal practice is to adopt the less risky alternative which would then amount to abandoning the project initially.

The maximum expected monetary value criterion, on the other hand, produces a strategy demonstrated in Figure 6.8. Reference should be made to the probabilities in Table 6.5 in order to check the EMV values in the figure.

The maximum EMV strategy differs now from the previous EMV strategy only in its instruction following a moderate efficacy outcome to the further laboratory trials. The firm is to proceed with the clinical trial, but will only launch the product if the outcome of the clinical trial is positive. Otherwise the project is abandoned. The overall

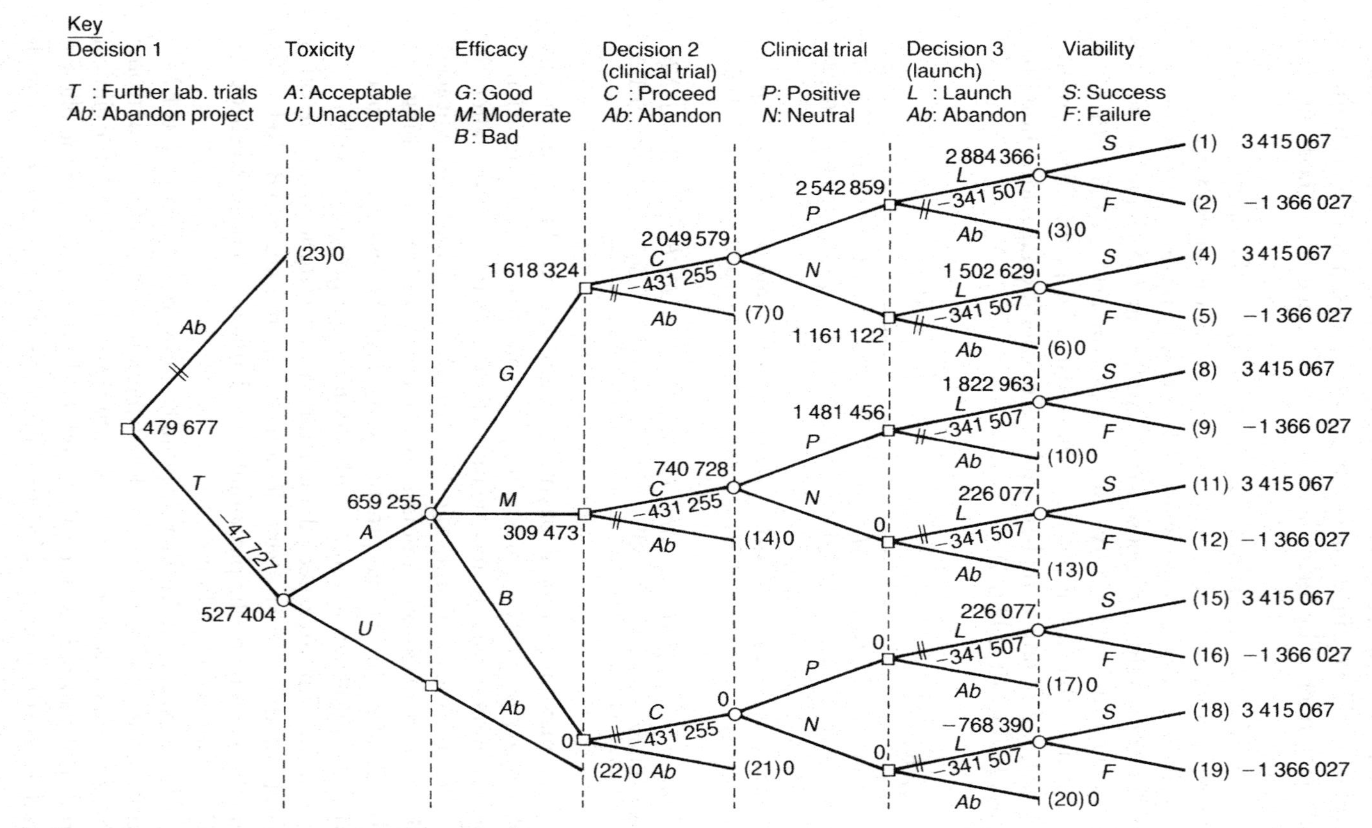

Fig. 6.8 Maximum EMV strategy with Pr(S) = 0.5

Table 6.6

Outcome	*Probability*	*NPV of profit or loss (£)*
Launch successfully	0.30	2 594 578
Launch unsuccessfully	0.10	−2 186 516
Abandon after clinical trial	0.12	−479 982
Abandon after further lab. trials	0.48	−47 727

EMV has been reduced from £979 136 to £479 677, a figure which is again made more meaningful by an examination of the feasible gains and losses when the strategy is used.

From Table 6.6 it can be seen that while a reduced prior probability of a successful project must make the project less attractive, the change in strategy has in this case kept the probability of the really substantial loss at approximately the same 10 per cent level, but the probability of the real gain now falls from 0.476 to 0.30. In addition, the extra possibility of losing nearly half a million pounds with probability 0.12 cannot be ignored.

If the prior probability of a successful launch is denoted by the general term π, the critical value of π at which the strategy changes can be determined. Our calculations so far show that this critical value must be between 0.5 and 0.7. We need only consider the branch of the tree following the 'proceed' action which in turn follows a moderate efficacy outcome (see Figure 6.9).

Using the notation established in Section 6.4, and Bayes' formula (2.4)

$$\Pr(S|PM) = \frac{\Pr(PM|S)\Pr(S)}{\Pr(PM|S)\Pr(S) + \Pr(PM|F)\Pr(F)}$$

$$= \frac{(0.20)\pi}{0.20\pi + 0.10(1 - \pi)} = \frac{2\pi}{\pi + 1}$$

Hence $\Pr(F|PM) = 1 - \Pr(S|PM) = \dfrac{1 - \pi}{\pi + 1}$.

Also $$\Pr(S|NM) = \frac{\Pr(NM|S)\Pr(S)}{\Pr(NM|S)\Pr(S) + \Pr(NM|F)\Pr(F)}$$

$$= \frac{(0.10)\pi}{(0.10)\pi + (0.20)(1 - \pi)} = \frac{\pi}{2 - \pi},$$

and hence $\Pr(F|NM) = 2(1 - \pi)/(2 - \pi)$.

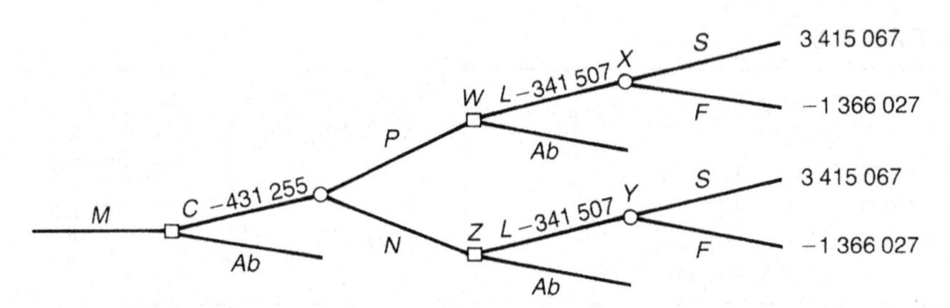

Fig. 6.9 Branch of the tree following moderate efficacy outcome

We can therefore calculate the EMV values at points X and Y on the tree in Figure 6.9, expressing them as functions of π. They are:

$$\frac{(3\,415\,067)(2\pi)}{(\pi + 1)} - \frac{(1\,366\,027)(1 - \pi)}{(\pi + 1)}$$

$$= (8\,196\,161\pi - 1\,366\,027)/(\pi + 1) \qquad (X)$$

$$\frac{(3\,415\,067)\pi}{(2 - \pi)} - \frac{(1\,366\,027)/2(1 - \pi)}{(2 - \pi)}$$

$$= (6\,147\,121\pi - 2\,732\,054)/(2 - \pi) \qquad (Y)$$

At node W, we launch if $(8\,196\,161\pi - 1\,366\,027)/(\pi + 1) > 341\,507$,
i.e. if $\pi > 0.217$
At node Z, we launch if $(6\,147\,121\pi - 2\,732\,054)/(2 - \pi) > 341\,507$,
i.e. if $\pi > 0.526$.

The critical value for the prior probability of success is therefore 0.526, assuming that all the other costs and probabilities are unchanged.

In the opposite direction, the strategy will presumably change if π is increased from 0.7. This can only increase the likelihood of a launch, which in turn will occur if we no longer abandon the project following a bad efficacy outcome. We need therefore only examine this branch of the tree as shown in Figure 6.10.

Following the same approach that was used above, we find

$$\Pr(S|PB) = \pi/(2 - \pi) \qquad \Pr(F|PB) = 2(1 - \pi)/(2 - \pi)$$
$$\Pr(S|NB) = \pi/(7 - 6\pi) \qquad \Pr(F|NB) = 7(1 - \pi)/(7 - 6\pi).$$

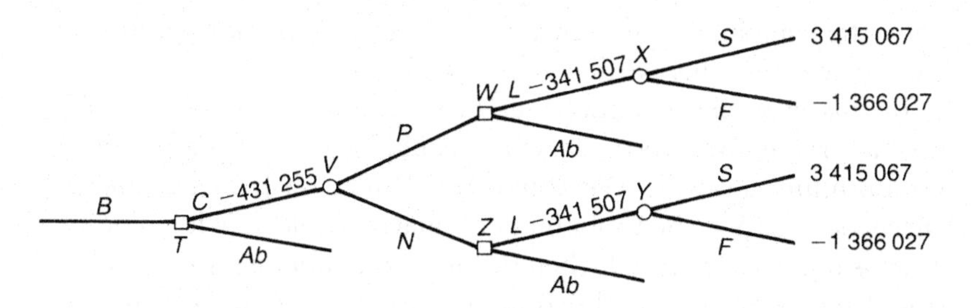

Fig. 6.10 Branch of the tree following bad efficacy outcome

Hence the EMV values at X and Y are

$$(X) \quad (6\,147\,121\pi - 2\,732\,054)/(2 - \pi) \tag{6.2}$$

$$(Y) \quad (12\,977\,256\pi - 9\,562\,189)/(7 - 6\pi). \tag{6.3}$$

The launch action is taken at X and Y if their EMVs are greater than the cost of the launch, £341 507. This requires that π should be greater than 0.526 and 0.795 respectively. Hence, for a value of π greater than 0.795 we still have to check that we do not abandon the project at node T in Figure 6.10. To evaluate the EMV at T we shall need $\Pr(P|B)$ and $\Pr(N|B)$ in terms of the prior probability π. Using the same approach to that used when these probabilities were calculated with a numerical value for π, we find

$$\Pr(P|B) = \frac{2(2-\pi)}{11-8\pi}, \qquad \Pr(N|B) = \frac{(7-6\pi)}{11-8\pi}.$$

The EMV at node V is found by multiplying the values in equations (6.2) and (6.3) by the corresponding conditional probabilities above, and adding. The inequality where the sum is greater than 341 507 is linear in π and reduces to $\pi > 0.748$. The implication of this analysis is that if $0.795 > \pi > 0.748$, the strategy derived previously for the case $\pi = 0.7$ is unaltered, except that if the efficacy is bad and the clinical trial is positive we launch; if the clinical trial is neutral the project is abandoned at this stage. However, if $\pi > 0.795$ the project is launched irrespective of the outcomes of the further laboratory and clinical trials (other than an unacceptable toxicity outcome, when, of course, the project is abandoned). For example, a prior probability of 0.8 that the additive is economically viable means that the belief in its success is so strong when linked to the possible gains and losses that the evidence from the laboratory trials and clinical trials should not deter the company from launching. There is clearly a need for great care in selecting this particular probability value.

In this example, attention has been confined to only one of the several probabilities and costs which could be assessed for its effect on the optimum strategy. We could in a similar fashion examine the effects of changing the probability of unacceptable toxicity. The situation is more complicated when the other probabilities are considered since these were given in the form of distributions of possibilities over six outcomes (see Table 6.1). Changing any individual probability would be compensated by a change in any set of the remaining five probability values in order to retain the necessary condition that they sum to one. There are several plausible ways in which this might be done. For example, we could change in pairs only – say PG and NG – thereby retaining the unconditional probability of G at its previous value. A detailed analysis is not presented here. In a project of any magnitude such an analysis would be justified, the aim being to highlight cases where the selection of values for the probabilities appears critical.

As far as the costs are concerned, the position is considerably simpler. Suppose that we are now prepared to accept the original values for the probabilities, but are concerned at the value (£5 000 000) assumed for the return from a successful launch of the project. Over what range of this return will our strategy remain optimal?

The actual return from a successful launch appears at six terminal nodes in the decision tree (see Figure 6.7). The procedure is to replace the numerical value used by some algebraic symbol, say R, and then to determine the conditions on R for which the given strategy remains optimal. This results in a set of simple inequalities for R. The critical inequalities occur at the decision nodes where, after acceptable toxicity, moderate efficacy and neutral clinical trial outcome, the launch/abandon decision is to be taken, and where, after acceptable toxicity and bad efficacy, the abandon/proceed with clinical trial decision is to be taken. The former requires a lower bound of £2 646 841 on the actual (i.e. before discounting to the NPV) gain from a successful launch, whereas the latter requires an upper bound of £5 086 771 on this gain if the original strategy is to be retained. Thus there is plenty of slack in one direction when the actual assumed value of £5 000 000 is taken, but the closeness of the upper bound may need some investigation. Certainly the difference is well within the error which might be expected in a case such as this. The reader will have noted that throughout this example figures have been quoted with an accuracy which is not really warranted. This has been done partly because it may have helped to recognize the appearance of the same figure from time to time during the discussion, and partly because, as we now see, there are occasions when close comparisons have to be made. There is also the reason that in rounding off values we are only substituting one set of figures for another which happens to be all zeros. The change in strategy caused by increasing the gain marginally above the five million mark consists of proceeding to the clinical trial even if the efficacy verdict is bad. The project is abandoned only if the clinical trial gives a neutral outcome following a bad efficacy outcome; otherwise it is launched. This strategy is valid until the gain from a successful launch reaches £8 000 000, at which point we launch irrespective of any outcome other than unacceptable toxicity.

Finally, it must be emphasized that in all cases we are measuring the sensitivity of the solution to changes in just *one* variable. It is unfortunately true that lesser simultaneous changes in two or more variables can sometimes cause a shift in the optimum strategy. If this type of analysis is contemplated, it will probably be worth programming the logic of the decision tree for a computer, so that many combinations of changes can be speedily and accurately assessed. Alternatively, we might turn to the approach of a stochastic analysis of the decision tree.

6.7 Stochastic analysis

In the previous section the possibility was discussed of allowing the parameters of the decision problem to vary, where the amounts by which they are varied would be controlled by the decision-maker. In a problem of any magnitude this could be rather a formidable task, even allowing for the fact that individual analyses would be carried out by computer. The result could be a wide-ranging set of outcomes for the same decision strategy and the decision-maker would be left with the task of sorting the results out, and particularly of apportioning weight to the extreme values.

One way of reducing the difficulty would be to try to fit probability distributions to the parameters. This may be feasible for the costs involved, where instead of asking the 'relevant experts' for a single estimate of a cost or gain, we pose further questions about the likely extremes that this value may take. These further questions have to be posed rather carefully. For example, if we feel that the concept of probability is understood, we could ask for the value which the experts feel should only be exceeded with a chance of one in twenty. A lower limit could be similarly defined. If these values are symmetrically placed around the single estimate, it is reasonable to assume a normal distribution for the parameter under consideration. The distribution would, in this case, be fully defined. If the limit values are not symmetrically placed, we have the choice of using a number of distributions, such as the log-normal, beta or gamma. A full discussion of this issue will not be attempted here, but reference may be made to Hampton *et al.* (1973) or Hull (1980).

Suppose, therefore, that probability distributions have been found for all gains and losses in the decision tree, and also that single estimates of the various probabilities will be used. Any decision strategy will therefore result in a set of outcomes whose probabilities can be derived. Moreover, each outcome will be the sum of a number of the gains and losses, and therefore either its distribution will be explicitly derivable or a good approximation by a normal distribution will be possible. The overall distribution of outcomes will therefore be a weighted sum of the individual distributions.

To carry out this type of analysis on all possible strategies would be prohibitively time-consuming. It is suggested that the EMV criterion be used to determine the optimum strategy under this form of assessment, and then a simple form of sensitivity analysis be used to determine the 'runners up'. Comparisons can then be made with these and also, where appropriate, with any initial terminal action strategy, such as abandon at the start. We shall return to the example to illustrate the point being made.

There are five gains and losses in the problem. We now refer to each as a random variable as shown in Table 6.7. Note that we have

Table 6.7

Description	*Letter*	*Mean value*
Further laboratory trials cost	X	47 727
Clinical trial cost	Y	431 255
Launch cost	Z	341 507
Return from successful launch	V	3 415 067
Loss from unsuccessful launch	W	1 366 027

taken the previously used single estimates as the mean values of the distributions. For simplicity we shall assume that the random variables are independently and normally distributed, and that each has a coefficient of variation equal to 0.1, so that each standard deviation is the corresponding mean value divided by 10.

The maximum EMV strategy has three possible outcomes, as was seen in Section 6.5. These can now be expressed more generally, as in Table 6.8. We use the well-known formula for the standard deviation of the sum or difference of a number of random variables, namely that it is the square root of the sum of the individual standard deviations squared. The distribution of the outcome is now the weighted outcome of the three normal distributions. In this case they are effectively separate, as can be seen from Figure 6.11.

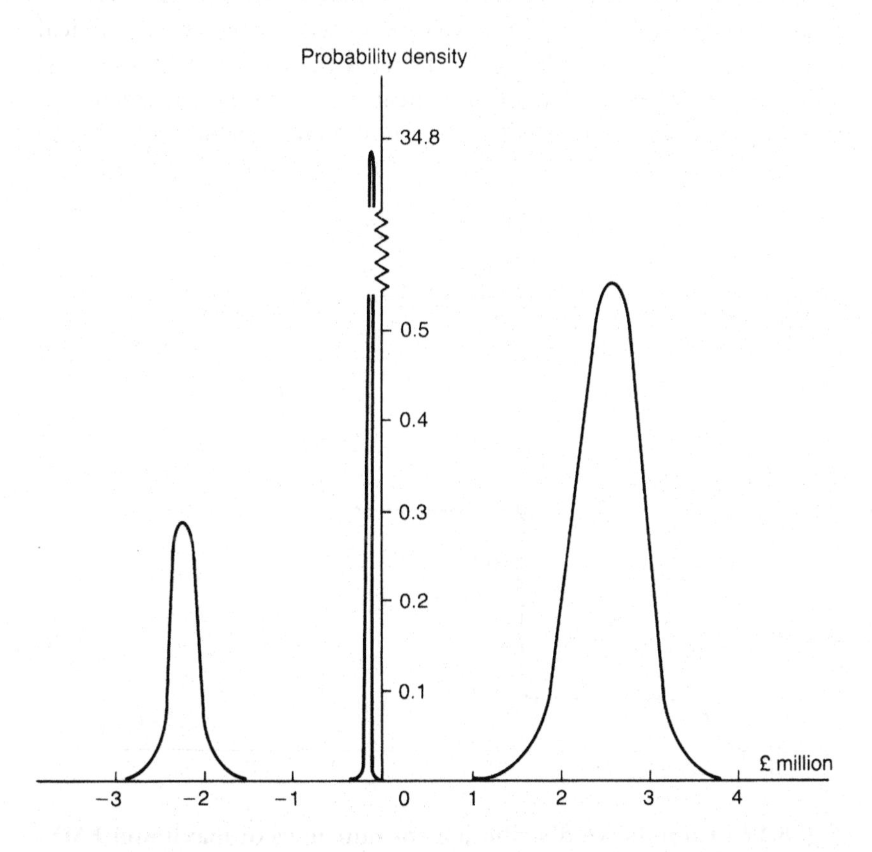

Fig. 6.11 Distribution of outcomes for maximum EMV strategy

Table 6.8

Outcome	*Probability*	*Random variable*	*Mean*	*Standard deviation*
Launch successfully	0.476	$V-Z-Y-X$	2 594 578	345 942
Launch unsuccessfully	0.108	$-W-Z-Y-X$	−2 186 516	147 340
Abandon after further lab. trials	0.416	$-X$	−47 727	4 773

An alternative way of graphing this result is to show it in a cumulative form, plotting the probability that the return is less than or equal to any particular value against the value itself. The cumulative graph, which like Figure 6.11 can readily be derived from tables of the cumulative normal distribution, is shown in Figure 6.12.

If a similar exercise is now carried out for the situation where the strategy is as above, except that we now always proceed to the clinical trial (assuming acceptable toxicity) unless the efficacy has been bad and this is followed by a neutral clinical trial outcome, in which case the project is abandoned. There are now four possible outcomes to

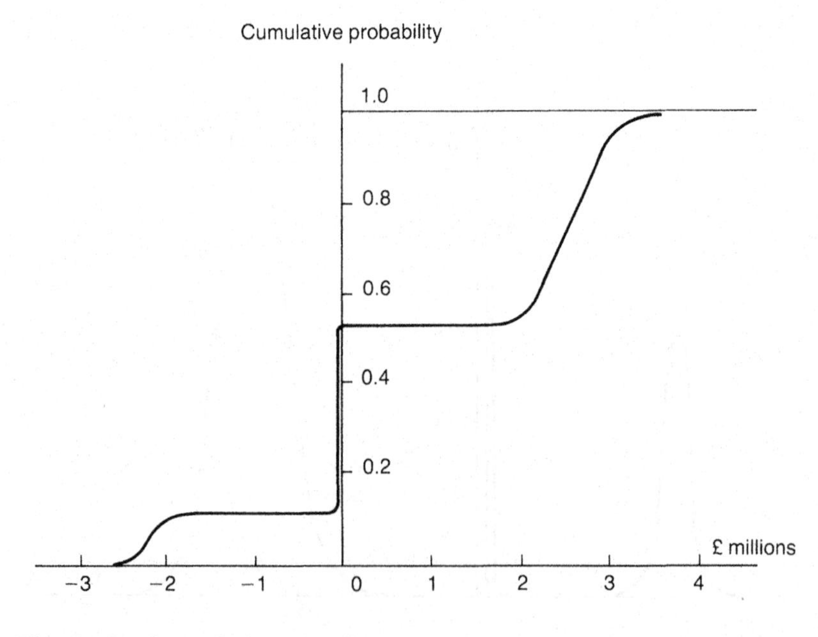

Fig. 6.12 Cumulative distribution for outcomes of maximum EMV strategy

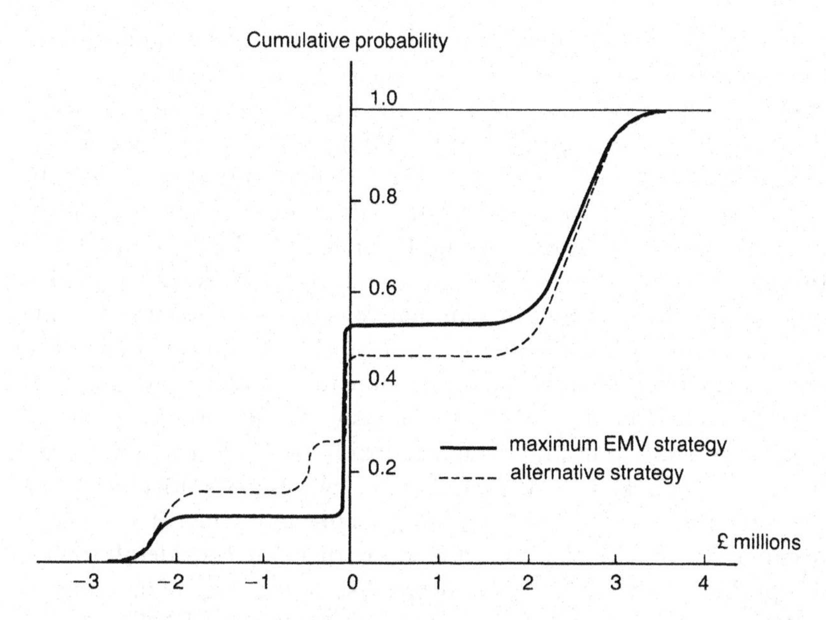

Fig. 6.13 Comparison of cumulative probabilities

this strategy (Table 6.9). Figure 6.13 matches the cumulative curves for these two strategies on the same diagram.

Figures 6.12 and 6.13 demonstrate the way in which the actual outcome is an observation from one of either three or four virtually distinct distributions. In other cases there will probably be overlapping distributions. In previous sections we had based our choice between strategies on the EMV or expected value of these distributions. Here the original strategy has an EMV of £979 136 compared with an EMV of £976 028 for the alternative now being considered. The

Table 6.9

Outcome	*Probability*	*Random variable*	*Mean*	*Standard deviation*
Launch successfully	0.532	*V–Z–Y–X*	2 594 578	345 942
Launch unsuccessfully	0.156	*–W–Z–Y–X*	−2 186 516	147 340
Abandon after clinical trials	0.112	*–Y–X*	−478 982	43 389
Abandon after further lab. trials	0.200	*–X*	−47 727	4 773

difference is very marginal, amounting to 0.3 per cent. Comparison of the cumulative curves in Figure 6.13 affords an alternative means of choice. If one curve were wholly to the right of the other, this would mean that for the strategy corresponding to the former curve, the probability that the outcome is greater than any value would be consistently greater than that for the latter curve. Thus, it could be argued that such a strategy would be preferable. In the example the two curves cross very close to the zero return value. A contemplation of Figure 6.13 shows that the alternative is relatively riskier, showing a higher probability of a positive return between approximately two and three million pounds, but at the same time a higher probability of substantial and moderate losses. The issue is thus resolved to one of risk. Converting to utilities would probably help to clarify the matter by (assuming risk aversion) bringing the curves closer together when the return is positive, but separating them more when the return is negative. In any decision situation where there are considerable risks, it is probably advisable to take the analysis to the stage of drawing up the cumulative risk curves for the strategies which the EMV criterion has shown to be in contention. Usually there will be a 'do nothing' strategy whose cumulative risk curve is simply the horizontal and vertical lines shown in Figure 6.14. This should not be forgotten.

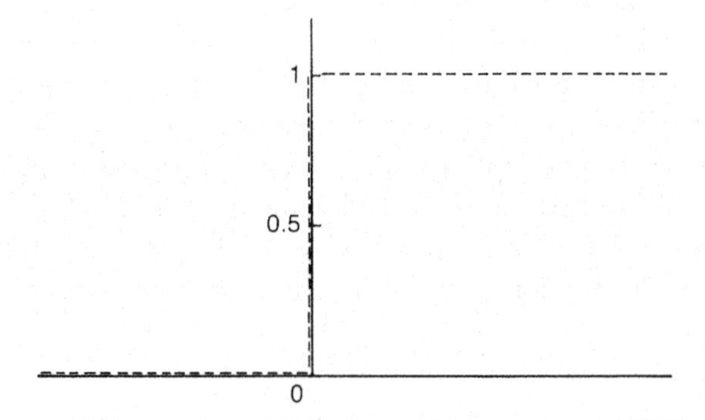

Fig. 6.14 Cumulative probability curve for 'do nothing' strategy

6.8 Further reading

Bunn, D. W., *Analysis for Optimal Decisions*, John Wiley, Chichester (1982), Sections 4.5, 4.6.

Hampton, J. M., Moore, P. G. and Thomas, H., 'Subjective Prob-

ability and its Measurement', *J. Roy. Stat. Soc.*, A, **136**, Part 1, 21–42 (1973).
Hertz, D. B. and Thomas, H., *Risk Analysis and its Applications*, John Wiley, Chichester (1983), Chapter 3.
Hertz, D. B. and Thomas, H., *Practical Risk Analysis*, John Wiley, Chichester (1984), Chapter 3.
Hull, J. C., *The Evaluation of Risk in Business Investment*, Pergamon, Oxford (1980).
Lindley, D. V., *Making Decisions*, John Wiley, London (1971), Chapter 8.
Moore, P. G. and Thomas, H., *The Anatomy of Decisions*, Penguin, Harmondsworth, Middlesex (1976), Chapters 4, 11.
Schlaiffer, R., *Analysis of Decisions under Uncertainty*, McGraw-Hill, New York (1969), Parts 1 and 2.
Weinstein, M., 'Risky Choices in Medical Decision Making: A Survey', *Geneva Papers on Risk and Insurance*, **11**, 40 (1986).

6.9 Problems and practical exercises

1 Berry Gilders is a company of building contractors. Their normal practice is to work for local authorities, and although there is no shortage of such work, the company feels that it could improve on the profit of £50 000 per month yielded by such work. To this end they have investigated two larger projects. One is to build an extension to an airport terminal and the other is to build a new high street branch for the Rempstone Building Society. Both projects are due to start in six months time and both will last a year, although bids for the airport terminal must be in before the end of this month, and the successful applicant will be announced at the end of the following month. Bids for the Building Society project have to be in by the end of the third month. It would therefore be possible to bid for the airport terminal extension, and, if unsuccessful, to bid then for the Building Society. Berry Gilders have resources for only one of the two jobs; moreover if they bid successfully for the airport terminal contract, they could not then back out should they also bid successfully for the Building Society contract.

Berry Gilders are considering three levels of tender for each project. Associated with each is the expected profit and also their assessment of the probability that the tender will be accepted (Table 6.10).

Table 6.10

	Profit (£ million)		
	High tender	*Medium tender*	*Low tender*
Airport project	1.5	1.0	0.8
Building Society project	2.0	1.5	1.0
Probability of success (either project)	0.3	0.5	0.7

Express the problem in the form of a decision tree and determine the maximum EMV strategy.

2 Thurlbee Knitwear have received a large order for next Spring's sales of children's sweaters from the large chain store Fergus and Aitkenhead. This is the first time that they have received an order from F & A and although it could cause some disruption to their production line, they are keen to accept it.

The order involves some technical problems which they are confident they can solve in time. There is therefore the danger that the order will be supplied late and also that the quality will not be up to the high standards demanded by F & A. If the order is delivered on time and is of satisfactory quality the profit to Thurlbee will be £50 000. A late delivery will incur a penalty of £30 000, and poor quality (the term is used relatively) a penalty of £20 000. If the order is both late and of poor quality it will be rejected, resulting in a loss of £40 000.

Thurlbee have three basic options. They can fit the order into their regular time (option R), they can work overtime (option O) or they can subcontract the order (option S). Option S could be used alongside options R or O. A surplus order made in this way (whether late or on time) could be sold at a profit of £10 000 if the quality is satisfactory or at a loss of £10 000 if the quality is unsatisfactory. The firm

Table 6.11

	Lateness	*Poor quality*
Option R	0.3	0
Option O	0	0.1
Option S	0.1	0.1

does not have the expertise to operate both O and R simultaneously. Use of option O reduces the profit by £10 000 and use of option S reduces it by £15 000.

Thurlbee estimate the probabilities shown in Table 6.11. Assume that all the probabilities are independent.

What would you advise?

3 Pargetter Parts plc is a manufacturer of mining pumps. The firm has a government contract which necessitates buying 200 castings, which he normally buys in batches of 200 for a price of £1000 per batch. In six months time the contract is to be renewed and Pargetter may or may not be awarded it. The renewed contract is for a further 400 pumps requiring 400 of the same castings.

His supplier now offers larger batches at a discount. He can buy a batch of 400 for £1600 or 600 for £2000. The offer is for a limited time only and certainly will not be available at the time the contract is to be renewed. Any castings not used on these contracts have no value.

List the manufacturer's strategies, and determine his maximin and minimax regret strategies. Comment on the appropriateness of these criteria.

If π is the probability that Pargetter receives the contract on renewal, for what range of values of π are his various strategies optimal under the maximum EMV criterion?

4 It has recently been announced that the Royal Shepshed Yacht Club is to issue a challenge for the next Americas Cup race. If the challenge is successful the following races three years later will be held on Rutland Water.

Acol Developers have the opportunity to buy a site for a hotel at Empingham on Rutland Water. If they do they must now make the initial decision about the size of hotel to build. A large hotel would cost £10 million inclusive of the land cost. If the Shepshed challenge is successful this size of hotel will produce a return whose present value is £14 million. On the other hand if the challenge is unsuccessful the return will have a present value of only £9 million. Both of these calculations do not include the cost of the hotel.

Alternatively they can build a smaller hotel at a cost of £7 million. Irrespective of the outcome of the race challenge this will produce a return whose present value is £9 million. Again this calculation does not include the cost of the hotel. There is also the opportunity to extend the hotel to the large size after the result of the challenge is known. This will cost an extra £4 million and produce the same returns as the initially planned large hotel.

Local experts assess the chance that Royal Shepshed will win at 0.3. An opportunity has arisen which may improve this chance. Dr Wally Luftkopf of the Department of Marine Engineering at Nanpantan University of Technological Science is developing a special keel in the shape of a handbag. Needing money to test this idea Luftkopf approaches Acol with the suggestion that if they will sponsor this research with £1 million, they, Acol, will have the exclusive knowledge on the outcome of the project in time to decide on the size of the hotel to be built. An independent assessor estimates that the project has a 50–50 chance of success, but that if it is successful, Royal Shepshed's chance of winning will be increased to 0.8.

Express the problem in the form of a decision tree and deduce the optimum EMV strategy. Is Dr Luftkopf's research really worth the money he is asking? If not what would you offer?

5 A textile designer has produced a design of a special tee shirt to commemorate the three hundredth anniversary of the birth of Thomas Bayes. He now has the choice of marketing the garment himself or of selling the design to a large manufacturer.

The market for Thomas Bayes tee shirts is unpredictable. If the idea catches on, sales will be high and he will make a profit of £100 000 if he markets the garments himself. On the other hand, if sales are low, he will make a loss of £10 000. In selling the design he can negotiate a contract whereby even if sales are low he will make a profit of £20 000, but if sales are high the profit will only be £50 000. The designer estimates that the probability of high sales is 0.6, and therefore of low sales is 0.4.

A firm of market research consultants specializes in commemorative tee shirts. They claim to conduct a survey whose record is such that, if the sales are high, they will predict 'high' with probability 0.7, and therefore predict 'low' with probability 0.3. Alternatively, if the sales are low, they will predict 'low' with probability 0.8 and 'high' with probability 0.2.

If the survey costs £400 what strategy would the designer adopt? How would this strategy change if the survey cost £800?

7

Market research and the value of information

OBJECTIVES

The basic objective of this chapter is to show the value of market research information in the reduction of errors in decision making. All readers should be able to appreciate the principles involved, which hinge on the use of Bayes' theorem to adjust prior beliefs in the light of further information. The actual mechanics are demonstrated for the guidance of the more mathematically adept.

7.1 Market research

In the example used in the previous chapter, there were a number of occasions when it appeared that the relevant experts were to be asked for their opinions on the outcomes of certain of Nature's actions. These concerned matters such as a successful clinical trial or an economically viable product launch, and it was required that the opinions be expressed in terms of probabilities. It was tacitly assumed that this information could be gained without any real cost, apparently relying on the experience and intuition of the individuals approached. This is not always the case. There are occasions when some form of commissioned investigation has to take place. At the one extreme this may take the form of a rigorously conducted survey of the market, involving taking a representative sample and the consequent statistical analysis. A full treatment of this aspect of market research is beyond the scope of this book but reference may be made to the books by Stuart (1976) and by Scheaffer *et al.* (1979) listed at the end of this chapter. A simple example of this type of investigation is given in Section 7.7.

Alternatively, it may be appropriate to obtain information in a relatively subjective form, but using expertise from outside the firm making the decision. For example, when the viability of a project depends on the state of a country's economy, advice may be sought on such a matter and interpreted to decide whether or not to open an export agency within that country. Another example may be a firm contemplating entering the 'wholefood' market on a large scale. The question for them is how to make a projection of a hitherto increasing market. Will it continue to increase or is it due to reach a saturation level in the near future? The answer to this question depends on many factors, such as the spread of social class of the main consumers of wholefoods, attitudes of retail chain stores, the views of the medical profession, etc. To the firm itself, an increasing market would then have to be translated into a profitable venture, whereas a stable market could result in a project which is no longer economically viable.

Two features of the concepts introduced in this section should be emphasized. The advice gained, which we shall term broadly as market research, is subject to error. Any form of sampling which is less than a complete census may by chance produce an unrepresentative sample. The risk of this can be controlled but not eliminated. Moreover, advice of a more subjective nature can also be wrong, since no expert is infallible. In the treatment of market research within the decision tree, account will be taken of the errors made when actions are based on the outcomes of the research.

The second point about market research is that there is a cost involved. This is the fee paid for outside expertise, and it does not depend on the use made of the information. Part of the study made in the following sections is concerned with the evaluation of the information offered, so that if the track record is known in terms of the probability of correct or incorrect information, a limit can be set on the economic fee. This concept can be extended to a wider limit where no track record is available. A formal structure for these ideas will now be introduced by means of an illustrative example.

7.2 Interproximal Cleaners plc

Interproximal Cleaners plc is a national firm of dry cleaners with high street shops in most of the larger towns and cities in the United Kingdom. They have the opportunity to open a branch in Melchester, a town where they have not previously operated. The shop under consideration already offers a privately run dry cleaning business, and Interproximal can readily calculate the net present value of this venture, assuming that the amount of business remains at its past level.

Melchester already has two other shops offering a dry cleaning service, one a local enterprise and the other a branch of a national company of similar standing to Interproximal. Interproximal are, therefore, concerned about the reaction of customers and potential customers to the opening of a branch under their name. Will the burghers of Melchester take their custom away in favour of the local firm, or will they be attracted by Interproximal's reputation for quality service and technical expertise? Clearly their behaviour will affect the viability of Interproximal's investment, and it is therefore in the interests of the firm to try to assess the consequences and their relative likelihoods.

Interproximal decide to carry out their analysis on the simple basis that, if they do open a branch in Melchester, the outcome will be one of three possibilities, namely an increased market share (I), no change from the present operations (P), or a reduced market share (R). It is assumed that no other outcome can occur. They then associate an NPV with each of these outcomes as in Table 7.1 (units of £100 000).

Table 7.1

Outcome	*NPV*
I	+3
P	+1
R	−3

The next problem is to assign probabilities to the outcomes. After much deliberation and with considerable apprehension, they arrive at the figures in Table 7.2. The corresponding decision tree is of a very simple structure, shown in Figure 7.1 together with its associated EMV pattern.

Table 7.2

Outcome	*Probability*
I	0.2
P	0.5
R	0.3

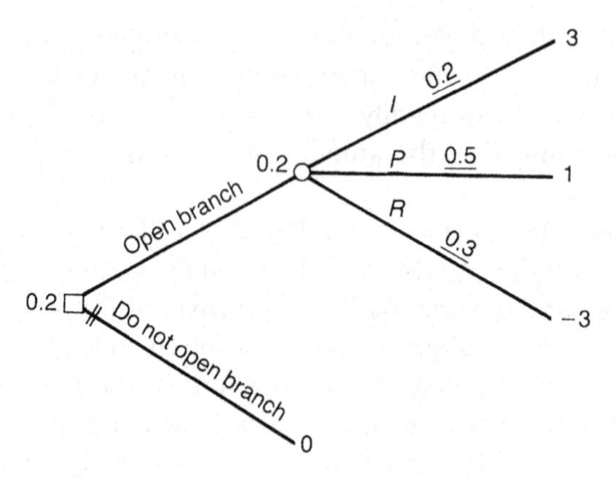

Fig. 7.1 Decision tree for Interproximal Cleaners

The resultant optimum EMV strategy is therefore to open the branch, and the optimum EMV itself is 0.2, or £20 000.

The management of Interproximal are conscious of the fact that, in coming to their decision by this criterion, the comparison was close. In fact a slightly more pessimistic view of the venture, which was expressed by one or two of the people involved, would have put the probabilities at

I 0.2; P 0.4; R 0.4.

The resultant EMV is then −0.2 for the 'open branch' action and therefore the optimum decision would be reversed. Because of the sensitivity of the optimum strategy to the assumed probabilities, Interproximal decide to seek outside professional help in assessing the viability of this project.

7.3 The form of the market research

The firm of Maxilla Enterprises plc has a great deal of experience of consumer surveys. When approached by Interproximal to carry out an investigation of the viability of opening a branch at Melchester, they agree to adapt their procedures in the following way. Firstly they will not simply carry out a survey of the shop's existing customers. Although Interproximal accept the impossibility of keeping secret their interest in the shop they do not want to advertise their intentions too widely by asking questions about the possibility of continuing custom if and when they take over. Instead, Maxilla intend to carry out a general survey of uses made of dry cleaning shops in Melchester, and by applying this information to similar towns where Interproximal have a branch, infer the likely impact of Interproximal on Melchester.

Much of the analyses will be subjective, and because of this, Maxilla are only prepared to present their findings as the simple dichotomy, favourable or unfavourable. Asked for the probabilities of each of these verdicts, Maxilla made the obvious reply that they could not answer this question until their investigation had started. This is reasonable, and it will later be shown how these probabilities depend not only on Maxilla's performance, but also on Interproximal's own prior probabilities of the three outcomes I, P and R.

A more convincing reply was obtained when Interproximal explained the three scenarios corresponding to their outcomes I, P and R. Asked then what were the probabilities of favourable and unfavourable verdicts, they were able to respond with the information in Table 7.3. Note that these conditional probabilities must sum across each row to unity, but no similar constraint applies to the column totals.

All this information has been gleaned at the stage when a possible consultancy arrangement with Maxilla Enterprises is being considered. No actual investigation has yet taken place, and Interproximal are still of the view that their best estimate of the probability distribution over the outcomes I, P, R is 0.2, 0.5, 0.3. The only further piece of information required is the fee to be paid for the investigation, and this is put at £10 000. Interproximal now have to decide whether or not to go ahead with the investigation and, if so, how they will use the resulting information.

7.4 Derivation of probabilities

The augmented decision tree is now shown in Figure 7.2. Denoting the outcomes from Maxilla's investigations by F (favourable) or U

Table 7.3

	Probability of verdict	
Scenario	*Favourable*	*Unfavourable*
Increased market share (I)	0.8	0.2
No change in market share (P)	0.5	0.5
Reduced market share (R)	0.1	0.9

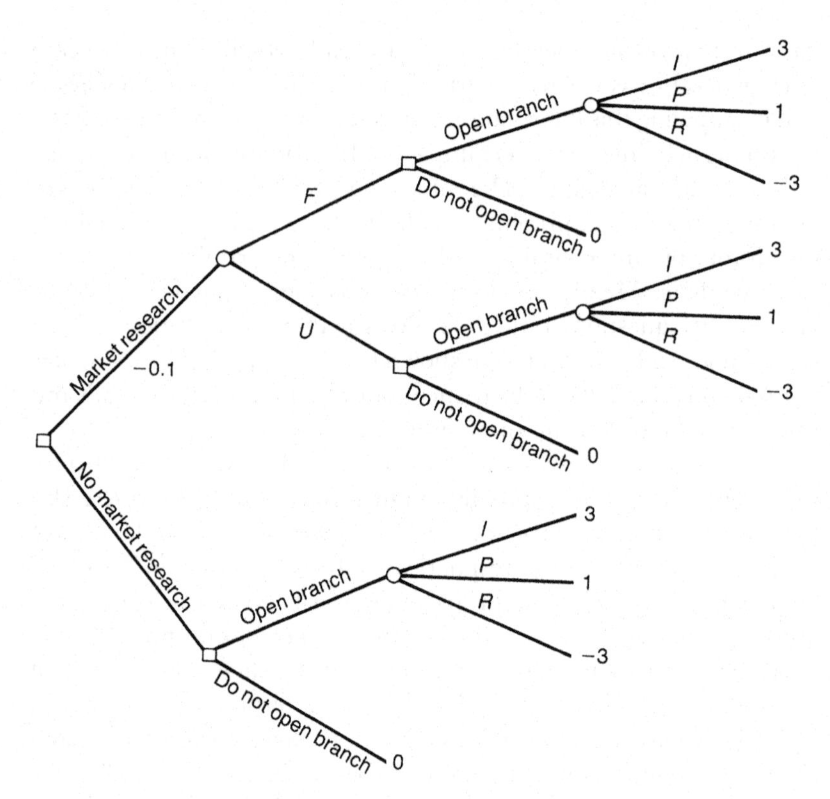

Fig. 7.2 Augmented decision tree, Interproximal Cleaners

(unfavourable), there is now the need to determine the probabilities $\Pr(F)$, $\Pr(U)$; $\Pr(I|F)$, $\Pr(P|F)$, $\Pr(R|F)$, $\Pr(I|U)$, $\Pr(P|U)$, $\Pr(R|U)$. The latter set of six conditional probabilities can be derived from the given data using Bayes' formula (equation (2.4)). The outcomes I, P and R are mutually exclusive and exhaustive and we therefore apply the formula in the following way:

$$\Pr(I|F) = \Pr(F|I)\ \Pr(I)/[\Pr(F|I)\Pr(I) + \Pr(F|P)\Pr(P) + \Pr(F|R)\ \Pr(R)].$$

$\Pr(F|I) = 0.8$, $\Pr(F|P) = 0.5$ and $\Pr(F|R) = 0.1$ from the table of information supplied by Maxilla. The prior probabilities $\Pr(I) = 0.2$, $\Pr(P) = 0.5$ and $\Pr(R) = 0.3$ are also used in this formula, as representations of the current state of knowledge.

Application of this formula in this case and for the remaining probabilities conditional on a favourable report yields the results:

$$\Pr(I|F) = \tfrac{16}{44},\ \Pr(P|F) = \tfrac{25}{44},\ \Pr(R|F) = \tfrac{3}{44}.$$

(The probabilities have been expressed as fractions with a common denominator for convenience in subsequent computations.)

Similar applications for the situation where the verdict is unfavourable produce the following probabilities:

$$\Pr(I|U) = \tfrac{4}{56},\ \Pr(P|U) = \tfrac{25}{56},\ \Pr(R|U) = \tfrac{27}{56}.$$

These, therefore, present the appropriate probabilities for the final stages of the branches of the tree when the initial decision to employ the market research has been taken. There remain the intermediate branches corresponding to outcomes of favourable or unfavourable for the market research. To determine each of these we refer to a further probability law derived in Chapter 2. This is the third law given in equation (2.3), and once again using the property that I, P and R are mutually exclusive and exhaustive,

$$\Pr(F) = \Pr(F|I)\Pr(I) + \Pr(F|P)\Pr(P) + \Pr(F|R)\Pr(R),$$

so that $\Pr(F) = (0.8)(0.2) + (0.5)(0.5) + (0.1)(0.3) = 0.44$. Similarly, or noting that $\Pr(U) = 1 - \Pr(F)$, $\Pr(U) = 0.56$.

This completes the probability requirements of the augmented decision tree (Figure 7.2) and we can now proceed to carry out the EMV analysis to determine the optimum strategy under this criterion. To end this section let us refer back to the way in which the conditional probabilities were requested from the market research consultants. They were asked for the probabilities of their verdicts given the three scenarios (I, P and R) specified by Interproximal. Combined with the prior probabilities of the scenarios this led to unique values for the reversed conditional probabilities and for the unconditional probabilities of the verdicts. Had the probabilities not been requested in this way, problems would almost certainly have arisen. For example, if we had asked directly for the conditional probabilities required in the decision tree, namely for $\Pr(I|F)$ etc., it is unlikely that these would match the given prior probabilities. Suppose in this case we had been given what are apparently quite reasonable values in the following way:

$$\Pr(I|F) = 0.80 \qquad \Pr(P|F) = 0.15 \qquad \Pr(R|F) = 0.05$$
$$\Pr(I|U) = 0.15 \qquad \Pr(P|U) = 0.25 \qquad \Pr(R|U) = 0.60.$$

$$\begin{aligned}\text{Then } \Pr(I) &= \Pr(I|F)\Pr(F) + \Pr(I|U)\Pr(U)\\ &= 0.80\ \Pr(F) + 0.15\ [1 - \Pr(F)]\\ &= 0.65\ \Pr(F) + 0.15.\end{aligned}$$

We originally decided that $\Pr(I) = 0.2$, and from this relationship it then follows that $\Pr(F) = 0.077$. Similar arguments for $\Pr(P)$ and $\Pr(R)$ produced values of -2.5 and 0.545 for $\Pr(F)$! Unless, therefore, the consultants have had access to the firm's prior probabilities and carried out calculations similar to those outlined in the main part of the section, this contradictory situation is always likely to arise. The fault lies in the question itself. We have seen that it can only be answered indirectly and with additional information which the consultant alone will not possess. It is important, therefore, not to ask questions requiring a fairly sophisticated level of reasoning, but to confine the needs to probabilities of clearly defined outcomes.

7.5 The augmented decision tree

The decision tree augmented by the market research decision and with the derived probabilities is shown in Figure 7.3. Also on this tree are the calculated EMVs for each node, assuming that the optimal EMV policy is adopted.

The optimal EMV strategy is therefore to use the market research, and if their report is favourable, to open a branch; if their report is unfavourable the project is abandoned. The EMV for this strategy is 0.54, or £54 000 in the original monetary units.

7.6 The value of information

The original statement of Interproximal's problem, before the possibility of market research was introduced, resulted in an EMV of 0.20. Market research in this particular case increased the EMV to 0.64 before the actual cost of the research was deducted. The difference between these figures, $0.64 - 0.20 = 0.44$, can therefore be considered as the value of the information supplied by the market research. This difference is usually known as the Expected Value of Sample Information (EVSI) and is defined as the difference between the optimal EMV when market research is used and when it is not. Two points should be made about this measure. Firstly, although as we have seen, it can be derived in advance of taking the decision about adopting market research, it relates specifically to the assumed efficiency for this problem, measured in terms of the conditional probabilities. Any

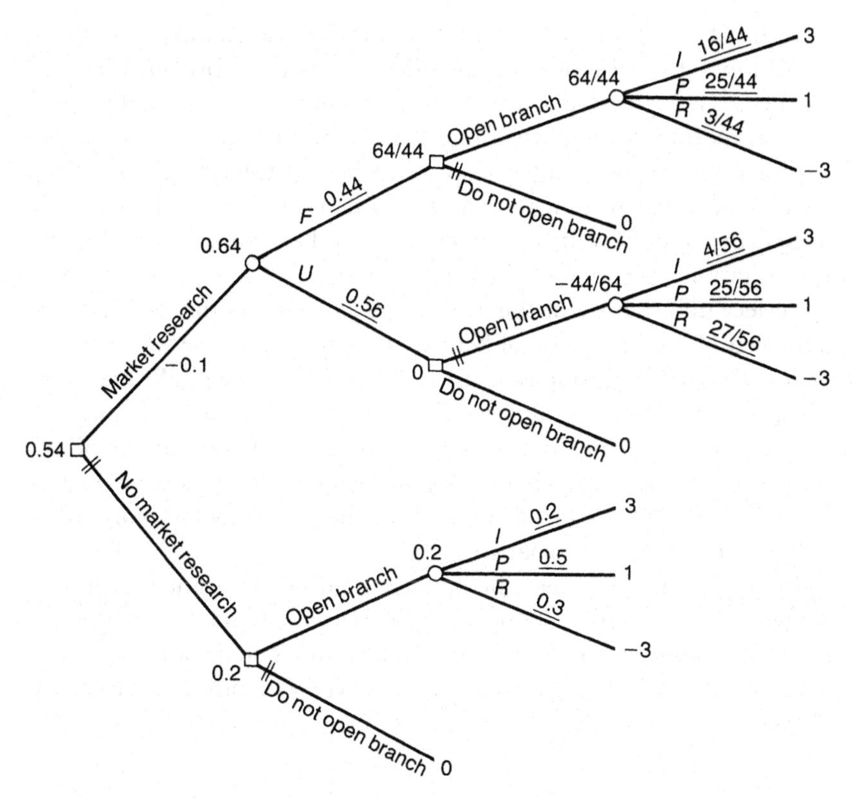

Fig. 7.3 EMV analysis for Interproximal Cleaners

change in these probabilities could result in a different EMV and would therefore change the EVSI. See, for example, Exercise 1 at the end of this chapter. The second point refers to the fact that once again a criterion is based on expected values. It measures the benefit in the long run if similar exercises are being carried out. In this case the benefit is that of making better decisions on the basis of market research information. Market research does not change the market, and it is of course true that the same action will always be less beneficial by the cost of the market research when an investigation is carried out. The EVSI is therefore a measure of the improvement so gained.

The EVSI is the maximum amount which should be paid for market research under these particular circumstances. The difference between EVSI and the cost of the market research is called the Expected Net Gain from Sample Information (ENGSI) which should therefore always be positive for the market research to be beneficial.

It may be that we lack information on the performance (in terms of conditional probabilities) of the market research organization. In such cases it is possible to argue towards a limit which should be paid for market information, assuming that the information is perfect. No organization can guarantee to provide perfect information, and the concept only has meaning as a target for market research, for example measuring the benefit which could still be gained from further investigation.

Perfect information in this rather special sense is the facility to be informed of the final state of Nature, that is in this case whether I, P or R will occur. Its value is measured before the actual information is disclosed. In some instances, of course, the information may be unwelcome, resulting in a loss or at best an abandonment of the project under consideration. Since the information has a price, this will add to the loss incurred. We cannot choose to pay for information only when it suits the balance sheet.

In Interproximal's case, the management would argue that if they knew that their market share would increase or remain at its present level, they would open the branch. If the market share was to reduce, they would abandon the project. The corresponding returns are therefore:

$$I, 3; \quad P, 1; \quad R, 0.$$

The best estimates available for the probabilities of these outcomes are the prior values 0.2, 0.5 and 0.3 respectively. Hence, weighting the returns by their probabilities results in an expected value of the return using this information of

$$3(0.2) + 1(0.5) + 0(0.3) = 1.1.$$

This is known as the Expected Monetary Value Under Certainty (EMVUC). The difference between this and the optimum EMV when no market research is used measures the value of using this information and is known as the Expected Value of Perfect Information (EVPI). This is the limit of the amount which should be paid for market research, no matter how efficient it is. In the example, this would be $1.1 - 0.2 = 0.9$, a sum of £90 000. The difference between this and the EVSI of £44 000 is a measure of the amount of improvement which could be gained if the market research could reduce the risk of error – for example, if Maxilla could improve their probabilities

Table 7.4

	Favourable (F)	*Unfavourable (U)*
Improved (*I*)	0.9	0.1
Present (*P*)	0.75	0.25
Reduced (*R*)	0.05	0.95

of favourable and unfavourable verdicts to the levels shown in Table 7.4.

At this level of efficiency, the EMV is 0.87 and the EVSI is therefore $0.87 - 0.20 = 0.67$. A value exactly half way between the previous EVSI and the EVPI has been achieved. Usually this is at the expense of disproportionately costly market research as the EVPI is approached.

7.7 Pinstripe Panels plc

In the example considered in the previous sections it was assumed that the market research was detached from the organization making the decision, in the sense that a package of investigation was negotiated, where the price and the efficiency (in terms of conditional probabilities) were given. It was then a matter of either accepting or rejecting the package. There are, however, occasions when the level of investigation can be adjusted virtually to any desired level, as for example is the case when a sampling survey is to be taken, whose cost depends on the sample size. The diminishing returns from equal increments of the sample suggest that there must be an optimal size for the cost parameters of a decision problem. In the example following a demonstration is given of the steps necessary to determine the optimal level of market research where the market consists of a well-defined set of individuals, any of whom may be included in the sample.

Pinstripe Panels are manufacturers of glass fibre sheets and panels, principally for the building and motor vehicle trades. They are currently contemplating launching a product which consists of a glass fibre panel which is specifically designed to give a particular vehicle, the Barkby Boxie, a more streamlined and sporty appearance. These would be fitted to any vehicle of this model. There are currently 100 000 registered owners of Boxies.

To set up a production line to manufacture Boxie panels would cost £500 000. Each panel sold would return a net contribution to profits of £60, giving a breakeven point of 500 000/60, or about 8333 panels. If, therefore, the number of Boxie owners who decide to buy the panel is Z, the decision tree is as shown in Figure 7.4.

At the end of this section we shall return to the particular values assumed in this example, but for clarity the argument will be developed using general algebraic symbols:

Let N = population size of potential customers
C = fixed cost of launching the product
m = contribution to profit of each unit sold
θ = proportion of the population buying the product.

Thus, if the product is launched, the total profit (or loss, if negative) will be

$$P = mN\theta - C. \tag{7.1}$$

This equation assumes that the demand is exactly met. Panels are manufactured only to customer orders.

The parameter θ must be between 0 and 1; also $C < mN$, since otherwise the product would make a loss even if the entire population bought one. (It is further assumed that nobody will buy more than one.)

Ideally we would launch if $P > 0$, so that from (7.1) $\theta > C/mN$.

In general, θ is unknown, although we shall usually have some idea of where θ is likely to be located. Suppose that we can express this in the form of a probability density function $f(\theta)$, ranging over $0 \leqslant \theta \leqslant 1$. Here we are using the concept of the probability density function to express our uncertainty about the value taken by what is really a parameter. It is a useful concept, particularly when the uncertainty can be refined as information in the form of data becomes

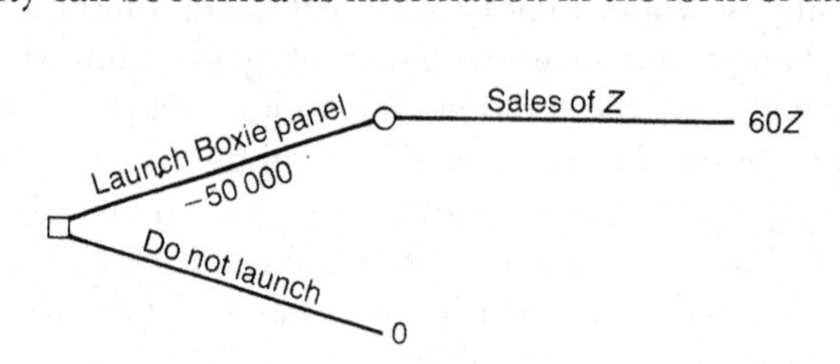

Fig. 7.4 Decision tree, Pinstripe Panels

available. Typically, the density might be a beta distribution (see Figure 2.13):

$$f(\theta) = \frac{(\alpha + \beta + 1)!}{\alpha!\ \beta!} \quad \theta^{\alpha}(1 - \theta)^{\beta}$$

A particular example of this distribution where $\alpha = 0$, $\beta = 0$ is the rectangular or uniform distribution, indicating that it is believed that all values between 0 and 1 are equally likely for θ. This is quite specific and certainly not the same as professing total ignorance of where θ is. Total ignorance is, in fact, a state of mind which is unacceptable for this procedure. For general values of α and β (which must be greater than -1) the distribution can be made skewed and concentrated to fit, say, a given

$$\text{mode} \quad \frac{\alpha}{\alpha + \beta}, \quad \text{or} \quad \text{mean} \quad \left(\frac{\alpha + 1}{\alpha + \beta + 2}\right),$$

and a second measure such as the variance

$$\frac{(\alpha + 1)(\beta + 1)}{(\alpha + \beta + 3)(\alpha + \beta + 2)^2}.$$

The distribution is the *prior* distribution of θ.

The expected monetary value if the launch decision is taken is the expected value of $mN\theta - C$, or $mN\mu - C$, where μ is the expected value of θ, defined using the prior density function of $f(\theta)$. Hence, we would launch the product if

$$mN\mu - C > 0,$$

and otherwise abandon it.

Now suppose that we have the possibility of carrying out market research in the form of taking a sample of size n from the population. The cost of the market research is

$$A + Bn,$$

where A is the cost of setting up the survey, and B is the marginal cost of each individual sampled. Our problem now amounts to finding the optimum value for n, optimum being defined as the value which maximizes the EMV for the whole decision tree. This includes the possibility of no market research, where not only will n be zero, but also the fixed cost A of sampling will not be incurred.

Suppose, therefore, that in the sample of size n, X customers report (truthfully) a willingness to purchase. With a population of large size N, the distribution of X can be assumed binomial (see Section 2.3.1). We now write the form of the distribution (equation (2.5)) as a conditional probability for a given value of θ, the proportion of purchasing customers, or equivalently the probability that any randomly selected customer will purchase:

$$\Pr(X|\theta) = \frac{n!}{X!(n-X)!}\,\theta^X(1-\theta)^{n-X}, \quad X = 0, 1, \ldots, n. \tag{7.2}$$

With a given prior distribution $f(\theta)$ for θ we can now adapt (7.2) applying Bayes' formula in the manner outlined in Section 2.5:

$$\begin{aligned} f(\theta|X) &= \Pr(X|\theta)f(\theta)\Big/\int_0^1 \Pr(X|\theta)f(\theta)\,d\theta \\ &= K\,\theta^X(1-\theta)^{n-X}f(\theta) \end{aligned} \tag{7.3}$$

$$\text{where } K^{-1} = \int_0^1 \theta^X(1-\theta)^{n-X}f(\theta)\,d\theta.$$

The probability density function $f(\theta|X)$ is known as the posterior distribution. It is a conditional distribution for the given value of X.

Note that although K depends on the X value realized, it does not depend on θ.

For any θ value, the return if the product is launched is $mN\theta - C$. To find the EMV we must now weight this return by the posterior probability $f(\theta|X)$ and sum – or in this case, since θ is a continuous variable, integrate.

Hence, if the product is to be launched, the EMV is

$$\int_0^1 (mN\theta - C)\,f(\theta|X)\,d\theta.$$

The EMV policy is therefore to launch if this expression is positive, which is equivalent to saying that the conditional or posterior mean value for θ should be greater than c/mN. This in turn gives a decision rule in terms of the observed value X.

For example, suppose that it is accepted that a beta distribution is appropriate for the prior distribution $f(\theta)$

$$f(\theta) = \frac{(\alpha+\beta+1)!}{\alpha!\beta!}\,\theta^\alpha(1-\theta)^\beta, \qquad 0 \leqslant \theta \leqslant 1.$$

The posterior distribution is therefore

$$f(\theta|X) = K \quad \theta^{\alpha+X}(1-\theta)^{\beta+n-X}, \qquad 0 \leqslant \theta \leqslant 1.$$

In this case there is no need to evaluate K as an integral, since the variable part of the function is recognizable as a further beta distribution with parameters $\alpha + X$, $\beta + n - X$. The complete expression is therefore

$$f(\theta|X) = \frac{(\alpha+\beta+n+1)!}{(\alpha+X)!\,(\beta+n-X)!}\,\theta^{\alpha+X}(1-\theta)^{\beta+n-X}, \quad 0 \leqslant \theta \leqslant 1. \tag{7.4}$$

Note that there is no difficulty with negative parameters since

$$0 \leqslant X \leqslant n.$$

The mean value of this distribution is

$$\mu(X) = (\alpha + X + 1)/(\alpha + \beta + n + 2),$$

and hence the product is launched if

$$(\alpha + X + 1)/(\alpha + \beta + n + 2) > C/mN, \text{ or}$$

$$X > \frac{C(\alpha+\beta+n+2)}{mN} - (\alpha+1) \tag{7.5}$$

Thus if X_0 is the smallest integer greater than $\dfrac{C(\alpha+\beta+n+2)}{mN} - (\alpha+1)$, the product is launched if $X \geqslant X_0$. The complete decision tree is shown in Figure 7.5.

It now remains to find the probability distribution of X. This can be determined by the property that $\Pr(X) = \int_0^1 \Pr(X|\theta)f(\theta)\,d\theta$. This is the extension of formula (2.3) for discrete outcomes, when the outcomes on which X is conditional are values of the continuous variable θ.

$$\Pr(X) = \int_0^1 \frac{n!}{X!(n-X)!}\,\theta^X(1-\theta)^{n-X}\,\frac{(\alpha+\beta+1)!}{\alpha!\beta!}\,\theta^\alpha(1-\theta)^\beta\,d\theta, \tag{7.6}$$

applying (7.2) to the example. The integral is that of the probability density function of a beta distribution, and since this is integrated over its full range ($0 \leqslant \theta \leqslant 1$) the integral can be evaluated directly as:

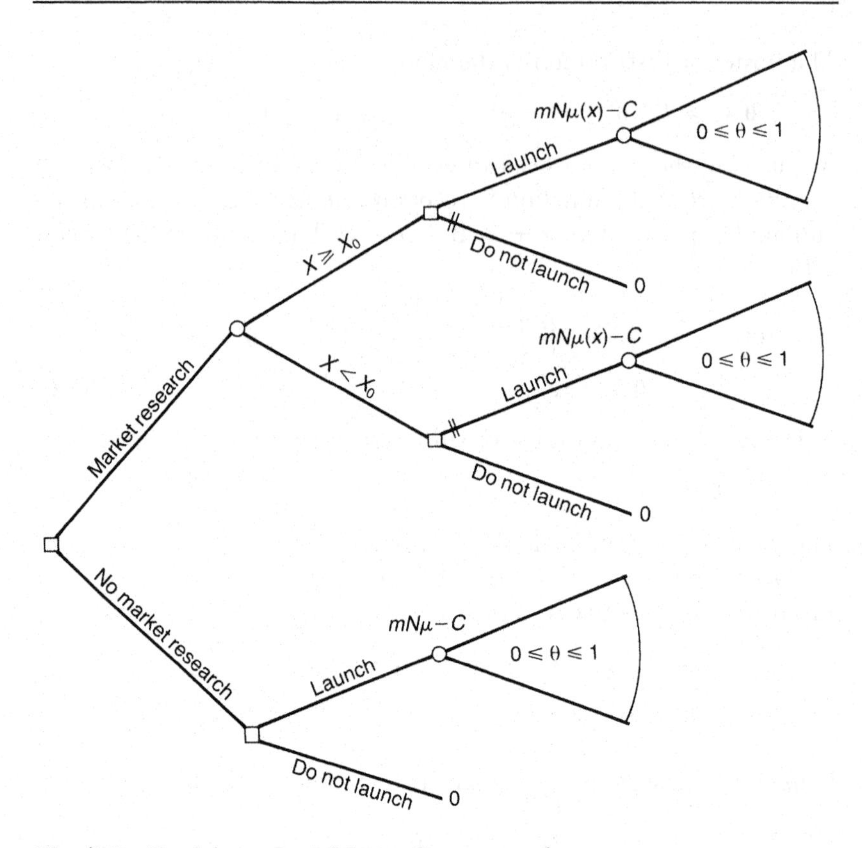

Fig. 7.5 Decision tree with market research

$$\Pr(X) = \frac{n!}{X!(n-X)!} \frac{(\alpha+\beta+1)!}{\alpha!\beta!} \Bigg/ \frac{(\alpha+\beta+n+1)!}{(\alpha+X)!(\beta+n-X)!}$$

This can be re-written in the form

$$\Pr(X) = \binom{\alpha+X}{X}\binom{\beta+n-X}{n-X} \Bigg/ \binom{\alpha+\beta+n+1}{n},$$

$$X = 0, 1, \ldots, n. \tag{7.7}$$

As such when α and β are integers it can also be derived as the probability of taking a sample of size $\alpha + X + 1$ from a population of size $\alpha + \beta + n + 1$ which contains $\alpha + \beta + 1$ defectives *and* n non-defectives, the sampling ceasing when $\alpha + 1$ defectives have been found. There is no obvious connection between the explanation and the sampling situation under consideration. The distribution of X is known as the negative hypergeometric.

The next stage of the EMV analysis is to derive the EMV at the node when the market research action has been taken. This is therefore the average EMV following each X value, weighted by the probability of X, or

$$\sum_{X = X_0}^{n} \Pr(X)\, [mN\mu(X) - C] \tag{7.8}$$

Note that the EMVs following values of X less than X_0 are all zero, since in these cases the project would be abandoned. The EMV of the branch corresponding to the adoption of market research is the expression (7.8) less the cost of the sampling, or

$$\sum_{X = X_0}^{n} \Pr(X)\, [mN\mu(X) - C] - A - Bn. \tag{7.9}$$

This is the expression which is maximized over n, and the maximum value is then compared with the EMV with no market research to ascertain the action to be taken. The difference between the two expressions is the Expected Net Gain from Sample Information (ENGSI).

The expected monetary value under certainty (EMVUC) can also be calculated. The venture is profitable for Pinstripe Panels if θ, the proportion ordering the Boxie panel, is greater than C/mN, in which case the profit is $mN\theta - C$. The optimum policy is therefore to launch the product if θ is greater than C/mN, but not otherwise. The expected monetary value is therefore

$$\int_{C/mN}^{1} (mN\theta - C) f(\theta)\, d\theta. \tag{7.10}$$

When θ has a beta distribution, this expression is the difference of two beta functions which, when the parameters of the distribution are integers, can be evaluated numerically. Alternatively, reference may be made to tables of this function, for example to the publication by Pearson and Johnson (1968).

We now demonstrate this approach by returning to the example and using numerical values. The population is of size 100 000 (N), the fixed cost of launching the project is £500 000 (C), and the contribution to profit per panel sold is £60 (m). Suppose, in addition, that the prior distribution of θ, the proportion of the population buying the panel, is beta with parameters $\alpha = 2$ $\beta = 26$. The mean of the prior distribution is therefore $(\alpha + 1)/(\alpha + \beta + 2)$, or 0.1, and the general form of the distribution is shown in Figure 7.6.

Note that this implies a proportion likely to be in the range 0.03 to 0.13, and very unlikely to be greater than 0.40. The sampling costs are assumed to be an amount of £10 000 (A) to set up the survey, after which each individual sampled is expected to cost £300 (B). The profit from a launched project (without market research) is therefore

$$(60)(100\,000)\theta - 500\,000$$

and the expected profit or EMV of the launch, assuming the prior distribution for θ, is the expected value of this expression,

$$(60)(100\,000)(0.1) - 500\,000 = 100\,000.$$

Since this is greater than zero, the EMV of 'no launch', the optimal action would be to go ahead with the project.

Although strict adherence to the EMV strategy produces the 'launch' action, the relative difference between the marginal profit of £600 000 and the fixed cost of £500 000 suggests that the decision is a close one. To be more specific, it appears that a departure in the actual value of θ, the proportion ordering, below the mean value of 0.1 could result in a loss to the firm. The actual break-even sales is, as we have already seen, 8333 panels corresponding to a θ value of

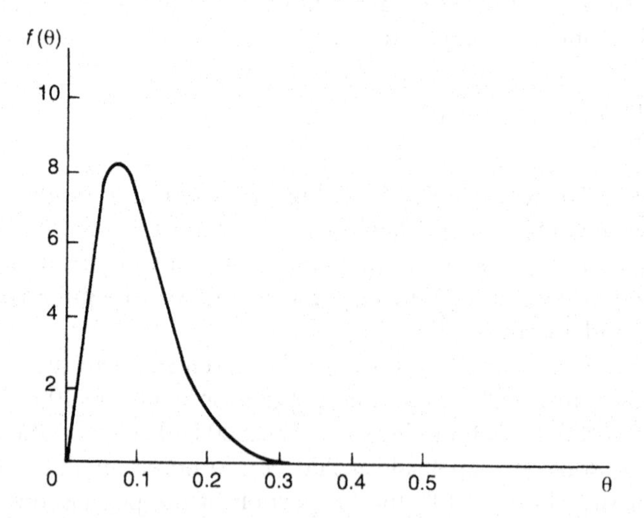

Fig. 7.6 Beta distribution, $\alpha = 2$, $\beta = 26$

0.0833 (or $\frac{1}{12}$), and hence any value of θ less than this critical value would mean an unprofitable venture. Using the given prior distribution for θ we can assess the chance of a loss by:

$$\Pr(\theta < 0.0833) = \int_0^{0.0833} f(\theta)\, d\theta.$$

This can be evaluated from tables of the incomplete beta function, or in this case, since the parameters α and β of the distribution are integers – one of which is the manageable value of 2 – the probability can be evaluated as the integral of a simple polynomial. The value here is 0.44. This relatively high chance of failure is one important incentive towards market research. Can the uncertainty in the situation expressed by the variability in the distribution of θ be refined by means of a sample of the population, so that a less variable posterior distribution will resolve more confidently the launch/no launch dilemma?

Following the procedure outlined in this section, suppose that a sample of size n is taken from the population, and X individuals in the sample report a willingness to purchase a panel. The posterior distribution of θ derived in equation (7.4) is therefore

$$f(\theta|X) = \frac{(n+29)!}{(X+2)!(26+n-X)!} \theta^{X+2} (1-\theta)^{26+n-X}, \quad 0 \leq \theta \leq 1.$$

This distribution has a mean value of

$$\mu(X) = (3+X)/(30+n),$$

and thus the product is launched if

$$(3+X)/(30+n) > 0.0833,$$

$$\text{or} \quad X > 0.0833n - 0.5 \qquad \text{(see inequality (7.5))}.$$

The critical value X_0 is the smallest integer greater than $0.0833n - 0.5$. It therefore increases in unit steps at values of n equal to 18, 30, 42, 54, etc. For example, a typical plan might be to take a sample of size 36 and to launch the product if at least three individuals indicate a willingness to buy.

Following equation (7.7), the probability of X is

$$\Pr(X) = \binom{2+X}{X} \binom{26+n-X}{n-X} \Big/ \binom{29+n}{n},$$

$$X = 0, 1, \ldots, n.$$

This reduces to

$$\Pr(X) = \frac{(X+2)(X+1)\; n(n-1)(n-2)\ldots(n-X-1)\; 29.28.27}{2(29+n)(28+n)(27+n)\;\ldots\;(26+n-X+1)}$$

Combining this with the profit

$$mN\mu(X) - C = 6 \times 10^6 \left(\frac{3+X}{30+n}\right) - 5 \times 10^5$$

gives the EMV for the particular sampling scheme as

$$\sum_{X=X_0}^{n} [6 \times 10^6 \left(\frac{3+X}{30+n}\right) - 5 \times 10^5] \Pr(X),$$

the expression corresponding to (7.8).

This can now be evaluated numerically, preferably using a computer. A plot of the EMV against the sample size n is shown in Figure 7.7. The EMV now has to be off-set against the cost of the survey. This is assumed to be linear in form with the given coefficients, and is therefore

$$10\,000 + 300n.$$

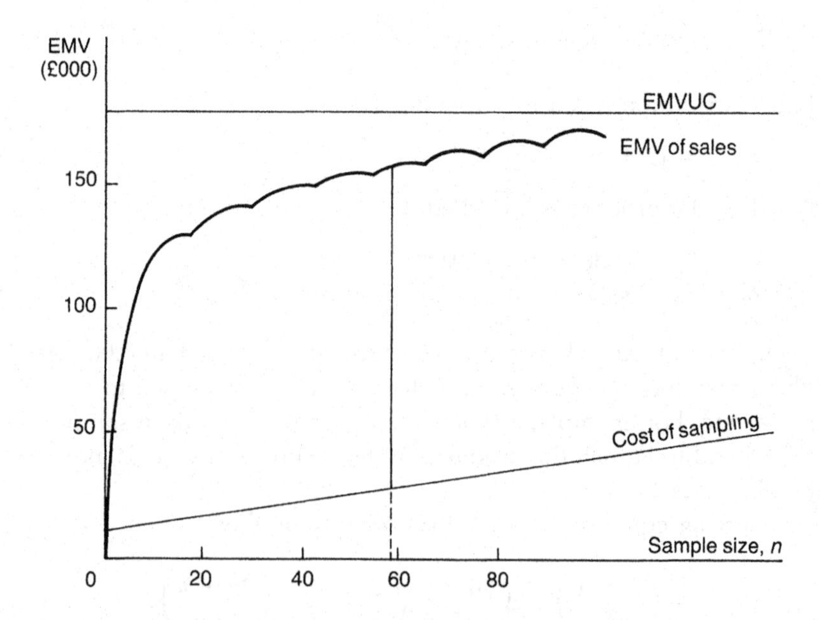

Fig. 7.7 The effect of sample size n on EMV

A plot of this relationship is also shown in Figure 7.7. The total EMV when any sample size is used is the difference between the curve and the straight line. This has a maximum of 127 407 when $n = 59$, and the plan, therefore, consists of taking a sample of 59 and launching the product if 5 or more individuals in the sample indicate a willingness to buy. The complete decision tree resulting from inserting numerical values in the tree of Figure 7.5 is shown in Figure 7.8.

Also plotted on the graph in Figure 7.7 is the expected monetary value under certainty (EMVUC). This is the EMV when the launch action is taken if $\theta > 0.083\,33$, but otherwise the project is abandoned. It is, therefore, calculated from the prior distribution as

$$\int_{0.083\,33}^{1} (6\,000\,000\theta - 500\,000)\, f(\theta)\, d\theta.$$

With $f(\theta)$ a beta function this is then the difference between two incomplete beta functions and can be evaluated as 177 152. Had the

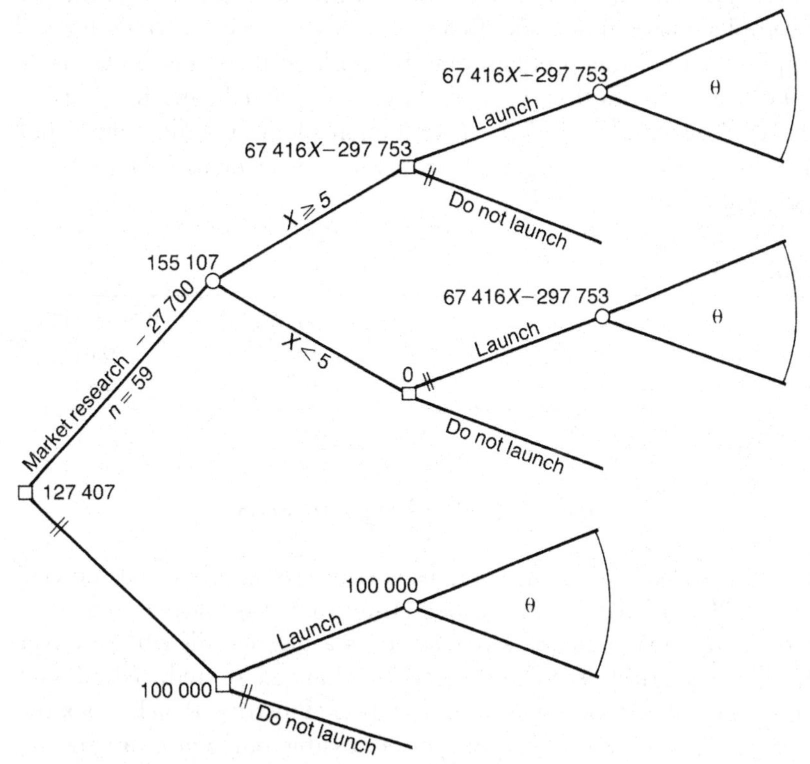

Fig. 7.8 EMV analysis for Pinstripe Panels plc

curve in Figure 7.7 been extended further it would have approached this value asymptotically, but the cost of sampling would have exceeded the EMV return well before the EMVUC was approached.

Finally we can measure the effectiveness of the strategy in terms of its ability to take the correct decision. With no market survey we saw that the optimal strategy was to launch the product. There was then a probability of 0.44 that θ would be less than 0.083 33 and the product would make a loss. With the optimal action now it is possible to make two types of error. When, in the sample, X, the number indicating a willingness to buy is at least 5, we launch the product. It is still possible that θ is less than 0.083 33 and a loss is made. Conversely, if X is 4 or less we abandon the project. Although we shall now never realize the outcome, it may be that θ turns out to be more than 0.083 33 and we abandon a project which would have been profitable. The two types of error are quite distinct; both can be derived from the sampling plan applied to the given prior probability distribution. The actual calculations (which will be omitted here) require the evaluation of further incomplete beta functions. Reference to this point and to further amplification of this problem may be made in the paper by Gregory (1984). It turns out that the probabilities associated with this decision are as shown in Table 7.5. Introduction of the market survey has therefore reduced the original probability of error of 0.44 to about half of its value.

Table 7.5

	$\theta < 0.08333$	$\theta > 0.08333$
$X \geqslant 5$ (launch)	0.19	0.81
$X \leqslant 4$ (abandon)	0.79	0.21

7.8 Concluding remarks

Two illustrations of the use of market research data have been given. They differ in that in the former (Interproximal Cleaners) the unknown state of Nature was described as a discrete distribution over the three possibilities of increased, unchanged or reduced market share. The model for the latter illustration (Pinstripe Panels) took the more realistic view of a continuum of possible outcomes, ranging the market share over any value between 0 and 1. This introduced a more complicated analysis, but the effort is surely justified for a decision of any real consequence. Readers who found the details of the analysis beyond their mathematical understanding should concentrate on the concepts involved and on the plausibility of the results.

7.9 Further reading

General

Gregory, G., 'Market Research and the Value of Information', Working Paper No. 86, Department of Management Studies, Loughborough University of Technology (1984).

Lindley, D. V., *Making Decisions*, John Wiley, London (1971), Chapter 7.

Moore, P. G. and Thomas, H., *The Anatomy of Decisions*, Penguin, Harmondsworth, Middlesex (1976), Chapters 5, 6.

Schlaiffer, R., *Analysis of Decisions Under Uncertainty*, McGraw-Hill, New York (1969), Chapters 10–14.

Survey sampling

Scheaffer, R. L., Mendenhall, W. and Ott, L., *Elementary Survey Sampling* (2nd edn) Duxbury Press, North Scituate, Mass. (1979).

Stuart, A., *Basic Ideas of Scientific Sampling* (2nd edn) Griffin's Statistical Monographs and Courses, No. 4, Griffin, High Wycombe (1976).

Statistical tables

Pearson, E. S. and Johnson, N. L., *Tables of the Incomplete Beta Function* (2nd edn) Cambridge University Press (1968).

7.10 Problems and practical exercises

1 Using the data from the Interproximal Cleaners example, suppose that the market research firm Maxilla Enterprises expresses its efficiency by conditional probabilities in the general form shown in Table 7.6.

Table 7.6

Scenario	*Probability of verdict*	
	Favourable (F)	*Unfavourable (U)*
Increased market share (*I*)	α_1	$1 - \alpha_1$
No change in market share (*P*)	α_2	$1 - \alpha_2$
Reduced market share (*R*)	α_3	$1 - \alpha_3$

Show that if $0 < 0.6\alpha_1 + 0.5\alpha_2 - 0.9\alpha_3 < 0.2$, there can be no benefit from market research according to the EMV criterion.

2 Find the EVSI in the problem of Exercise 1 when $\alpha_1 = 0.9$, $\alpha_2 = 0.25$, $\alpha_3 = 0.05$. Explain why the EVSI is marginally more than it was with the original probabilities of $\alpha_1 = 0.8$, $\alpha_2 = 0.5$, $\alpha_3 = 0.1$.

3 Refer to Exercise 1 of Chapter 6. Berry Gilders now decide that they must have one of the contracts. They are therefore prepared to put in a bid for the Building Society contract which will yield a profit of £700 000, but which will be certain to be accepted. How would this affect their bid for the airport contract?

4 Refer to Exercise 3 of Chapter 6. Suppose now that a firm offers to advise Pargetter on whether or not they will receive the contract. What is the maximum amount that they should pay for this information, assuming that *a priori* Pargetter believe that there is a 0.6 probability that they will be successful? The advising firm bases its recommendation to Pargetter on its knowledge of the view of the chairman of the awarding committee. If the chairman does not favour Pargetter there is absolutely no hope that they will receive the contract. On the other hand if the chairman favours Pargetter, he has a probability of 0.9 of persuading the committee to recommend Pargetter. What is the value of the information of the chairman's view?

5 Refer to Exercise 5 of Chapter 6. What is the maximum that the textile designer should pay for

(a) market research with the given prediction probabilities?

(b) perfect information?

Why do these answers differ substantially?

8
Linear programming

OBJECTIVES

By the end of this chapter the reader should be able to

(a) understand the formulation of the linear programming model;
(b) solve a two-variable LP problem by the graphical method;
(c) carry out a sensitivity analysis on costs and constraints for a two-variable problem;
(d) understand the concept of shadow prices;
(e) know what can be expected from a linear programming computer package;
(f) understand and solve the transportation problem;
(g) carry out sensitivity analysis on the transportation problem.

8.1 Optimization

In this chapter we depart from the theme of the book so far where decisions have been taken in the face of uncertain outcomes. Here it is assumed (at least in the basic model to be considered) that all elements of the model are known exactly and that there are therefore no chance or random features to be considered. Logically it follows that in such circumstances taking a decision is a matter of finding the best strategy, or in another word of optimizing. Later on we shall see how uncertainty in the model can be catered for by sensitivity analysis.

Although, therefore, we are now concerned with optimization, in many practical situations finding the mathematically optimum strategy is of less importance than finding a good strategy which works. There are a number of reasons for this. Firstly, there is the mathematical one that the expression which is to be minimized or maximized can be rather 'flat', so that significant departures from the optimizing strategy produce little effect on the value of the expression. This is not the main reason. Very often the problem area being modelled is highly complex and has to be broken down into a number of sub-areas each of which is then to be optimized. Sub-optimization in this way will not generally produce a global optimum, although with careful choice of the partitioning it should not be far off. Clearly, in such a situation there is little merit in expending considerable resources to determine the absolute sub-optima when good approximations are available at low costs. Furthermore, it is often difficult to relate the global objective which is to be optimized to the operation of the sub-systems. For example, a company may decide that it will plan its operations to maximize its profit after tax and depreciation. How does this objective guide it in the way it operates the fleet of vehicles used for the distribution of its products? What maintenance and replacement policies are implied by the global objective? In practice we usually argue in such a case that if our vehicles can provide the service required by the system at a minimum cost, a high level of profit should result. Optimization of the company's profit will not necessarily follow, largely because we have imposed as constraints on the sub-problem the service requirements which should be examined in any global study. Is it, for example, really in the cause of maximizing profit when the company makes daily deliveries to all of its depots?

Consider also the commonly met problem of production planning, and in particular the case of the tufted carpet manufacturer planning his production for the following four weeks. He requires a plan which will tell him how many yards of the various qualities and widths he should produce in this period. His plan is constrained by a number of factors. These are the total time he has available, the number of looms and other ancillary production units such as dye kettles, backing and shearing machines, the amounts of the different yarns both currently available and on immediate order, the skills of the work force, etc. Within these constraints and allowing for the time lost when a production facility is changed from one style of carpet to another, there will be a large number of feasible plans. The aim of the production manager is to find the best plan within the constraints, where 'best' is defined by some agreed criterion, such as maximizing the match with known and forecast customer demand or replenishing the finished goods stock to the best level of an agreed pattern. Optimization is scarcely relevant here. Not only is the objective difficult to agree unequivocally, but the constraints themselves are in many cases based on estimates, as would be the case with the actual time to needle a certain length of carpet where allowance is made for interruptions owing to breakdowns. A good plan would be expected to meet the constraints and also to meet the requirements of the sales, accounting and plant maintenance functions; it would also be flexible enough to cope with extra or reduced total output when the estimated constraints do not materialize precisely.

Despite the misgiving expressed above, we shall in this chapter be concerned with optimization. The reason for this is that there are circumstances where optimization is a relatively simple process, and it is almost as easy to find the optimum decision as it is to establish a good decision. A further justification is that a complex situation may be simplified to the point where an optimum can readily be found. In such situations it is reasonable to assume that the simplified optimum will, if it is feasible, be a good solution to the original complex problem

In formal mathematical terms the problem is one of selecting values for a number of variables, say $X_1, X_2, \ldots, X_m$ which optimize an objective function $g(X_1, X_2, \ldots, X_m)$. The variables are subject to a number of constraints which may be expressed as

$$h_1(X_1, X_2, \ldots, X_m) = 0, h_2(X_1, X_2, \ldots, X_m) = 0, \ldots h_n(X_1, X_2, \ldots, X_m) = 0.$$

The variables in the case of the multi-product production planning problem denote the amounts of each of the m products to be made in a particular period of time. Alternatively, they may be the amounts of the constituent parts of a mixture where the mixture has to satisfy certain constraints on specified properties and the aim is to form a blend at minimum cost of constituent parts. The blending of oil industry products, the mixing of cattle food and the blending of scrap metals are typical examples of this type of application. In theory the method is applicable to any form of optimization, but unfortunately the mathematical complications rule out all but the simplest forms of the functions g and h_i.

As far as solving the problem is concerned it might appear that all we have to do is to use each of the h_i equations to express one of the variables in terms of the others and we can thereby eliminate n of the variables in the objective function. The standard methods of differential calculus can then be used to find the optimizing values of the remaining variables. The mathematical difficulties of this approach can be appreciated if we try a very simple algebraic problem, namely that of minimizing $X + 3Y$ subject to the constraint that $X^3 + 2X^2Y + XY^2 + Y^3 - 1 = 0$. Simple elimination of either X or Y is now impossible, and the only possibility is to eliminate one of a pair of transformed variables. The complications when there are, say, twenty variables and perhaps a dozen constraints can well be imagined.

An alternative approach is to use the method of Lagrange multipliers. In principle this amounts to augmenting the expression to be optimized by adding multiples of the constraint expressions. The augmented function is then either maximized or minimized, treating the multiples introduced as further variables. Finding the optimum is then a matter of solving a set of equations resulting from equating the partial derivatives of the augmented expression to zero. Note that the constraints themselves are amongst these equations. This is still a procedure with considerable computational difficulties when the number of variables or constraints is large. In the simple problem quoted in the previous paragraph, the procedure evaluates the minimum at 1.739 when $X = -1.581$ and $Y = 1.107$. A full treatment of this approach is beyond the scope of this book. Many of the standard texts on mathematical analysis contain an account of Lagrange multipliers. For example, the book by Lucking (1980) has a useful introductory section, and a more rigorous approach can be found in the publications by Ledermann and Vajda (1981), and Bunn (1982). Further use of the Lagrange multiplier approach will not be made in the illustrations to follow, since alternatives which are superior in terms of practicability can be demonstrated.

The intention in this chapter is to direct most of the reader's attention to linear programming, the simplest of the mathematical programming models. Many problem situations can be modelled in the linear form and for others the form will be a good approximation. The book *Quantitative Approaches in Business Studies* by Clare Morris (1983) has (in Chapter 15) an excellent introduction to the linear programming technique.

As its name implies, linear programming (LP) is the special case of mathematical programming where the objective function $g(X_1, X_2, \ldots, X_m)$ and the constraints $h_i(X_1, X_2, \ldots, X_m)$ are all linear functions in the variables $X_1, X_2, \ldots, X_m$. A linear function is of the form

$$a_0 + a_1X_1 + a_2X_2 + \ldots + a_mX_m,$$

where $a_0, a_1, \ldots, a_m$ are all constants. Hence, not only does this rule out powers of the variables such as X_1^2, X_4^3, $\sqrt{X_5}$, but also cross-products X_1X_2 and reciprocals $1/X_3$ are not permitted. Care must be exercised with situations where there is a constant set-up cost a_0 and a constant marginal cost of production a_1. If X items are produced, the cost is therefore $a_0 + a_1X$, a linear function in X. It is linear, however, only if $X \geqslant 1$. The condition $X = 0$ corresponding to no production does not incur the set-up cost, and the linearity condition does not, therefore, hold for all feasible values of X. Consequently, the LP approach cannot be used without further modification.

An additional condition which is imposed on the model is that all the variables should be non-negative. This is useful, since in practice most of the variables we deal with cannot take negative values. When this is not the case for a variable X, it can be written as

$$X = Y - Z,$$

where both Y and Z are non-negative and at least one of them is zero.

Finally, although the problem as stated so far takes constraints as equations $h_i(X_1, X_2, \ldots, X_m) = 0$, the technique will cope equally well with inequalities of the 'less than or equal to' or 'greater than or equal to' variety. Again this is useful because often we are dealing with bounds on the amount of time available on a machine, or on, say, the octane level of aviation fuel. In the linear case we might, therefore, start with a constraint of the form

$$b_1X_1 + b_2X_2 + \ldots + b_mX_m \leqslant b_0.$$

This is converted into an equation by introducing the extra variable S_1, so that

$$b_1X_1 + b_2X_2 + \ldots + b_mX_m + S_1 = b_0.$$

The variable S_1 is called a *slack* variable. It measures the extent by which the bound is not reached by any solution $X_1, X_2, \ldots, X_m$. It must, therefore, behave like the other variables and be non-negative. Note that it is important that it should be introduced with the coefficient $+1$, since had the inequality been in the opposite direction,

$$C_1X_1 + C_2X_2 + \ldots + C_mX_m \geqslant C_0,$$

the slack variable would then have the coefficient -1,

$$C_1X_1 + C_2X_2 + \ldots + C_mX_m - S_2 = C_0,$$

where again $S_2 \geqslant 0$.

In both cases the slack variables can be treated as one of the original variables of the model. It will not normally enter the objective function, which does not contravene the linearity condition, since it can be taken as having a multiplying coefficient of zero. In summary, therefore, the linear programming problem consists of maximizing (or minimizing) a function

$$a_0 + a_1X_1 + a_2X_2 + \ldots + a_mX_m,$$

subject to constraints

$$b_{1j}X_1 + b_{2j}X_2 + \ldots + b_{mj}X_m = b_{0j}$$

for $j = 1, 2, \ldots, n$. In general n will be less than m (recall that the X variables include slacks).

The technique used to solve this problem, the simplex method, will not be demonstrated here. Reference to most texts on operational research (e.g. Ackoff and Sasieni (1968), Daellenbach and George (1978), Harper and Lim (1982), Moskowitz and Wright (1979), Wilkes (1980), Johnson (1986) or French *et al.* (1986)), or to texts devoted to LP or mathematical programming (e.g. Gass (1964), Hadley (1962), Vajda (1961)) will provide the information.

The simplex method can be carried out 'by hand'. It is an iterative procedure, requiring only the use of a few simple arithmetic rules to reach an optimum solution. The same procedure, or a variant of it, is also used in the computer packages now available for this technique. With the ready accessibility of computing facilities, it is doubtful if any practical LP problem would be solved manually nowadays. Certainly the effort would not be justified with more than, say, five or six variables. With only two variables it is possible to demonstrate the technique graphically. Thus, although few practical situations of this limited scope will arise, the lessons taken from such a demonstration can be carried over to the larger and more realistic problems, particularly when it comes to the interpretation of the computer package.

Much of the effort in any LP application goes into the collection of data. Before that we have to agree on an objective and to translate the objective into a mathematical form. Maximization of profit may, amongst others, require the assessment of the raw materials costs of a unit of each product made. We also have to discover the constraints – technical, policy and possibly ethical – on the solution we can offer, and again translate these into mathematical forms. The discipline of this exercise may itself prove invaluable in understanding the problem. Solving the problem is then relatively quick and cheap, but further mileage can be obtained by examining the sensitivity of the solution to the assumptions made about the coefficients such as costs, bounds, etc. in the model formulation. It is hoped to demonstrate these points in the example which follows.

8.2 The Delluge Camp Park

The owner of Delluge Camp Park in the Mediterranean state of Cote d'Effluence has been approached by Flybynite Holidays plc to have a number of camping places on his site allocated throughout the season to Flybynite's customers. Flybynite will pay the owner a fee of 50 Rz (Razu – the local currency) for any space on his site measuring 50 square metres. The usual space allocation for casual bookings on the Delluge Park is in units of 60 square metres and each such space realizes on average an annual income of 40 Rz. The total area of the Park is 15 000 square metres, but there are also other restrictions on the way that the Park can be managed.

Firstly, if he is to maintain his registration with the Cote d'Effluence Tourist Board, the owner must provide at least 30 places for casual hire. In addition the demand on the toilet and washing facilities depends very strongly on the number of tents in the Park, and the owner reckons that his facilities can cater for a total of no more than 280 tents. Finally, he provides recreational facilities for children and estimates that these can cope with a demand of up to 650. Beyond this figure he tends to have difficulties with noisy and frustrated children, and also with irritated parents. He finds with his normal casual hirings he gets on average 1.5 children per tent (one tent per place). Previously he has had only casual bookings and his facilities both for toilet and washing and for children have more than coped with his limit of 15 000/60 = 250 tents.

He now makes further enquiries with the Flybynite company and discovers that their publicity and charge structure are designed to attract families. Their tents will take a maximum of four children, although the firm's records show that an average number of 2.5 children per booking has been recorded.

The Park manager is attracted by this proposal, particularly when he realizes that the average income per square metre under the proposal is 1 Rz, whereas for casual bookings it is only $\frac{2}{3}$ Rz. Although he does not wish to lose his acceptance by the Tourist Board, he still has

15 000 – (30)(60) = 13 200 square metres to offer, sufficient for 264 Flybynite places. He is about to accept Flybynite's offer to this extent when he remembers his other restrictions. Not only would the total number of tents (264 + 30 = 294) exceed the limit (280) for his toilet and washing facilities, but such a change in the management of the Park could be expected to attract (264) (2.5) + (30)(1.5) = 705 children, again well above the Park's capacity of 650. The manager realizes that, as under the present conditions of only casual bookings, he has spare capacity in both the toilet and the children's facilities, he ought to be able to take up some of Flybynite's offer. The initial question is – how much?

8.3 Formulation of the problem

The first stage in the solution is to formulate a mathematical model of the problem area, or in other words to translate the text of the previous section into mathematical relationships. To do this we must first ask ourselves about the form of the action to be taken. In this case the manager wants two numbers, namely the number of places to be allocated to Flybynite and the number to be retained for casual bookings. Call these X and Y respectively. Certainly he will be interested in other values such as the total income which this action will produce and the load that the action will place on the Park's facilities, but these can be readily derived from the given X and Y values.

We now take each of the constraints in turn and express them mathematically. Take first the total area limitation. Since each Flybynite place requires 50 square metres and each casual place 60 square metres, the total requirement of this action is $50X + 60Y$. This must be less than or equal to the total amount available of 15 000 square metres, and thus the constraint can be expressed precisely as

$$50X + 60Y \leqslant 15\,000. \tag{8.1}$$

Next there is the Tourist Board's requirement that there should be at least 30 places for casual bookings. This is simply expressed as

$$Y \geqslant 30. \tag{8.2}$$

The demand on the toilet and washing facilities requires that the total number of tents (i.e. places used) should be no more than 280.

$$X + Y \leqslant 280. \tag{8.3}$$

Finally, there is the limit on the facilities provided for children. With his general (X,Y) action, the expected number of children would be $2.5X + 1.5Y$. The constraint is therefore:

$$2.5X + 1.5Y \leqslant 650. \tag{8.4}$$

There is also the necessity to consider only non-negative values of X and Y, so that

$$X \geqslant 0,\ Y \geqslant 0. \tag{8.5}$$

It is possible that some of the constraints may be redundant in the sense that if one is true, another must necessarily be so. Here the $Y \geqslant 0$ constraint in (8.5) is redundant when compared with (8.2). Obvious examples like this can be eliminated from the problem, although no harm is done by keeping them in, particularly when a computer package is used. Comparisons between constraints in several variables looking for redundancies are almost certainly not worth the effort.

Any pair of values (X,Y) satisfying the constraints (8.1)–(8.5) would be a feasible action for the manager to take. His problem now is to find the best action, which in this case is interpreted as the action which will maximize his return R, where

$$R = 50X + 40Y. \tag{8.6}$$

The linearity of the model is now apparent. All of the constraints are linear, and in addition, the objective function R in equation (8.6) is also a linear function of the variables. The problem has been reduced to a mathematical exercise in optimization.

With only two variables the problem can be represented geometrically. Any action (X,Y) is equivalent to a point with these coordinates in a standard two-dimensional plane. The constraints restrict the feasible actions to a region within this plane, and the aim now is to find the point or points within the region of feasible actions which maximizes R. Each constraint divides the whole space into two by a straight line, one side of which (and the line itself) consists of feasible actions, the other side being the infeasible actions. The dividing line is simply the constraint expression replacing the inequality sign by an equality. The simplest way of determining which side of the line corresponds to the feasible actions is to take a point which is obviously either feasible or infeasible and to locate it in relation to the line. Once one point on a particular side has been identified as either feasible or infeasible, the same label may be attached to the entire side. For example, for the first constraint (8.1), the line $50X + 60Y = 15\,000$ is drawn. The origin (0,0) then identifies the side of the line corresponding to feasible actions, since $X = 0$, $Y = 0$ satisfies the constraint (see Figure 8.1). The infeasible side is denoted by hatch marks along the appropriate side of the line.

When all the constraints are treated in this way each in turn will slice off part of the remaining region of feasible actions. Note that constraints (8.5) restrict the region to the positive quadrant. In theory it is possible to eliminate completely the region, but in practice this usually means that somebody has used incorrect information about a constraint. Normally we end up with a *region of feasible actions* which is bounded by straight lines. It also has the important property of being *convex*. This means that if any two points A and B within the region are taken, the whole line connecting them is also within the region. Examples of regions which are convex and which are not convex are given in Figures 8.2(a) and (b).

A region such as that shown in Figure 8.2(b) could not possibly arise using the approach under discussion here, since all the boundary lines continue indefinitely.

The complete diagram for the Delluge Park example is shown in

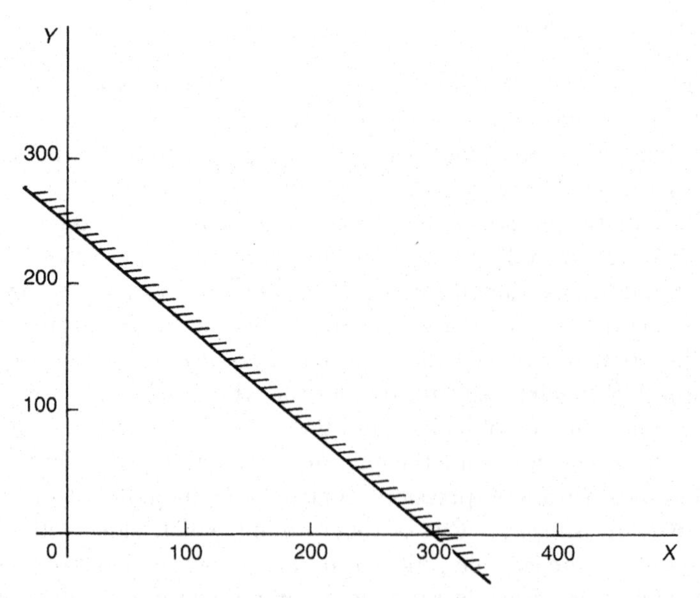

Fig. 8.1 Graphical representation of the total area constraint

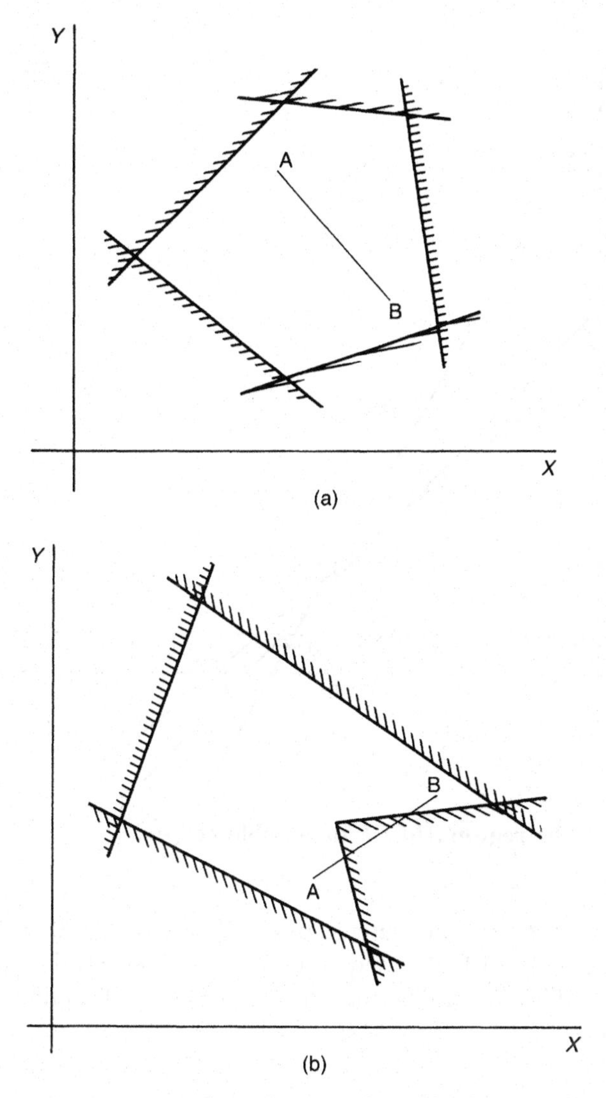

Fig. 8.2 (a) A convex region; (b) a region which is not convex

Figure 8.3, where the region of feasible actions is the pentagon *ABCDE*. Note that it satisfies the conditions of convexity.

When such figures are drawn in a freehand manner it is useful to check the accuracy of the diagram by checking the configuration of closely placed intersections such as *B, C* and *F*. In this case the co-ordinates of *F*, found by solving the simultaneous equations corres-

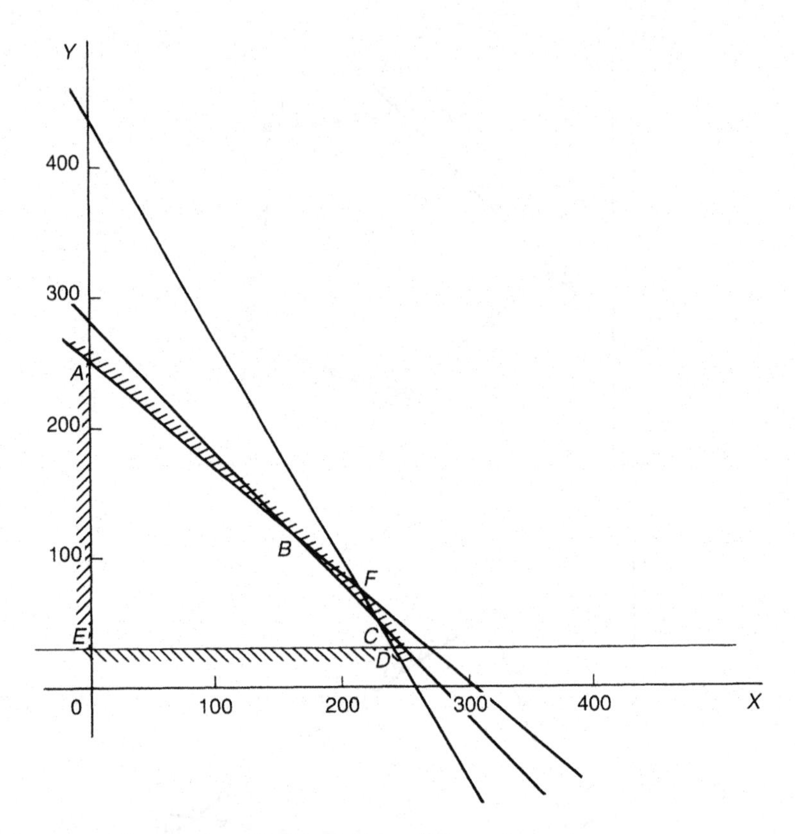

Fig. 8.3 The region *ABCDE* of feasible actions

ponding to the two lines intersecting at F, are $X = 220$, $Y = \frac{200}{3}$. This point does not satisfy the inequality $X + Y \leqslant 280$ and therefore lies on the side of this line remote from the region of feasible actions, as shown in the figure.

8.4 Derivation of the optimal solution

Before being approached by Flybynite, the Delluge Park manager operated with 250 places for casual use. This used all the space available, and also was well within the limitations of facilities for toilets and washing and for children's recreation. It produced a total income of (40) (250) = 10 000 Rz.

As a starting point in his present deliberations, the manager would like to be assured that he could achieve this income from the more general (X,Y) action. To do this any pair of values X and Y would be satisfactory, provided that $R = 50X + 40Y = 10\ 000$. The stipulation therefore amounts to constraining the coordinates (X,Y) of the action to lie on the straight line defined by this equation. Plotted on the diagram of the feasible actions, the line appears as in Figure 8.4. The important point to note about this line is that it intersects the region of feasible actions in the chord AK, and thus any point within this chord is both feasible and productive of a return of 10 000 Rz. Moreover, no other action has these two properties simultaneously.

Now suppose that the manager, encouraged by progress so far, goes for the more ambitious income of 10 500 Rz. To achieve this he would need an action (X,Y) for which $R = 50X + 40Y = 10\ 500$. When this line is plotted on the diagram of feasible actions (see Figure 8.5) it will be seen that the line is parallel to the line $50X + 40Y = 10\ 000$ and also that it still intersects the region. Further additions to required income produce similar parallel lines, all of which

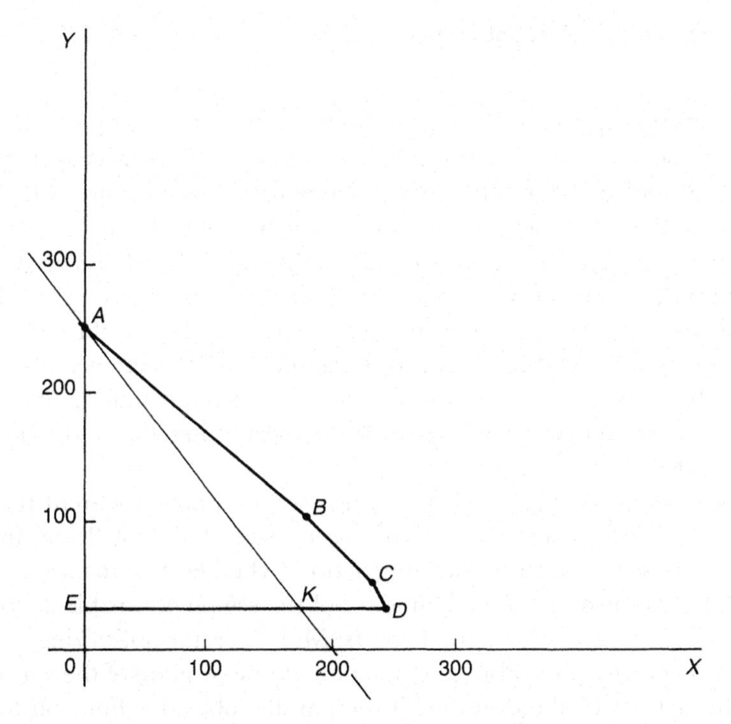

Fig. 8.4 Feasible strategies for a profit of 10 000 Rz

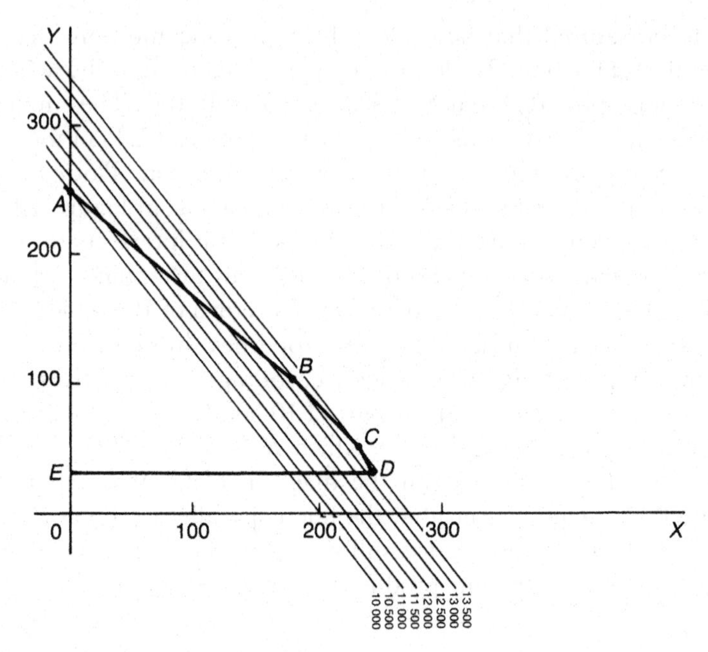

Fig. 8.5 Lines of equal income

intersect the region until at the value of 13 500 the line just touches the region at the single point C. From then on the lines miss the region entirely and therefore although the actions would produce a higher income the corresponding (X,Y) values are outside the region of feasible actions, and therefore cannot be used. It follows that the maximum return occurs when $X = 230$, $Y = 50$ and is 13 500 Rz. Had there not been a line in the family considered through the boundary we would have had to refine the scale of increments to the right-hand side of the objective function, R, until the limiting line was found. Also, moving the lines in the opposite direction reduces the value of the objective function.

Having solved the problem in this way, it is now apparent that a much quicker method could have been used. The first lesson from this exercise is that the optimizing action (whether it be maximum or minimum) must lie on the boundary of the region of feasible actions. We can discard all the internal points. Moreover, the optimizing value must be taken by the objective function at one or more of the vertices of the region. In the example, if the parallel objective function lines happened also to be parallel to a boundary on the appropriate side of the region of feasible actions, all values on this side would give the same optimal value of the objective function, and this of course includes the two vertices at the extremes of this side. In such cases we have multiple optimal values, any of which could be chosen to maximize the return.

Because of these properties (which generalize to the problem with any number of variables and constraints) the optimization problem is reduced from considering the countless infinity of points within the region to a comparison of the values of the objective function at the finite number of vertices of the region. In this case there are only five candidates for the optimum, namely the coordinates at the points A, B, C, D and E. By solving the simultaneous equations of the two lines intersecting at each of these points, we can find the coordinates and therefore the value of the objective function corresponding to each point (Table 8.1). The optimizing property of C(X = 230, Y = 50) is again demonstrated. This approach still depends on the drawing of a reasonably accurate diagram, since it is important that non-feasible intersections of constraint lines (such as the point F in Figure 8.3) should not be considered in the above table.

Table 8.1

Vertex	X	Y	$R = 50X + 40Y$
A	0	250	10 000
B	180	100	13 000
C	230	50	13 500
D	242	30	13 300
E	0	30	1 200

The general problem with any number of variables and constraints is solved by what is known as the simplex method. Although this particular approach will not be discussed here in detail, it depends heavily on the result that only vertices need to be considered. Unfortunately, despite the fact that there are only a finite number of such points, this number can in practice be very large, and there is still a need for a systematic approach to examining the vertices. In essence the simplex approach starts with one vertex – often a rather poor one as far as the objective function is concerned – and then moves to an adjacent vertex always in the direction of improving (or, to be precise, not deteriorating) the objective function. Because of the convexity of the region of feasible actions, this will lead to the optimizing vertex. More importantly it only uses a manageable fraction of the total number of vertices, thereby reducing the computational effort to an economic level.

8.5 Sensitivity analysis – objective function

The solution found assumes that the incomes of 50 Rz and 40 Rz for Flybynite and casual spaces respectively were known precisely. Sometimes this is not the case. For example, the manager's estimate of an income of 40 Rz for casual spaces depends on assumptions about the weather, money exchange rates, alternative attractions, local inflation rates, etc. How wrong would the manager have to be before he should change his optimal strategy? This is largely a matter of thinking of the order of magnitude of the errors involved and comparing these with the change in the cost which would lead him to change his strategy. Alternatively, the Flybynite cost, which was a specific offer and therefore not subject to error, may, however, be open to further negotiation. Suppose for example that Flybynite are anxious to take as many places at Delluge Park as they can. They have done some investigation and have come to the same conclusion, namely that the manager can count on an income of some 40 Rz for each of his places. Innocent of the LP model, they had therefore assumed that an offer of 50 Rz would lead to a contract for all 280 places. Even when informed of the need to retain 30 places to keep the Tourist Board happy and also of the constraint on the facilities for children, they had to be persuaded by the LP model that their offer was insufficient to obtain the action which would be in their best interests, namely at point D of the region where $X = 242$, $Y = 30$. By how much will their offer have to be increased to persuade the manager to move to this action? Alternatively, they might have thoughts along the lines that their offer of 50 Rz produced an allocation of 230 places. Could they have achieved the same allocation with a smaller offer?

Behind the thoughts expressed above is the knowledge that in this example the optimal action can only be one of five possibilities corresponding to the vertices of the region. It follows that there must be a range of values of the incomes from Flybynite and casual spaces for which any of these vertices is optimal, and that we need to identify the ranges in order to answer the questions raised. Geometrically this can be seen, since the income coefficients in the objective function determine the slope of the objective function line, and this in turn determines the vertex of the region of feasible actions at which the family of parallel lines finally just touches the region. More objectively the range can be determined by a simple algebraic procedure.

Suppose that the incomes are f and c for Flybynite and casual spaces respectively. Then the total income for any action (X,Y) is $fX + cY$. Applying this to the five candidates for optimality we have Table 8.2.

Table 8.2

Vertex	X	Y	$fX + cY$
A	0	250	$250c$
B	180	100	$180f + 100c$
C	230	50	$230f + 50c$
D	242	30	$242f + 30c$
E	0	30	$30c$

Hence action C will be optimal if $230f + 50c$ is greater than or equal to the income values corresponding to the other vertices, namely $250c$, $180f + 100c$, $242f + 30c$, and $30c$. Of the four inequalities here, two are redundant and the operative ones which remain correspond to comparisons with the incomes from adjacent vertices. These reduce to the basic inequalities

$$c \leqslant f \quad \text{and} \quad c \geqslant 0.6f \tag{8.7}$$

Note that the original values $c = 40, f = 50$ satisfy these inequalities, verifying once more that vertex C is the optimal action for these incomes.

We can now turn to the questions raised in the introductory paragraph of this section. Here we take Flybynite's offer of 50 Rz as fixed, and determine the range of values for c for which the action remains optimal. From the inequalities (8.7) we now see that the action is optimal if c is more than 30 Rz but less than 50 Rz. The actual value used in the model lies well inside these limits and hence in the circumstances the answer is robust to changes in the assumed value within an error of -10 to $+10$. Turning now to Flybynite's problem, if they assume a value of 40 for c, inequalities (8.7) now present a range of values for f from 40 to $66\frac{2}{3}$, over which the manager would use the same optimal strategy. From this it is clear that Flybynite would have to increase their offer to more than $66\frac{2}{3}$ Rz before additional places would be allocated; conversely their offer could be reduced to just over 40 Rz before they would lose allocated places.

Similar inequalities can be drawn up for the costs appropriate to each vertex being optimal. The actual manipulations, left as an exercise for the reader, result in Table 8.3.

Table 8.3

Vertex	*X*	*Y*	*Optimizing inequalities*	
A	0	250	$c \geq 0$	$c \geq 1.2f$
B	180	100	$c \leq 1.2f$	$c \geq f$
C	230	50	$c \leq f$	$c \geq 0.6f$
D	242	30	$c \leq 0.6f$	$f \geq 0$
E	0	30	$c \leq 0$	$f \leq 0$

Again with only two unknowns (f and c) under consideration these inequalities can be plotted, so that the whole space of the (f,c) co-ordinates is divided up into a number of regions, each of which corresponds to the optimality of a particular vertex. Note that the regions must cover the whole space, and that there are no overlaps. The plot is shown in Figure 8.6.

With such a diagram the optimum action for any income figures could be directly found simply by plotting the (f,c) value. The current value of $f = 50$, $c = 40$ is seen at point Z to be well within the vertex C region. The limits on the two values for which the vertex remains optimal are found by drawing vertical and horizontal lines through Z. Care should be exercised when the sensitivity analysis is carried out simultaneously on the two variables. For example, although vertex C remains optimal if $40 \leq f \leq 66\frac{2}{3}$ and if $30 \leq c \leq 50$, both of these ranges assume that the other variable is at its original given value, that is at $c = 40$ and at $f = 50$ respectively. The values of $f = 60$, $c = 35$, although satisfying the inequalities individually do not, when plotted, produce a point within the region C. The point lies just across the boundary in region D (at the location W in Figure 8.6) and therefore an action $X = 242$, $Y = 30$ would then be optimal. It is particularly important to bear this fact in mind when sensitivity analysis on problems of a higher dimension is carried out, and the limits are produced in the form of a computer output. The figure also shows that if all the incomes (or costs) are changed by the same multiplicative factor, the

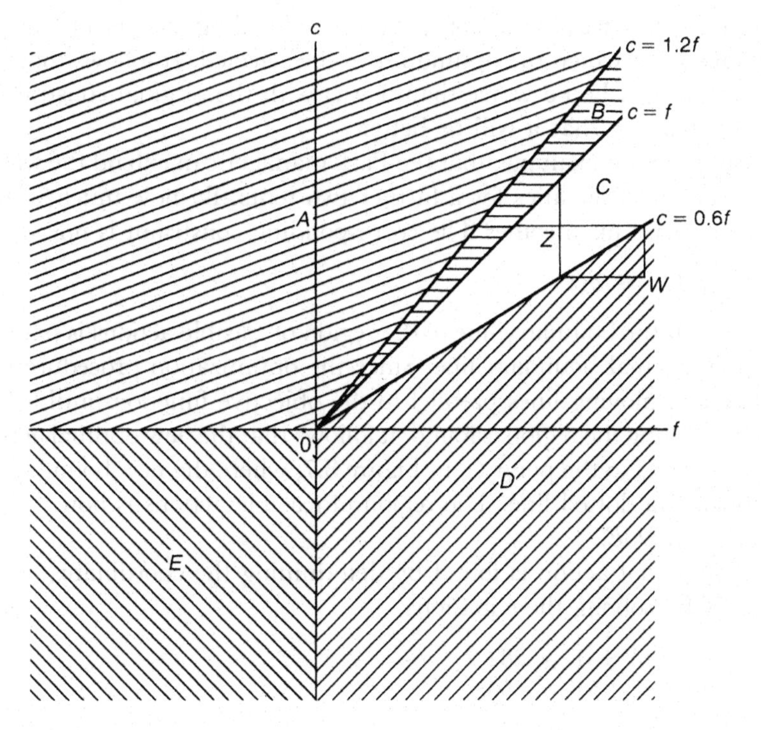

Fig. 8.6 Division of the cost space into optimal strategy regions

optimizing action is unchanged. The plotted point moves radially from the origin and therefore radial boundaries cannot be crossed. This is also clear from the linearity of the objective function. The same property does not necessarily hold for additive changes.

8.6 Sensitivity analysis – constraints. Shadow prices

In the treatment of the optimization problem so far, we have assumed that the constraints are absolute limitations on the feasible actions. Any action which satisfies the constraints is permissible, but even the slightest excess over a single constraint rules out the action from consideration. Although this concept of an absolute division between the acceptable and the unacceptable will be retained, there will be occasions when it is possible to relax one or more of the constraints, usually at a price. The effect of such relaxations is naturally to enlarge the region of feasible actions, thereby extending the set of feasible actions from which the optimum is to be selected. It follows that the optimum can be no worse than it was with the original constraints, and it is likely that it will be better.

Suppose, for example, that the Park manager can extend the washing and toilet facilities for a further two tents. By how much would this increase his income? The corresponding constraint is now

$$X + Y \leqslant 282,$$

for which the boundary line of the region of feasible actions is parallel to the original boundary line. Since the optimization procedure is a matter of moving lines parallel to the objective function until a line just touching the region is found, it follows that the new optimal action corresponds to the point C_1 in Figure 8.7, where the new boundary is shown as a dotted line. The region of feasible actions is now A, B_1, C_1, D, E.

The actual action is found by determining the coordinates of C_1, that is by solving the equations:

$$\begin{aligned} X + Y &= 282 \\ 2.5X + 1.5Y &= 650. \end{aligned}$$

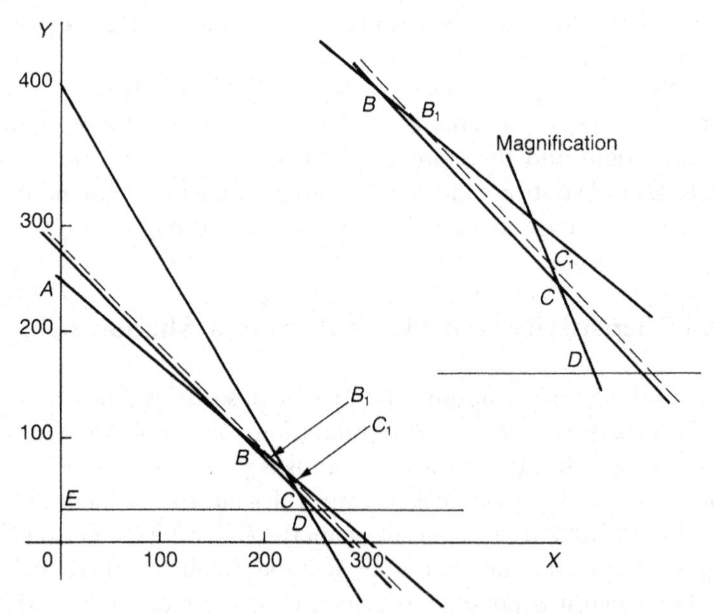

Fig. 8.7 Effect of extending toilet and washing facilities

The action is, therefore, $X = 227$, $Y = 55$ and the corresponding income is $50X + 40Y = 13\ 550$. There is therefore an increase in income of 50 Rz, and this may well persuade the manager that this is a satisfactory return on the investment required to provide the facilities. Note that it implies a reduction of 3 in the places offered to Flybynite, but an increase of 5 in places available for casual bookings.

Encouraged by these results, the manager proposes a further extension of the facilities for two more tents and discovers that this results in an action $X = 224$, $Y = 60$ and a further increase in income of 50 Rz. Similarly for an extension to 286 tents, the action is $X = 221$, $Y = 65$ and an income of 13 650 Rz. The pattern is now established. Each unit by which this constraint is relaxed increases the income by 25 Rz, and in so doing reduces the number of Flybynite places (X) by 1.5 but increases the number of casual places (Y) by 2.5. (Note that in the argument leading up to this result we actually relaxed the constraint in steps of two units in order to avoid consideration of fractional numbers in the actions. We shall return to the question of integer solutions in Chapter 10.)

If, therefore, the project is viable on a marginal basis of a return of 25 Rz for each extra tent catered for, the manager will naturally wish to pursue this idea to the limit. Clearly there is a limit to which the argument can go and this can best be seen in Figure 8.8, which is a magnification of Figure 8.7 in the region of the vertices B and C. As the constraint is relaxed, the corresponding boundary lines move parallel to themselves, giving successive optimal actions at C, C_1, C_2, C_3. If we pursue the relaxation to a limit of 288, the corresponding action is C_4, which is beyond the other constraint $50X + 60Y \leqslant 15\ 000$. The actual coordinates of C_4 are $X = 218$, $Y = 70$, yielding a value of 15 100 for $50X + 60Y$. Clearly the procedure must stop when the progression of C points reaches F. In this particular case the coordinates of F are fractional ($X = 220$, $Y = 66\frac{2}{3}$), so the actual limit may not be attainable in practice. Notwithstanding this matter, the extent to which the increase in income of 25 Rz may be achieved for every unit increment of the right-hand side of this constraint is thus circumscribed. The actual limit for this constraint occurs when the right-hand side is $286\frac{2}{3}$, an increase of $6\frac{2}{3}$. Beyond this the constraints on total area and on facilities for children become operative, the washing and toilet constraint being redundant.

The value 25 is said to be the *shadow price* for the constraint. It is the amount by which a unit relaxation in the constraint will increase the

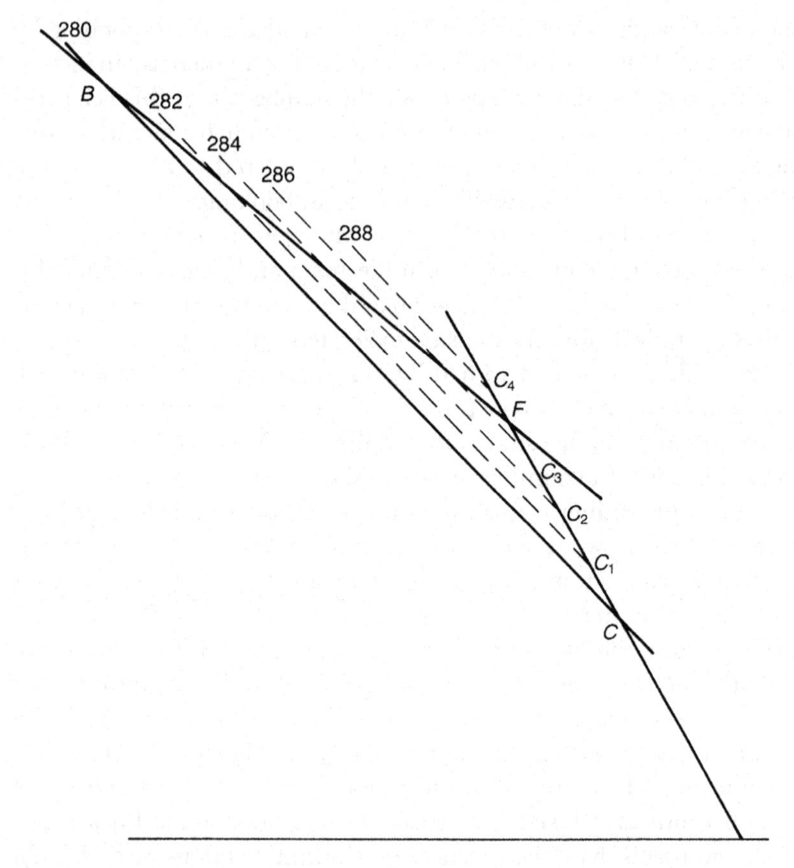

Fig. 8.8 Sensitivity analysis on washing and toilet facilities

objective function, or alternatively it is the maximum price which should be paid to have the constraint relaxed by one unit. The terms *marginal price* and *opportunity cost* have also been used for this concept. Most importantly, as we have seen above, it is only appropriate for changes in a limited range in the constraint. Had the inequality in the constraint been in the 'greater than or equal to' direction, the relaxation would have taken the form of a reduction in the right-hand side of the constraint.

A similar argument applied to the constraint

$$2.5X + 1.5Y \leqslant 650$$

produces a shadow price of 10 (Rz). Moreover, each unit increase in the right-hand side produces an increase of one in the X value and a decrease of one in the Y value in the optimum action. The limit on the range of validity for the shadow price occurs when the line parallel to CD goes through G (see Figure 8.9), that is when $X = 250$, $Y = 30$ and the right-hand side is 670, an increase of 20.

Note that shadow prices are expressed in units of the constraints as fed into the problem. In the above these were 25 Rz for each extra tent's washing and toilet facilities and 10 Rz for each extra child's recreational facilities. It is sometimes tempting to simplify the inequalities by multiplying through by a constant. For example, the second case considered above could be expressed equivalently as

$$5X + 3Y \leqslant 1300,$$

thereby removing the decimal numbers. The shadow price would, if this inequality were used, be halved at 5 (Rz) since we are now working in units of recreational facilities for half a child.

So far shadow prices have been calculated for the constraints whose lines meet at the optimal vertex. Applying the same idea to the other constraints produces shadow prices of zero, for the obvious reason that if there is a positive slack in a constraint there can be no benefit in extending the slack by relaxing the constraint. For example, the optimal action of $X = 230$, $Y = 50$ requires a total space of 14 500 square metres. This means that 500 square metres in the park will not be used, and it would therefore be pointless (in terms of income) to try to extend the size of the park. Similarly, there would be no merit in trying to persuade the Tourist Board to lower their requirement

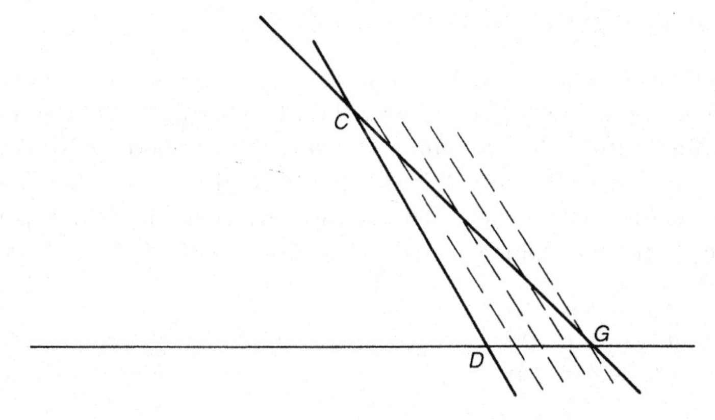

Fig. 8.9 Relaxation of the constraint on recreational facilities for children

from 30 casual places. One major asset from this analysis is, in fact, that it points out which constraints should be attacked to improve the profitability and which should, for the time being, be left alone.

We have stressed the point that shadow prices are operative over only a limited range. This is true if an individual constraint is changed and the remainder are unaltered. If, however, two constraints with positive shadow prices are simultaneously changed, the range for each constraint can be extended considerably. The point can be demonstrated in the case of the example, since we are only dealing with two constraints. In the more general situation there will be rather more constraints and we are therefore working in a space of more than two dimensions. Thus, the lesson from this is still valid, although the precise values cannot be so easily shown.

Suppose, therefore, that the right-hand side of the toilet and washing constraint (8.3) and the children's facilities constraint (8.4) are augmented by the positive quantities a and b respectively. (Note that we could permit negative values corresponding to a tightening of the constraints. In some examples this could have a beneficial effect on the objective function, but not here.) The constraints are now

$$X + Y \leqslant 280 + a, \; 2.5X + 1.5Y \leqslant 650 + b. \tag{8.8}$$

The intersection of the boundary lines found by replacing the inequalities by equality signs lies at

$$X = 230 - 1.5a + b, \qquad Y = 50 + 2.5a - b,$$

and the corresponding value of the objective function is

$$13\,500 + 25a + 10b. \tag{8.9}$$

From this it can be seen that increasing a and b improves the objective function, the only restriction being that the coordinates of the action used must satisfy the remaining constraints (8.1) and (8.2), namely $50X + 60Y \leqslant 15\,000$ and $Y \geqslant 30$. We must also ensure that $X \geqslant 0$. Note that the coefficients in (8.9) are the corresponding shadow prices. These restrictions reduce to

$$7.5a - b \leqslant 50 \tag{8.10}$$
$$-2.5a + b \leqslant 20 \tag{8.11}$$
$$1.5a - b \leqslant 230 \tag{8.12}$$

Plotted, they define a region $OLMN$ as shown in Figure 8.10. Maximizing the increase in the income is now a further linear programming

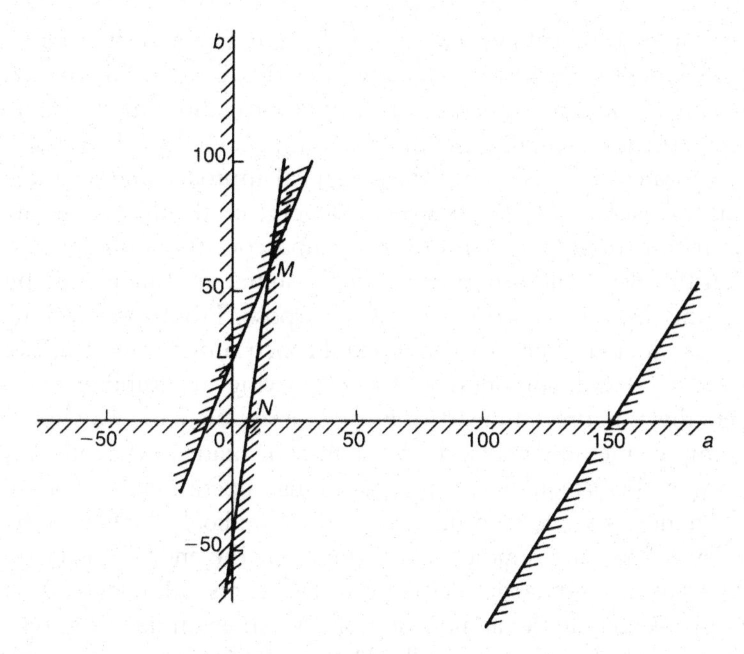

Fig. 8.10 Region of feasible constraint changes

problem, although it is, in this case, clear that the maximum must occur at M ($a = 14$, $b = 55$) with an income of 14 400 Rz.

In summary, therefore, we observe that the shadow price is zero for any constraint where there is already a positive slack, which is only a rather elaborate way of saying that there is no benefit from loosening a constraint which already has slack in it. For the constraints with zero slack, we saw that we could measure the shadow prices, and we determined limits on the amounts by which each constraint could be relaxed for this shadow price, assuming that all other constraints were unaltered. Finally, we derived the region over which the amounts by which the constraints could jointly be relaxed, again for the given shadow prices.

8.7 Solution by computer package

In practice most optimization problems are much larger than the one used for illustrative purposes in this chapter. They are larger in the sense of involving more decision variables and also more constraints.

Clearly the graphical approach is inadequate for such situations. An iterative method devised by Dantzig in 1951 will cope with any size of problem, the limitation being only that of the computing device used. This method is generally known as the simplex method. It is based on an algebraic generalization of the graphical approach and operates by means of a series of tableaux which converge on the optimal solution. The arithmetic application of the method consists of operations for which the electronic computer is highly suited, and many computer programs have been written for this purpose. Problems of a limited size – say up to five or six variables and constraints – could feasibly be tackled by hand, although with the growing accessibility of computers this is not recommended. For a description of the way in which the simplex method works, reference may be made to virtually any of the many books on linear programming, operational research or management science, and in particular to the books by Gass (1964), Hadley (1966) and Vajda (1961) quoted at the end of this chapter.

Most of the standard LP computer programs will offer sensitivity analysis as well as the actual solution. As an example of the type of output which is encountered, the Delluge Park problem was run on IMPS (Interactive Mathematical Programming System) a package devised by E. C. Love and J. Stringer and adapted by D. G. Johnson at the Australian Graduate School of Management, University of New South Wales.

The output on pp. 208–9 starts by listing and labelling the variables and constraints. The optimum income is given as line (8.17), and the optimal action in the preceding block under OPTIMUM solution basis. Lines (8.14) and (8.16) indicate the action to be taken, the amounts given in (8.13) and (8.15) showing by how much the constraint right-hand sides exceeded the requirement of the optimal action. For example, the $x(4)$ slack of 500 is identified from the preceding block as referring to the space constraints. Similarly, the surplus of 20 for $x(3)$ refers to the Tourist Board requirement. By inference the remaining constraints are at their limits.

The shadow price analysis for the binding constraints is self-explanatory, except that we had previously confined our attention to the benefits we would receive by relaxing the constraints. Should we be obliged to move in the opposite direction, the same price would operate initially in the sense that it would reduce the optimal value. The lower limits in lines (8.18) and (8.19) are the limits at which the reduction per unit change in the right-hand side would itself change.

```
Contents of linear programming data file  delluge


Variable names
                 x             y



Constraint    1
'space        ' 50.0000       60.0000
<= 15000.0


Constraint    2
'touristboard'0.000000D+00  1.00000
>= 30.0000


Constraint    3
'toilet.wash ' 1.00000       1.00000
<= 280.000


Constraint    4
'children     ' 2.50000       1.50000
<= 650.000


Objective function (income       ) to be MAXIMIZED
                 50.0000       40.0000


> lpge


List of additional simplex variables

                                              Simplex
Constraint      Name            Type          Variable        Label
----------      ----            ----          --------        -----
     1          space            <=           x(  4)          slack
     2          touristboard     >=           x(  3)          surplus
     2          touristboard     >=           x(  7)          artificial
     3          toilet.wash      <=           x(  5)          slack
     4          children         <=           x(  6)          slack

> simp
none


OPTIMUM solution basis

                    x(  4)    'slack          '        =      500.000 (8.13)

                    x(  2)    'y              '        =      50.0000 (8.14)

                    x(  3)    'surplus        '        =      20.0000 (8.15)

                    x(  1)    'x              '        =      230.000 (8.16)

MAXIMUM value of the function 'income         '        =      13500.0 (8.17)
```

```
Type 'y' if you wish to see the shadow prices ? y


SHADOW PRICE ANALYSIS

--Binding Constraints

                                                   Right Hand Side Range
Constraint    Name          Shadow Price     Lower Limit    Upper Limit
---------     ----          ------------     --------------------------
     3        toilet.wash    25.0000          272.000        286.667         (8.18)
     4        children       10.0000          600.000        670.000         (8.19)


--Non-Binding Constraints

                                                   Right Hand Side Range
Constraint    Name          Shadow Price     Lower Limit    Upper Limit
---------     ----          ------------     --------------------------
     1        space           0.0            14500.0        0.10000 0D+08    (8.20)
     2        touristboard    0.0            0.0000         50.0000          (8.21)

Type 'y' if you wish to see a cost analysis of basic variables ? y


COST RANGING ON BASIC VARIABLES

--Explicit Variables

                             Objective       Cost Coefficient Range
Variable      Name           Coefficient     Lower Limit    Upper Limit
--------      ----           -----------     --------------------------
 X(  2)       y              40.0000         30.0000        50.0000          (8.22)
 X(  1)       x              50.0000         40.0000        66.6667          (8.23)

--Augmented Variables

                             Objective       Cost Coefficient Range
Variable      Name           Coefficient     Lower Limit    Upper Limit
--------      ----           -----------     --------------------------
 X(  4)       slack           0.0            -1.00000       .333333          (8.24)
 X(  3)       surplus         0.0            -10.0000       10.0000          (8.25)
```

A similar argument can be applied to the non-binding constraints, which of course have zero shadow prices. The upper and lower limits for lines (8.20) and (8.21) are the extremes before which the shadow prices become positive, or in this case, before the corresponding line, moving parallel to itself is absorbed into a coordinate axis or passes through the current optimum point or vanishes off the diagram. (Note that the figure $0.100000D + 08$ represents the computer's attempt to reach plus infinity.)

In the final section of the output on cost ranging we have in the first tableau (lines (8.22) and (8.23)) the limits of the costs for which the solution remains optimal. These agree with the values found in

Section 8.5. The last tableau refers to what are termed *augmented* as opposed to *explicit* variables. The explicit variables are the genuine original variables of the problem, being in this case the number of Flybynite and casual places. The augmented variables are slack variables introduced into the inequalities to convert them into equations, which in addition have positive values in the final solution. Usually such variables do not enter the objective function, or equivalently enter with a zero coefficient (as is the case here). If, however, there is a possibility of using the slack profitably or conversely of paying a fee if the slack is not used, then this should be included in the objective function. For example, with the space constraint there are 500 square metres spare in the optimal solution. The Park Manager could rent this space to a hot dog concession, he could grow tomatoes on it or turn it into a tennis court, from any of which he could show a profit. Alternatively, it could be that any space not used will need more maintenance in terms of grass-cutting, litter collection, etc. and therefore there will be a cost. The range -1 to $+\frac{1}{3}$ indicated in line (8.24) gives the range of a cost of 1 Rz per square metre to a profit of $\frac{1}{3}$ Rz per square metre within which the solution remains optimal. The same argument would hold for the Tourist Board constraint (line 8.25) although in this case it is more difficult to conceive of plausible interpretations.

8.8 The transportation problem – problem formulation

A special case of the linear programming problem arises when a commodity is available at a number of sources and required at a number of destinations; where the cost of transportation is the product of the cost per unit for each route, and the amount sent along the route. This latter proviso is important – it requires, for example, that ten units cost ten times as much to send as one unit. There is no fixed cost in using any route and also no saving when larger amounts are transported. The problem now is to find the allocation of the commodity to the different routes which minimizes the total cost.

In mathematical terms, suppose that we have m sources and n destinations and that

A_i = amount available, source i,
R_j = amount required, destination j.

We impose the further restrictions that the total amount required equals the total amount available, that is that $\Sigma A_i = \Sigma R_j$, although this requirement can be lifted.

Suppose also that the cost of transporting one unit of the commodity from source i to destination j is C_{ij}. Then, if X_{ij} is the amount transported from source i to destination j, the total cost is

$$Z = \Sigma\Sigma\, C_{ij} X_{ij}, \tag{8.26}$$

which is to be minimized subject to

$$\sum_{j=1}^{n} X_{ij} = A_i, \qquad i = 1, 2, \ldots, m \tag{8.27}$$

$$\text{and} \quad \sum_{i=1}^{m} X_{ij} = R_j, \qquad j = 1, 2, \ldots, n \tag{8.28}$$

Also $X_{ij} \geqslant 0$ for all i, j.

Equation (8.27) expresses the constraint that everything at source i is sent to some destination, and conversely equation (8.28) is the constraint that the total arrivals at destination j equals the requirement for that destination.

Hence in equation (8.26) a linear function of non-negative variables is being optimized subject to constraints (8.27) and (8.28), all of which are also linear, and the problem can therefore be solved by the standard approach of LP. The special structure of this problem, however, permits the use of a different method which allows reasonably large and practical problems to be solved by a manual approach. This will be illustrated in the following section.

8.9 The transportation problem – methods of solution

The Brakeless Car Hire Company has a number of centres in the United Kingdom where its customers may either pick up or surrender its vehicles. On the following day it can match the demands with available vehicles at the centres apart from customers arriving at Folkestone (8 customers), Gatwick (31 customers), Harwich (3 customers) and Southampton (10 customers). The firm has suitable vehicles available only at the following centres: Bath (2 vehicles), Birmingham (24 vehicles), Canterbury (5 vehicles), London (15 vehicles), Oxford (2 vehicles) and Nottingham (4 vehicles). They have sufficient of their own drivers to take vehicles from any centre to any other centre, but as they estimate that any of these journeys costs 20p per vehicle mile, they wish to find an allocation which minimizes the overall cost. Table 8.4 summarizes the distances (miles), followed by the corresponding costs (*p*) in brackets, the availabilities and requirements.

The method of solution to be demonstrated here is usually known as 'stepping stone'. It starts off with a feasible solution which is also *basic* (a term to be defined) and then carries out iterations to improve the solution until the optimum is reached. It can start with any basic feasible solution, although it will generally save iterations if a good starting solution can be found. The term 'basic' refers to a particular form of efficient solution. If a problem has m rows (sources) and n columns (destinations) it will always be possible to find a solution which only uses $m + n - 1$ routes between sources and destinations. It may, because of degeneracy (a further term which requires definition) be possible to find a solution with less than this number. Because of this a basic solution must also have the property that it is impossible to find a solution which uses only a subset of its used routes. To illustrate the point, suppose that a solution included the following used routes:

Birmingham–Folkestone	2 cars
Birmingham–Gatwick	2 cars
London–Folkestone	5 cars
London–Gatwick	10 cars

This particular group of allocations could be amended by using

Birmingham–Gatwick	4 cars
London–Folkestone	7 cars
London–Gatwick	8 cars

The remaining parts (unstated) in the solution are unaltered, but within the quoted parts the number of cars leaving Birmingham and London and arriving at Folkestone and Gatwick are unchanged. The omission now of the Birmingham–Folkestone route has reduced the total number of used routes by one. The original solution could therefore not be basic.

We need now to find a good first basic feasible solution. Although we might therefore look favourably on the lower-cost routes, it is important to realize that it is the relative costs in either a single row or

Table 8.4

From	Cars available	To: Folkestone	To: Gatwick	To: Harwich	To: Southampton
	Cars required	8	31	3	10
Bath	2	174 (3480)	124 (2480)	188 (3760)	64 (1280)
Birmingham	24	179 (3580)	136 (2720)	159 (3180)	130 (2600)
Canterbury	5	16 (320)	74 (1480)	96 (1920)	129 (2580)
London	15	71 (1420)	28 (560)	73 (1460)	78 (1560)
Oxford	2	122 (2440)	83 (1660)	125 (2500)	66 (1320)
Nottingham	4	192 (3840)	153 (3060)	151 (3020)	161 (3220)

a single column that should influence us here. For example, any car going from Nottingham will cost at least 3020p, and since all four cars at this source must go to a destination, we could subtract out this minimum cost to obtain an equivalent but more informative matrix of costs. We could apply a similar argument to the costs in any row and, furthermore, having applied this argument to each row, we could then apply it to the reduced costs in each column. There would then be at least one zero cost in each row and column. In this particular example reducing the row costs in turn produces a zero value in each column and further reduction by columns is therefore impossible. The tableau of reduced costs is therefore as shown in Table 8.5.

This problem is equivalent to the original one. Costs have been reduced by row minima, and therefore the actual cost is the cost for this reduced problem plus the sum of each row minimum multiplied by the number of cars available. This addition is therefore

$$2(1280) + 24(2600) + 5(320) + 15(560) + 2(1320) + 4(3020) = 89\,680.$$

The first solution is found by selecting a zero-cost route which looks promising, in the sense that all the other costs in its row and column are relatively large. This is not an explicit rule, and in fact any starting point will work; a little thought at this stage can save time taken with too many iterations. The Canterbury–Folkestone route appears to be a reasonable candidate. From now on the rules can be more explicit. Each route used must be loaded to its maximum defined by the smaller of the amounts available and required. When the allocation is made there will in general be a surplus at either the corresponding source or the corresponding destination. This is then allocated to the route remaining in the row or column which has the smallest cost. Again the loading is made to its maximum, and again the surplus is allocated to the smallest-cost route in the row or column. In this way a complete allocation will be effected. It is unlikely to be the minimum cost allocation, but with any luck it should be a good one. In this case it results in Table 8.6.

The cost of this solution is

$$89\,680 + 2(1200) + 12(120) + 2(580) + 10(0) + 5(0) + 15(0) \\ + 2(340) + 3(820) + 1(0) = 97\,820.$$

Note that there are 9 used routes, which equals $m + n - 1$, where $m = 6$, $n = 4$. Also a solution found in this way must be basic.

The iterative approach now proceeds in a standard way. We first

Table 8.5

From	Cars available	To: *Folkestone*	*Gatwick*	*Harwich*	*Southampton*
	Cars required	8	31	3	10
Bath	2	2200	1200	2480	0
Birmingham	24	980	120	580	0
Canterbury	5	0	1160	1600	2260
London	15	860	0	900	1000
Oxford	2	1120	340	1180	0
Nottingham	4	820	40	0	200

Table 8.6

From	Cars available	To: Folkestone	Gatwick	Harwich	Southampton
	Cars required	8	31	3	10
Bath	2	2200	(2) 1200	2480	0
Birmingham	24	980	(12) 120	(2) 580	(10) 0
Canterbury	5	(5) 0	1160	1600	2260
London	15	860	(15) 0	900	1000
Oxford	2	1120	(2) 340	1180	0
Nottingham	4	(3) 820	40	(1) 0	200

test our most recent solution for optimality. If it is not optimal, we then find a better solution.

Test for optimality. Allocate to each row and column a *shadow cost.* These must be such that for each *used route* the sum of the corresponding row and column shadow costs must equal the actual route cost. For a basic feasible solution this modest piece of analysis can always be carried out. In fact there are too many costs for them to be derived uniquely, and any shadow cost can arbitrarily be given any value. The value given and cost chosen does not affect the outcome of the test. Usually we take the first row shadow cost as zero. The analysis is most conveniently done around the right hand and lower perimeter of the tableau.

We now examine the routes not used in the current iteration and calculate the difference between the sum of the row and column shadow costs and the actual route cost. If none of these differences are positive, an optimal solution has been reached. Otherwise we have to proceed to a further iteration.

Improving the solution. In the tableaux shown in Tables 8.7–8.13 the differences between shadow and actual costs are given for all routes where the difference is positive. These routes are candidates for inclusion in the next solution, and in fact the inclusion of any of them would effect an improvement. It can be proved that the actual differences are rates of improvement, amounting to the reduction in the total cost for each car using the route. The actual number of cars which can be introduced into any hitherto unused routes depends on other, less obvious, constraints, but a good rule is nevertheless to use the route with the largest positive difference.

Having identified in this way the new route to be included in the next solution, we trace from it a path moving horizontally and vertically but only changing direction at *used route* cells. The path, which may cross itself at unused routes without changing direction, must return to its starting point at the new used route. Provided that the previous procedures have been followed, such a path always exists and it is unique. The next step is to trace round the path placing + and − signs alternately at the change of direction points, starting with a + at the new used route. Examining only the used routes with − signs, we find the minimum of the amounts allocated to these routes. The new iteration is found by taking this amount and adding or subtracting it, according to the sign, to all amounts in the change of direction routes.

The procedure is summarized in Table 8.7. Figures in the rectangular blocks are the differences between shadow and actual costs. The largest such figure occurs for the Bath–Southampton route, and the corresponding path is indicated by the dashed line. Note that it goes through but does not change direction at the Birmingham–Harwich and Bath–Harwich routes. Amounts 2 and 10 are sent along the routes with negative signs, and therefore the smaller of these, 2, is sent around the route. The revised solution, together with its shadow costs and cost differences is shown in Table 8.8.

Note that in going to this solution, one used route has been 'gained', another 'lost', preserving the nine routes required for a basic solution. Also the cost now is 89 680 + 5980 = 95 660, a reduction of 2160, and that the reduction 2160 = 2(1080) is the product of the amount 'sent along the path' and the cost difference.

There are now four positive differences in the costs, the largest being a value of 500 occurring in the Oxford–Folkestone route. Note that the path beginning and ending at this route is a little more complicated than the simple rectangle encountered at the first iteration. Nevertheless it satisfies the conditions required. The smallest amount at the routes with negative signs is 2, and this is sent around the route to yield Table 8.9.

The total cost is now 94 660, a reduction of 1000. Proceeding with the shadow costs in the usual way, we reach the stage where no further progress can be made and three rows and two columns remain unlabelled. This is because the solution now contains less than $m + n - 1 = 9$ used routes. In turn this has arisen because there was a tie in finding the smallest amount to be sent round the path, so that two routes were simultaneously removed (Birmingham–Harwich, Oxford–Gatwick) and only one introduced (Oxford–Folkestone). The term for this phenomenon is *degeneracy*, where a subset of the sources supplies exactly a subset of the destinations. Degeneracy is removed by introducing a small amount x available at some source and required at some destination in such a way that a further used route is brought into the solution in a place where we shall be able to proceed with the shadow cost determination. The way to achieve this would be to select one of the two routes in the tie, for example Oxford–Gatwick. The complete tableau may now be derived (Table 8.10). Had there been a triple tie, it would have been necessary to introduce two x quantities.

Table 8.7

From	Cars available	To: Folkestone	Gatwick	Harwich	Southampton	Shadow cost
	Cars required	8	31	3	10	
Bath	2	[280] 2200	(2 −) 1200	2480	[1080] + 0	0
Birmingham	24	[420] 980	(12+) 120	(2) 580	(10−) 0	−1080
Canterbury	5	(5) 0	1160	1600	2260	−2480
London	15	[420] 860	(15) 0	900	1000	−1200
Oxford	2	[500] 1120	(2) 340	1180	[220] 0	−860
Nottingham	4	(3) 820	40	(1) 0	200	−1660
Shadow cost		2480	1200	1660	1080	

Table 8.8

From	Cars available	To: Folkestone	Gatwick	Harwich	Southampton	Shadow cost
	Cars required	8	31	3	10	
Bath	2	2200	1200	2480	(2) 0	0
Birmingham	24	[420] 980	(14) + 120	(2) − 580	(8) 0	0
Canterbury	5	(5) 0	1160	1600	2260	−1400
London	15	[420] 860	(15) 0	900	1000	−120
Oxford	2	[500] + 1120	(2−) 340	1180	[220] 0	220
Nottingham	4	(3−) 820	40	(1+) 0	200	−580
Shadow cost		1400	120	580	0	

Table 8.9

From	Cars available	To: Folkestone	Gatwick	Harwich	Southampton	Shadow cost
	Cars required	8	31	3	10	
Bath	2	2200	1200	2480	(2) 0	0
Birmingham	24	980	(16) 120	580	(8) 0	0
Canterbury	5	(5) 0	1160	1600	2260	
London	15	860	(15) 0	900	1000	−120
Oxford	2	(2) 1120	340	1180	0	
Nottingham	4	(1) 820	40	(3) 0	200	
Shadow cost			120		0	

Table 8.10

From		Cars available	To: Folkestone	Gatwick	Harwich	Southampton	Shadow cost
		Cars required	8	31 + x	3	10	
From	Bath	2	2200	1200	2480	(2) 0	0
	Birmingham	24	980	+ (16) 120	580	− (8) 0	0
	Canterbury	5	(5) 0	1160	1600	2260	−900
	London	15	860	(15) 0	900	1000	−120
	Oxford	2 + x	(2) 1120	− (x) 340	1180	[220] + 0	220
	Nottingham	4	(1) 820	[0] 40	(3) 0	200	−80
Shadow cost			900	120	80	0	

One positive cost difference remains in this tableau, and the simple path shows that the amount x is taken round. Since x is small the procedure amounts to transferring the x value to the Oxford–Southampton route, producing no real saving in total. The fact that this is the only positive difference between shadow and actual costs does not necessarily imply that there will be no further positive differences and that the solution is therefore optimal. The point is made here where in Table 8.11 there are two further positive differences, and in this case a real saving can be made. The differences here are equal and we arbitrarily choose Birmingham–Folkestone as the starting point for the path.

An amount 2 is taken around the path. Since now we are in the fortunate position of having the x value at a positive sign on this path, the degeneracy has been removed and the x value can be dropped from the model. Table 8.12 shows that there are now no positive differences between shadow and actual costs and that therefore the corresponding solution is optimal. There is in fact one zero difference (London–Folkestone) and the significance of this will be discussed in the next section. Had the x value been retained in the final solution it would have been taken as zero in the interpretation. Its purpose was as a device to achieve the optimal solution.

The allocation is

Bath–Southampton	2 cars	London–Gatwick	15 cars
Birmingham–Folkestone	2 cars	Oxford–Southampton	2 cars
Birmingham–Gatwick	16 cars	Nottingham–Folkestone	1 car
Birmingham–Southampton	6 cars	Nottingham–Harwich	3 cars
Canterbury–Folkestone	5 cars		

The total (minimum) cost is 94 380p.

8.10 The transportation problem – multiple solutions

In the final tableau of the example in the previous section the differences between the sum of the shadow costs and the actual cost for each route were all non-positive. This meant that the allocation in this tableau was minimal cost. There was, however, one unused route (London–Folkestone) where the difference was zero. The implication of this is that there are several equivalent allocations, all of which give the same optimal total cost. They are found simply by adopting the

Table 8.11

From	Cars available	To: Folkestone (Cars required 8)	Gatwick (31)	Harwich (3)	Southampton (10 + x)	Shadow cost
Bath	2	2200	1200	2480	(2) 0	0
Birmingham	24	[140] + 980	(16) 120	580	(8 −) 0	0
Canterbury	5	(5) 0	1160	1600	2260	−1120
London	15	[140] 860	(15) 0	900	1000	−120
Oxford	2 + x	(2−) 1120	340	1180	(x+) 0	0
Nottingham	4	(1) 820	40	(3) 0	200	−300
Shadow cost		1120	120	300	0	

Table 8.12

From	Cars available	To: Folkestone	Gatwick	Harwich	Southampton	Shadow cost
	Cars required	8	31	3	10	
Bath	2	2200	1200	2480	(2) 0	0
Birmingham	24	(2) 980	(16) 120	580	(6) 0	0
Canterbury	5	(5) 0	1160	1600	2260	−980
London	15	860	(15) 0	900	1000	−120
Oxford	2	1120	340	1180	(2) 0	0
Nottingham	4	(1) 820	40	(3) 0	200	−160
Shadow cost		980	120	160	0	

Table 8.13

From	Cars available	To: Folkestone	Gatwick	Harwich	Southampton	Shadow cost
	Cars required	8	31	3	10	
Bath	2	2200	1200	2480	(2) 0	
Birmingham	24	(2 − y) − 980	(16 + y) + 120	580	(6) 0	
Canterbury	5	(5) 0	1160	1600	2260	
London	15	(y) + 860	(15 − y) − 0	900	1000	
Oxford	2	1120	340	1180	(2) 0	
Nottingham	4	(1) 820	40	(3) 0	200	
Shadow cost						

same procedure that would have been used had the route had a positive difference in the costs. The resultant path can then take any value at its positive and negative change of direction routes, consistent with all the amounts sent being non-negative. The actual procedure is summarized in Table 8.13.

In the table the amount y can be taken round the path, and in order that the amounts should be positive, y must be between 0 and 2. In fact *any* value over this range will give the same optimal answer, this being the one instance where a solution with more than '$m + n - 1$' used routes should be considered. The choice of a value for y allows options whose effects cannot be quantified. The method minimizes total cost, but there may be some other reason, perhaps as subjective as comparisons of scenery or driver preference, why the London–Folkestone route should be used.

8.11 The transportation problem – shortages and surpluses

In the original statement of the transportation problem the total number of cars required exactly balances the total number available. If the method of solution only applied in such fortunate circumstances, it would indeed be of limited use. Fortunately, there is a very simple device which permits the same approach to be used when there is either a shortage or a surplus.

Suppose, for example, that there had been 5 cars available at Bath, instead of 2. We have a surplus of 3 cars. The problem now is not only one of deciding the allocation to the destinations, but also of deciding which cars should be left at their present location (assuming that this is what the system would do in practice). If therefore the Brakeless Company are indifferent about where the 3 surplus cars are left, the equivalent model is to introduce a *dummy* destination requiring the 3 cars. Moreover, since the cost of not using a car, that is sending it to the dummy destination, is the same for each location, we can use a value zero for the transportation cost in each route for the dummy destination column. Note that if the cost is the same but not zero, we can still use the zero value by the 'subtraction of costs in the same row or column' argument of Section 8.9. If instead we wished to make some locations less attractive for the surplus than others, we might express this relative preference by differing costs in the dummy destination column.

The outcome of this argument is that the problem has been restored to the balancing availability and requirement model, and can therefore be solved in the same way. Readers are invited in Exercise 9 to solve the problem mentioned above.

A similar argument is applied to the situation where there is a shortage, except that in this case a dummy source is introduced to supply the shortage. In the final solution any destination unfortunate enough to be receiving a positive amount from the dummy source will in fact be that amount short of its requirements. Again, it can be argued that all the costs from the dummy source will be zero. Finally the reader is reminded that when zero costs are introduced for a dummy source or destination, they are introduced in the first tableau, before any cost reduction has been carried out. The implication of having a zero on every row (dummy destination problem) or a zero in every column (dummy source problem) is that the cost reductions are effective in only one direction.

8.12 The transportation problem – sensitivity analysis

Having found the optimal solution, we can now determine how the optimum is affected by changes in the parameters of the problem. Any change in an amount available must be accompanied by an identical change in an amount required, perhaps in a dummy. The simplest way to proceed is then to take the amended solution, test it for optimality and, if necessary, proceed as before. Of more interest is the effect of changes in the costs.

Sensitivity analysis on the costs is in principle very simple. The optimization criterion requires that the difference between the sum of the row and column shadow costs and the actual cost should not be positive. When this position has been achieved it is clear that if we then reduce any actual cost to the extent that a previous negative difference becomes positive, we have further iterations to carry out and, moreover, the route whose cost has been reduced then becomes a used route. It follows that the amount of each negative difference is the amount by which the route 'misses out' on being included in the solution. For example, referring back to the example of Section 8.9, in the optimum solution the Bath–Folkestone route had a difference of $0 + 980 - 2200 = -1220$. The original cost of 3480 would have to be reduced by at least 1220 in order that this route be considered for inclusion in the final solution. The smallest difference (apart from the zero difference of the multiple solution) occurs for the Nottingham–Gatwick route and is -80. Recalling that the original costs are the products of the distances in miles between locations and the assumed cost of 20p per mile driven, this Nottingham–Gatwick difference represents missing out by a mere 4 miles. Move Nottingham four miles closer to Gatwick, and the optimum solution changes! At a more sensible level, the original distances may have been taken from a table in a standard United Kingdom road atlas. They measure distances between town centres. It could now be rewarding to examine the more marginal cases in detail, adjusting the distances to take into account the actual location of the company's depot. The initial formulation of the problem requires costs for every possible route. Some of the routes are clearly too expensive to be used, and a very rough estimate of the costs will then suffice. After the first 'optimum' has been found, routes requiring cost refinement will become apparent.

A further use of sensitivity analysis arises when the costs are independently managed. Imagine a situation where competing car-hire firms are offering to provide the ferrying service. Each car-hire firm operates one route only, and the prices are their bids for each route. Solving the problem allocates car-hire firms to routes where the total cost is minimized. The difference in the costs for a non-used route is the amount by which the bid has to be reduced before it can be accepted.

8.13 The transportation problem – other applications

The transportation problem models the situation where a single commodity is to be carried at minimum cost from a number of sources to a number of destinations. Systems of factories, warehouses, depots and outlets can fit this description. An early application was the transportation of coal between pits and power stations. There are also situations where the allocation aspect of the model, rather than the actual transportation, prompts its use in quite unusual circumstances. An extension of the bidding model suggested in the previous section runs as follows. Suppose that a firm has a number of similar jobs (building retail outlets around the country, printing electoral rolls, etc.) which it wishes to have completed quickly. It offers them for tender. A numbr of contractors place bids for some or all of the jobs, and the firm now wishes to accept the minimum price bids. The real problem arises because each contractor has a limited capacity in the number of jobs that it can accept. As a transportation problem, the contractors are the sources with amounts available equal to their individual capacity limits; the jobs are the destinations each with a requirement of one contractor. The costs are the bids for each job from each contractor. Assuming that the total capacity of the contractors exceeds the number of jobs, there is a need for a dummy destination to take up all of the unused capacity. Also there is considerable degeneracy in this problem.

A further application arises in production planning, where a demand which is variable but assumed known has to be met from a limited production capacity. Units can be made and stored at a cost for future sales. Here the sources are the opening stock and the monthly capacities. The destinations are the monthly demands and the final stock. In general the costs are the penalties incurred by not selling in the month of production. An illustration of this application is given in Exercise 8.

Finally, there is the job assignment problem. Suppose that a consultancy firm has five jobs which it wishes to assign, one job going to each of five senior consultants. Because of their differing abilities and experiences, the consultants vary in their suitabilities for each job. If we can measure the suitabilities on a linear scale, we might then argue that the aim of the firm should be to maximize the total suitability when each consultant is assigned to one job. (*Note:* the linear scale for suitabilities means that we would be indifferent between an assignment which allowed suitabilities of 9 to job A and 6 to job B, and one which assigned suitabilities of 8 to job A and 7 to job B. They both add up to a total of 15.) Suppose that measured on a scale of 0–10, where the higher values denote greater suitability, the actual values are as shown in Table 8.14.

Table 8.14

		Jobs				
		A	*B*	*C*	*D*	*E*
Consultant	Brown	6	8	7	9	5
	Fletcher	2	6	4	7	2
	Harrison	3	4	3	8	2
	Naylor	5	7	5	9	6
	Wilkinson	7	10	9	10	7

The sources are the consultants, each of which has one unit, himself, available. The destinations are the jobs, each requiring one consultant. Since the transportation problem would minimize the total suitability, the values have to be adjusted to an unsuitability score, found by subtracting each from the maximum of 10. The usual transportation approach would involve a considerable amount of degeneracy, and hence this particular problem is usually approached by what is known as the Hungarian method. For a full account of this method, see for example Vajda (1961). The method relies on the fact that if any amount is added to or subtracted from the costs in the same row, the optimizing solution is unaltered. In this case it can be demonstrated that if amounts 1, 4, 2, 1 and 0 are added to the scores for the consultants in the order listed, and if 3, 0, 1, − 1 and 3 are added to the job scores again in the order listed, we have the equivalent problem shown in Table 8.15.

Table 8.15

		Jobs				
		A	*B*	*C*	*D*	*E*
	Brown	10	9	9	9	9
	Fletcher	9	10	9	10	9
Consultant	Harrison	9	7	7	10	8
	Naylor	9	8	7	9	10
	Wilkinson	10	10	10	9	10

As 10 is the maximum score possible, if we can find an assignment which uses only 10 scores, it must be the optimum for this problem and therefore for the original problem. Although there are nine scores of 10, we need just five of them with one in every row and column. There is, in fact, a unique solution:

Brown, A; Fletcher, B; Harrison, D; Naylor, E; Wilkinson, C.

Note that only Harrison, who because of his total score is arguably the worst of the consultants, is assigned to his most suitable job. Moreover only job C has assigned to it the most suitable consultant.

8.14 Further developments of linear programming

In this chapter we have examined the basic model of linear programming and the particular case of the transportation problem. The model demands linearity of the constraints and of the objective function; it also provides solutions where the variables can take any non-negative value. There are occasions when some of these demands can be relaxed, resulting in modification to the approach. The non-negativity requirements can be overcome by writing the variable

$$X = U - V$$

where U and V are both non-negative, and one of them zero. Fortunately most practical problems are concerned with non-negative variables and this modification is rarely needed. Another variation occurs when we are dealing with people, production units, vehicles or any commodity where the strategy has to be an integer. The standard LP does not recognize this necessity, and unfortunately it is often incorrect to round off the answer to the nearest whole number. It may be possible to find a better integer solution, or alternatively in rounding off we may have stepped out of the region of feasible solutions. One approach to this problem is given in Chapter 10.

A further variation occurs when the coefficients in the model are taken as random variables instead of constants. The general term for this is *stochastic programming*, and although the approach is beyond the scope of this book, reference may again be made to Vajda (1961).

Finally, there are techniques which will deal with the situation where the objective function is not linear. Some of these can guarantee to produce only a local optimum, that is a strategy which is better than any in its immediate vicinity, whereas the aim is of course for the global optimum. Reaching a local optimum is like reaching the summit of one of a range of mountain peaks. Doubts arise when the mountaineering is being done in thick fog, a situation which is analogous to much non-linear programming. One case where the optimum can be shown to be always global arises when the objective function is *convex*. The term convex has been defined in terms of a region in Section 9.3. The function $R = g(X_1, X_2, \ldots, X_m)$ is said to be convex if the region defined by $R \geqslant g(X_1, X_2, \ldots, X_m)$ in $(m + 1)$–dimensional space is convex. For example, the simple function $R = X^2$ is convex, since if (R_1, U) and (R_2, V) are any two points in the region defined by $R \geqslant X^2$, then

$$R_1 \geqslant U^2 \quad \text{and} \quad R_2 \geqslant V^2 \tag{8.29}$$

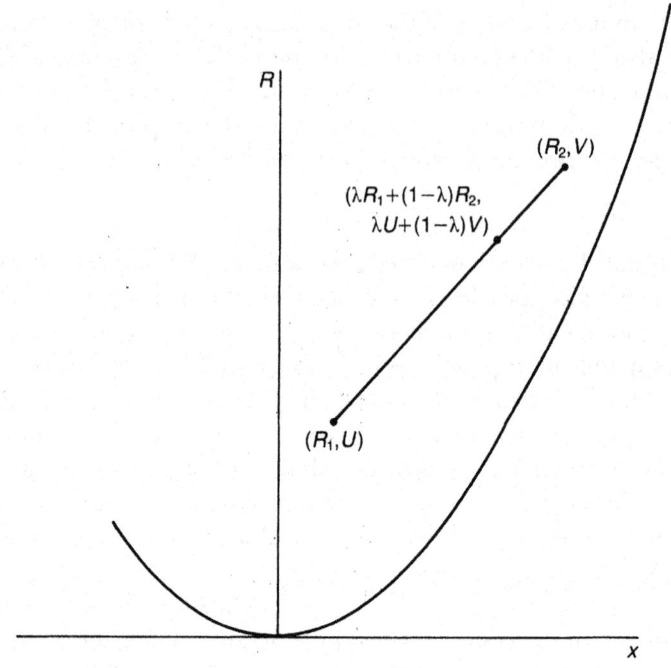

Fig. 8.11 The convex function $R = x^2$

The general point on the segment joining these two points is

$$(\lambda R_1 + (1 - \lambda)R_2, \lambda U + (1 - \lambda) V),$$

where $0 < \lambda < 1$. See Figure 8.11.

Since all the quantities are positive, it follows from (8.29) that

$$\begin{aligned}\lambda R_1 + (1 - \lambda)R_2 &\geqslant \lambda U^2 + (1 - \lambda)V^2 \\ &= [\lambda + (1 - \lambda)]\,[\lambda\, U^2 + (1 - \lambda)\, V^2] \\ &= \lambda^2\, U^2 + \lambda(1 - \lambda)\,(U^2 + V^2) + (1 - \lambda)^2\, V^2 \\ &= \lambda^2\, U^2 + \lambda(1 - \lambda)\,[(U - V)^2 + 2U\, V] + \\ &\quad (1 - \lambda)^2\, V^2 \\ &\geqslant \lambda^2\, U^2 + 2\,\lambda\,(1 - \lambda)U\, V + (1 - \lambda)^2\, V^2 \\ &\quad \text{since } \lambda(1 - \lambda)\,(U - V)^2 > 0 \\ \text{and hence } \lambda R_1 + (1 - \lambda)R_2 &\geqslant (\lambda\, U + (1 - \lambda)V)^2.\end{aligned}$$

This demonstrates that the intermediate point of the segment also lies in the region, and the region is therefore convex. Establishment of the convexity property for more complicated functions needs more careful analysis but follows this method in principle.

A further approach can be used when the objective function can be expressed as the sum of functions, each of which depends only on one variable. For example, an expression of the form $3X_1^2 + (2/X_2)$ is separable in this sense. The property can be achieved sometimes by a transformation of the variables. For example, the term X_1X_2 is not separable, but by writing

$$Y_1 = X_1 + X_2, \qquad Y_2 = X_1 - X_2 \tag{8.30}$$

we see that $Y_1^2 - Y_2^2 = 4X_1X_2$, and hence the Y variables have the property of separability. The equation (8.30) can be incorporated into the constraints, noting that although X_1 and X_2 are non-negative, the same condition does not hold for Y_2. The approach now is to approximate each of the functions of one variable by a set of linear segments and to optimize in a way requiring consideration of only one segment at any iteration. A fuller description of the method (known as *separable* programming) may be found in Williams (1978). A local, but not necessarily global, optimum can be found.

8.15 Further reading

Lagrange multipliers

Ledermann, W. and Vajda, S. (eds.), *Analysis*, Vol. IV of the *Handbook of Applicable Mathematics*, Wiley, Chichester (1981).

Lucking, R. C., *Mathematics for Management*, Wiley, Chichester (1980).

Linear programming

Ackoff, R. L. and Sasieni, M. W., *Fundamentals of Operations Research*, Wiley, New York (1968).

Bunn, D. W., *Analysis for Optimal Decisions*, Wiley, Chichester (1982).

Daellenbach, H. G. and George, J. A., *Introduction to Operations Research Techniques*, Allyn and Bacon, Boston (1978).

French, S., Hartley, R., Thomas, L. C. and White, D. J., *Operational Research Techniques*, Arnold, London (1986).

Gass, S. I., *Linear Programming* (2nd edn) McGraw-Hill, New York (1964).

Hadley, G., *Linear Programming*, Addison-Wesley, Reading, Mass. (1962).

Harper, W. M. and Lim, H. C., *Operational Research* (2nd edn) M & E Handbooks, Pitman, London (1982).

Johnson, D., *Quantitative Business Analysis*, Butterworths, London (1986).
Morris, Clare, *Quantitative Approaches in Business Studies*, Pitman, London (1983).
Moskowitz, H. and Wright, G. P., *Operations Research Techniques*, Prentice-Hall, Englewood Cliffs, N. J. (1979).
Vajda, S., *Mathematical Programming*, Addison-Wesley, Reading, Mass. (1961).
Wilkes, F. M., *Elements of Operational Research*, McGraw-Hill, London (1980).
Williams, H. P., *Model Building in Mathematical Programming*, Wiley, Chichester (1978).

8.16 Problems and practical exercises

1 A hotel has 110 similar bedrooms, each of which can be occupied by either one or two persons. A room rented to one person results in a contribution of £8 per day; to two persons the contribution is £10 per day. The dining room (which is used exclusively for hotel guests) has a maximum capacity of 160. Each room is serviced daily. If it has a single occupant the time taken is 15 minutes, but a room with two occupants takes 20 minutes to service. The hotel employs 10 maids for three hours daily on room service.

What mixture of single and double room bookings should the hotel aim for?

What would be your reactions to the alternative proposals:

(a) build an annex with another 5 bedrooms, or
(b) extend the dining room capacity to 180, or
(c) employ an extra maid for one hour per day?

What is the limit on the extra time that could profitably be used under (c)?

2 The Coweswater Ferry takes cars and cars plus caravans on scheduled trips to the Black Isle. The ferry has room for vehicles of total length 400 metres and can carry vehicles of total weight no more than 120 tonnes. The ferry operators require that the total number of cars and cars plus caravans should not exceed 75 on any trip. The charge for a car is £20 and for a car plus caravan £30.

The operators wish to accept bookings in a way which will maximize the total amount charged for a trip. Assuming that a car on average is 4 metres long and weighs 1.5 tonnes, and that a car plus caravan is 8 metres long and weighs 2 tonnes, what mixture of cars and cars plus caravans should they accept?

3 A sports hall is available for evening recreational use which can be broadly classified as 'noisy' (indoor football, country dancing, advanced bagpipe tuition, etc.) or 'quiet' (bridge, flower arranging, taxidermy, etc.). On any evening it is used only for one activity and the profit (hire plus bar takings) on a 'noisy' evening averages £25 whereas on a 'quiet' evening it is £10. To placate the neighbours it has been decided that there should be at least twice as many quiet as noisy evenings. The hall employs two managers, each of whom can put in a maximum of 220 evenings per year. A noisy evening requires both managers, but a quiet evening needs only one. The sports hall is available for 348 evenings in a year. Determine how the evenings should be divided between 'noisy' and 'quiet' activities to maximize the total profit.

Show how the optimum strategy would change if the hire fee for quiet evenings were increased, keeping the hire fee for noisy evenings unaltered and assuming that bar takings are unaltered.

4 A firm requires coal with phosphorus content no more than 0.05 per cent and ash impurity no more than 6 per cent. Three grades of coal, X, Y and Z are available at the prices shown in Table 8.16.

Table 8.16

Grade	*Price (£/ton)*	*% Phosphorus*	*% Ash*
X	25	0.07	3.0
Y	30	0.04	7.0
Z	38	0.03	1.0

Show how the grades should be mixed to meet the impurity restrictions at minimum cost.

(*Hint:* Consider the composition of one ton of blended coal. Eliminate one variable as the difference between unity and the sum of the other two. Introduce an inequality to ensure that the eliminated variable is non-negative.)

Over what range of price per ton for grade Z is the solution optimal, assuming that the prices of grades X and Y are unchanged?

5 The North-West Rutland Water Authority is contracted to supply water to six urban regions. Four reservoirs are available to provide water, although each is limited in its capacity. The supply network involves pumps and requires maintenance, so that the costs depend on the connections made. A summary of these costs (£ per million gallons supplied) together with the capacity limits and demands is given in Table 8.17. Any excess supply can be disposed of without cost.

Table 8.17

	Region						
	Jandon	*Veeton*	*Lipton*	*Teiby*	*Kenton*	*Lumham*	*Capacity Limit (m.g.d.)*
Reservoir							
Earbrook	31	50	42	40	65	27	30
Porcston	48	63	53	52	80	29	25
Tapton	37	55	52	49	71	31	20
Sandiland	53	72	57	62	84	45	35
Requirements: (m.g.d.)	14	11	18	13	21	15	

[m.g.d. = millions of gallons per day]

Find a minimum cost distribution pattern together with any alternative patterns with the same minimum cost.

An industrial development is proposed for Veeton which will increase its demand for water by 4 m.g.d. What will be the extra cost to the system of supplying this water? Why could Veeton feel that this was rather excessive?

6 The Wessex Education Authority has an agreement whereby children from its primary schools take swimming classes in the public baths at Melchester, Wintoncester, Casterbridge and Emminster. Table 8.18 lists for each school the number of classes requiring this weekly service and also the distances in miles from each public

Table 8.18

School	*Baths* Melchester	Wintoncester	Casterbridge	Emminster	*Number of Classes*
Weydon-Priors	4	13	12	11	3
Bulbarrow	7	12	8	9	3
Pilsdon	4	16	8	5	8
Sherton Abbas	6	17	5	4	3
Stourcastle	4	14	6	7	8
Shottsford	9	15	3	8	8
Maximum number of Classes	13	5	13	5	

baths. Each public baths also has a maximum number of classes per week which it can offer.

If each class requires a bus, what allocation of schools to baths will minimize the total distance travelled by the buses? Give full details of any alternative equivalent solutions.

7 Six firms put in bids for government contracts to build five new telephone exchanges. The buildings are similar but the firms vary in their capacities to carry out the contracts in the time available. The bids submitted (£000 000) and the limits on capacities are given in Table 8.19.

Which firms should have their bids accepted to minimize the total cost?

Table 8.19

Firm	*Capacity*	*Telephone exchange:* *V*	*W*	*X*	*Y*	*Z*
A	1	45	45	45	45	45
B	2	53	50	48	55	40
C	2	61	60	57	59	42
D	3	65	60	58	63	48
E	1	58	59	55	60	43
F	1	50	48	57	59	43

Firm E does not have a bid accepted. What is the least amount by which one of his bids should have been reduced in order that he should win one of the contracts?

8 A manufacturer of industrial cabinets has orders for the first four months of the year.

January 6; February 10; March 5; April 10.

At the beginning of January he has 3 cabinets in stock. His monthly output is 6 cabinets in regular time and, if necessary, 2 cabinets in overtime. Each cabinet made in overtime incurs a penalty cost of £200, and any cabinet stored has a penalty cost of £100 per month.

Show how the problem can be formulated as a transportation problem and find the minimum penalty production plan, assuming that there is to be no stock at the end of April.

9 Solve the problem of Section 8.9 where the number of cars available at Bath is 5. By how much is the total cost reduced?

(*Hint:* Start with the previous solution augmented by a dummy.)

9

Multiple objectives

OBJECTIVES

With the material in this chapter the reader should be able to reconcile the problem of having several objectives in a linear programming formulation. This he can do firstly by finding the set of non-inferior activities, which are strategies for which no other strategy is uniformly better in the sense that it is better for all objectives. The more specific approach of goal programming is then introduced. The reader should be able to handle simple two-variable problems, and to understand how larger problems can be programmed on a computer.

9.1 The problem

In problems considered so far we have assumed that there is a single objective to be optimized. A single figure has been attached to the outcome of each decision strategy denoting the return to the decision-maker resulting from actions taken. With a decision tree this was usually taken as the monetary return, so that under the EMV criterion, the optimum strategy was the maximum monetary return weighted by the probabilities of the respective outcomes. A modification of this was to substitute utility for monetary value when either it was felt that monetary returns were not linear or where the consequences of the outcome could not wholly be expressed in monetary terms.

It often happens that individuals and organizations are operating with several objectives. A firm may argue that it wishes to maximize profits and that it also wishes to minimize costs. These aims may appear to be similar, but they can produce radically different actions, particularly when they are expressed mathematically. The firm might also aim to maximize its output, using this as a more easily measured surrogate for profit. Such objectives are termed *acquisitive* by Daellenbach and George (1978). Alternatively, objectives may be *retentive*, where the aim is to keep some property of the system within certain bounds. For example a firm may wish to keep the number of its employees around a given level, or there may be a limit on its warehouse capacity for raw material or finished goods stock. Retentive objectives can generally be treated as constraints on the actions taken, and we saw in the previous chapter how such constraints could be dealt with in the particular case when they were linear functions of the decision variables. In this chapter attention is confined to acquisitive objectives.

Briefly the problem is to find an optimum action when there are several objective functions. Again in brief the answer to this is that in general it cannot be done. We would indeed be fortunate to find a single operating strategy which optimized more than one objective. Quite often the objectives work directly against each other. Maximizing output usually means the opposite of minimizing costs. In planning a journey across the Atlantic Ocean, minimizing the time taken certainly does not give the same answer as minimizing the cost of the fare, and it becomes even more complicated if we throw in considerations of passenger comfort.

In theory we might reduce all of the objectives to a single function if we could define a utility function in terms of the variables of each objective. In an industrial context, therefore, we might need to define a firm's utility as a function of its monthly costs, output, profit, employment level, etc. This would be a very complicated process, simplified somewhat if we could persuade ourselves that the utility function could be decomposed, either additively or multiplicatively, into functions of the individual variables. For a fuller account of the procedures reference should be made to Bunn (1982). It is doubtful if the approach will have many practical applications.

Two main approaches will be illustrated in this chapter. One amounts to the consideration of non-inferior alternatives and is a method of listing possible actions, none of which has an alternative which is superior for all of the different objectives. The other approach is called *goal programming*, where an attempt is made to quantify the effects of not being at the optimizing limit of each objective. Both approaches are based on the simplest linear programming model. Extensions to non-linear objectives or to integer solutions may be feasible in special cases. The Delluge Park example (Section 8.2) will be used to illustrate the approaches.

The general problem is to allocate values to a number of different variables ($X_1, X_2, \ldots, X_m$). These variables are subject to a set of linear constraints

$$b_{1i} X_1 + b_{2i} X_2 + \ldots + b_{mi} X_m \leq C_i, \qquad i = 1, 2, \ldots, n. \tag{9.1}$$

The constraints have been expressed in the 'less than or equal to' form. Equalities in the opposite direction can be transformed by multiplying by -1, and constraints in the form of equations can be converted into the form of (9.1) by writing them as both 'less than or equal to' and as 'greater than or equal to'.

Ideally the aim is to maximize (or minimize) a set of k linear functions of the X_i variables

$$Z_j = a_{j1} X_1 + a_{j2} X_2 + \ldots + a_{jm} X_m, \qquad j = 1, 2, \ldots, k. \tag{9.2}$$

Usually k is much smaller than m or n. Also it is unlikely that the same feasible values ($X_1, X_2, \ldots, X_m$) will optimize all of the objectives (9.2).

9.2 The method of non-inferior activities

The method of non-inferior activities argues that in the general case where no single optimum to all of the objectives exists, consideration should be confined to those actions which are not uniformly worse (in terms of the objective functions) than any other feasible action. If therefore $X_1^*, X_2^*, \ldots, X_m^*$ is a member of this set, and if

$$Z_j^* = a_{j1} X_1^* + a_{j2} X_2^* + \ldots + a_{jm} X_m^*, \qquad j = 1, \ldots, k,$$

are the objective functions, all of which are to be maximized, then it will not be possible to find another feasible action $X_1, X_2, \ldots, X_m$ for which the Z_j value is greater than Z_j^* for all j values. Any other feasible action must have at least one objective function where it is inferior to any member of the set. The concept is similar to that of dominance in the theory of games which was introduced in Chapter 3. No member of the set of non-inferior activities (or actions in our terminology) is dominated by any other feasible action.

The method of finding the set of non-inferior activities is quite simple, but the actual listing of the set could, in practice, be very tedious. The method is as follows. A convex linear combination of k objective functions is a weighted sum of these functions where the weights are constants λ_j, such that $\lambda_1 + \lambda_2 + \ldots + \lambda_k = 1$ and all $\lambda_j \geq 0$. (It follows that they all must be less than or equal to one.) We then maximize this expression

$$Z^* = \lambda_1 Z_1 + \lambda_2 Z_2 + \ldots + \lambda_k Z_k.$$

subject to the same constraints on the X values. Recall that the Z values are themselves linear functions of the X values and that therefore Z^* is also a linear function of the X values. The problem is a standard linear programming problem, producing an optimum which must be on the boundary of the region of feasible actions. It can easily be shown that any set of λ values satisfying the above convexity conditions will produce as the solution of its LP problem a non-inferior action. These will include the actions maximizing each objective function individually and ignoring the others (corresponding to one λ equal to one, and the remainder zero). Except where the objective function coincided in its 'slope' with the slope of a constraint, the actions so found will be at vertices of the region of feasible actions. Other vertices may also be included, as will be demonstrated in the illustration following.

One further important point is that there will be considerable interest in intermediate actions between the vertex solutions mentioned above. These represent compromises between actions when only one objective function is maximized. To find them requires some care. They correspond geometrically to the multiple solutions discussed in Section 8.4 of the previous chapter, and are themselves convex linear combinations of vertex solutions. However, it is not sufficient to take a convex linear combination of any set of vertex coordinates. This could easily produce a point which is in the interior of the region of feasible actions and therefore, as we have seen, not optimal under any linear objective function. The vertices included must be multiple solutions for a particular set of the λ coefficients, so that for this combined objective function they all realize the same value. This argument should be clearer when it is demonstrated in the next section.

The outcome of this analysis will be a listing of actions from which the decision-maker has to select. The selection criterion may be some subjective preference. Alternatively, the exercise of seeing the compromises available to him may lead the decision-maker to an objective weighting of the aims in the problem, which in turn reduces it to a single-objective problem (whose solution should be virtually identical with the action selected) or to a more rational use of the second approach, which is called goal programming.

9.3 The method of non-inferior activities – Delluge Park illustration

The proprietor of Delluge Park is still faced with the problem of how many sites (X) he should allocate to Flybynite and how many sites (Y) he should retain for casual letting. The constraints on X and Y introduced in Section 8.3 are unchanged. The proprietor still has the objective of maximizing his return, but it also occurs to him that this might be a little short-sighted. Much of his custom comes from repeat visits and also from people who hear of his reputation from friends who have stayed at the Park. It is therefore important for his long-term viability that as many adults as possible should stay at Delluge Park. He knows from his own records that on average there are 3.6 adults on each of his casual places. An enquiry to Flybynite elicits the information that their experience is that they are generally booked by a family singly, so that although more children tend to be booked, the number of adults per Flybynite place is only 2.4 on average. Assuming that all stays are of the same length (a different assumption could readily be allowed for) this means that if the proprietor wishes to maximize the Park's exposure to adults, he should maximize $2.4X + 3.6Y$. His complete problem, recalling equations (8.1)–(8.5) of Chapter 8, is:

$$\begin{aligned}
\text{maximize } Z_1 &= 50X + 40Y && (9.3)\\
\text{and } Z_2 &= 2.4X + 3.6Y && (9.4)\\
\text{subject to } \quad 50X + 60Y &\leqslant 15\,000 && (9.5)\\
Y &\geqslant 30 && (9.6)\\
X + Y &\leqslant 280 && (9.7)\\
2.5X + 1.5Y &\leqslant 650 && (9.8)\\
X \geqslant 0, \quad Y &\geqslant 0. && (9.9)
\end{aligned}$$

It has already been shown that maximizing Z_1 produces a strategy $X = 230$, $Y = 50$ and a return of 13 500 Rz. This is point C in Figure 9.1. A similar analysis for the second objective function Z_2 produces an optimal action $X = 0$, $Y = 250$ at point A.

The manager therefore has a problem. He cannot maximize both objective functions simultaneously. If he uses the approach of non-

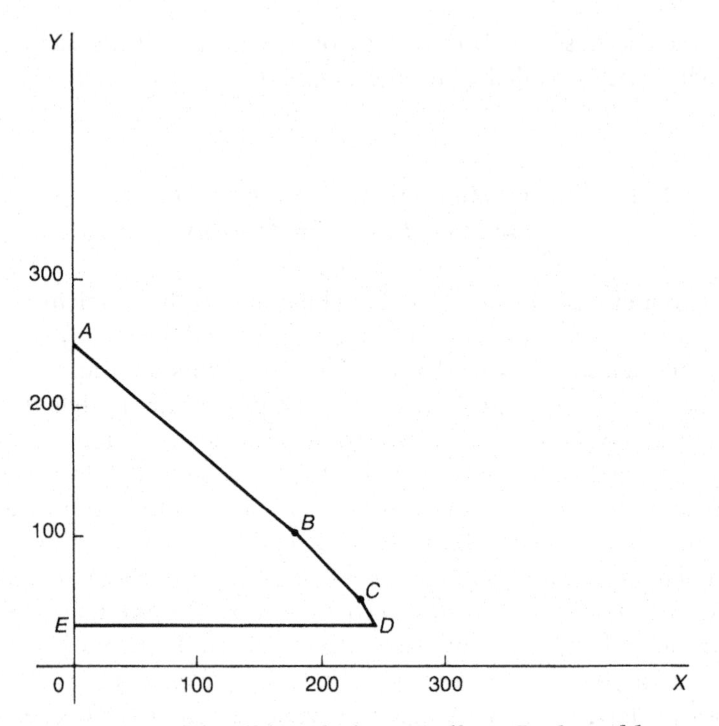

Fig. 9.1 Region of feasible solutions, Delluge Park problem

inferior activities, he has to solve a set of linear problems defined by taking values of λ between 0 and 1, where he is to maximize

$$\begin{aligned} Z^* &= \lambda(50X + 40Y) + (1 - \lambda)\,(2.4X + 3.6Y) \\ &= (47.6\lambda + 2.4)X + (36.4\lambda + 3.6)Y \end{aligned} \tag{9.10}$$

subject to the constraints (9.5)–(9.9).

A summary of the solutions to these problems can readily be obtained as shown in Table 9.1. When λ takes the values 0.034 75 and 0.107 14 the problem has multiple solutions corresponding to the boundary segments *AB* and *BC* respectively of the region of feasible solutions. Thus the set of non-inferior activities consists of these two lines. Mathematically speaking it is equivalent to convex linear combinations of the coordinates of *A* and *B* and of the coordinates of *B* and *C*. A similar combination of *A* and *C* or of all *A, B* and *C* would lead to interior points of the region and would not form part of the set of non-inferior activities. This is an illustration of the discussion in the penultimate paragraph of the previous section. In the two-dimensional case, the explanation can be accepted readily, but higher-

Table 9.1

Range of λ values	*Optimal point*	X	Y	Z_1	Z_2
0 – 0.03475	*A*	0	250	10 000	900
0.03475 – 0.10714	*B*	180	100	13 000	792
0.10714 – 1	*C*	230	50	13 500	732

order cases where it is impossible to draw the diagram require a more careful algebraic treatment.

The set of non-inferior activities is a form of *efficient frontier* for the region of feasible actions. No other action should be considered if Z_1 and Z_2 are retained as the objectives. In this case the actions can be characterized by the two parameters a and b, where

$$X = 180a, \qquad Y = 250 - 150a,$$

and as a increased from 0 to 1, this characterization accounts for all points on the line AB;

$$X = 180 + 50b, \qquad Y = 100 - 50b,$$

and again as b increases from 0 to 1, this accounts for the points on the line BC. The corresponding values of the objective functions are:

$$\begin{aligned} Z_1 &= 10\,000 + 3000a, & Z_1 &= 13\,000 + 500b \\ Z_2 &= 9000 - 108a, & Z_2 &= 792 - 60b \end{aligned}$$

Equivalently the results could be set out as in Table 9.2 choosing some convenient grid for the scale of either X or Y.

Note that as Flybynite places increase from 0 to 180 (and casual places decrease correspondingly from 250 to 100) we trade off a loss of 18 adult customers for every increase of 500 Rz return, but that as the

Table 9.2

X (Flybynite places)	0	30	60	90	120	150
Y (casual places)	250	225	200	175	150	125
Z_1 (return)	10 000	10 500	11 000	11 500	12 000	12 500
Z_2 (adults)	900	882	864	846	828	810
X (Flybynite places)	180	190	200	210	220	230
Y (casual places)	100	90	80	70	60	50
Z_1 (return)	13 000	13 100	13 200	13 300	13 400	13 500
Z_2 (adults)	792	780	768	756	744	732

Flybynite places increase further from 180 to 230 (casual decrease from 100 to 50) the trade off is at a rate of 60 adult customers for an increase of 500 Rz return.

The aim is to put all the evidence as clearly as possible before the caravan park proprietor, who is the real decision-maker. It is now up to him to make his choice. The alternative approach which now follows attempts to build the decision-maker's preference into the actual analysis, and thereby come to a specific recommendation. It demands a rather more enlightened attitude on the part of the decision-maker in that he has to become involved in the actual modelling of the decision process, but the eventual outcomes using both approaches should be the same. Effectively goal programming makes more objective the selection from the non-inferior activities set.

9.4 The method of goal programming

There are several variants on the method of goal programming, but the essential argument is that in cases where several objectives cannot be optimized simultaneously a compromise objective should be optimized. The compromise consists of a linearly weighted function of the departures from some agreed level of each objective. These 'agreed levels' are said to be goals.

In formulating the goal programming model two extra concepts are introduced. A desirable level or goal for each objective must be determined. It is possible to determine the optimum value of each objective function alone and treat these as the goals. This is done in the awareness that all of the goals will not be achieved, and that the departures when they occur will be on the negative side, that is reality will fall short of the goal. Alternatively the goals could be determined by the requirements of the decision-maker. He may be satisfied with a level of output below the maximum, a utilization of facilities below 100 per cent and even a profit level below the optimum. Having decided on appropriate levels or goals, he then finds a strategy which will match the goals as closely as possible.

The second new concept is that of weighting the departures from the objective functions. This first of all allows priorities to be given to the various goals, and these priorities effectively turn the problem into a single objective linear programming problem. Alternatively, they can be thought of as assigning a priority sequence to the objectives.

The highest priority goal is satisfied first, and thence in order until no further goals can be achieved. Each goal, therefore, has full priority over lower-order goals. In the event that all goals are fully achieved, the decision-maker should think once more about the levels chosen for the goals, since he may not have been sufficiently ambitious and better goals are in fact attainable. The further drawback to this approach is that since each goal is only considered when its turn in the sequence arrives, the actual levels of those lower-priority objectives, whose goals are not met, may be a very long way from the set goals.

Usually the consequences of departures from the goals will depend on the sign (+ or −) of the departure. Exceeding the profit goal by £1000 is certainly of more benefit than a shortfall of the same amount Again this can be allowed for in the weights assigned to the departures. Weights should also take into account the units used to measure the goals, since objectives are likely to be expressed in money, hours, weight, etc. One way of overcoming this could be to express departures as proportions of the goal value, with the possibility of weighting these proportions by the relative importance of the objectives. This and other points mentioned in this section will be illustrated using the Delluge Park example once more.

In general mathematical terms, we have k objectives

$$Z_j = a_{j1}X_1 + a_{j2}X_2 + \ldots + a_{jm}X_m, \qquad j = 1, 2, \ldots, k \tag{9.11}$$

to be maximized, subject to the m linear constraints

$$b_{1i}X_1 + b_{2i}X_2 + \ldots + b_{mi}X_m \leq C_i, \qquad i = 1, 2, \ldots, n. \tag{9.12}$$

Suppose that goals $G_1, G_2, \ldots, G_k$ have been set for the k objectives. The departure of objective Z_j from goal G_j can be expressed as

$$a_{j1}X_1 + a_{j2}X_2 + \ldots + a_{jm}X_m + d_j^- - d_j^+ = G_j. \tag{9.13}$$

The actual departure is $d_j^- - d_j^+$, where both of these terms are non-negative, but at least one is zero. Individually the terms are interpreted as the amount by which the goal is exceeded (d_j^+) and the amount of the shortfall (d_j^-). Equations (9.13) now augment inequalities (9.12) as constraints on the decision variables $X_1, X_2, \ldots, X_m$, which themselves have been augmented by the variables $d_1^+, d_1^-, d_2^+ \ldots, d_k^-$. The objective function to be minimized is a linearly weighted function of these departure variables, that is

$$w_1^+d_1^+ + w_1^-d_1^- + w_2^+d_2^+ + \ldots + w_k^-d_k^-.$$

Thus in this general form the problem has been formulated as a linear programming problem. With all of the weights non-negative, there is no possibility of an optimum with both d_j^+ and d_j^- positive. On the other hand, it may be tempting to encourage a departure in one direction, but not in the other. For example, if the goal refers to profit, most companies would welcome a value in excess of the set goal. This could be engineered by making w_j^+ negative. The method will adjust automatically provided that w_j^+ is less in absolute value than w_j^-, so that the benefit from a unit excess is less than the loss of a unit shortfall.

When the weights are treated as priorities, so that a sequence of problems is to be solved, a special variant of the simplex method of linear programming may be used. For further details see for example Hsiao and Cleaver (1982) or Moskowitz and Wright (1979).

If the goals are the individual optimum values, then for a maximizing problem there can be no excesses over the goals and therefore all the d_j^+ values must be zero. It follows that only the negative weights need be estimated.

9.5 The method of goal programming – Delluge Park illustration

The proprietor of Delluge Park has two objectives, namely profit and adult customers. Expressed in terms of Flybynite sites (X) and casual sites (Y) those objectives are

$$Z_1 = 50X + 40Y \tag{9.14}$$
$$Z_2 = 2.4X + 3.6Y. \tag{9.15}$$

He places a higher priority on profit than on the number of adult customers, and also sets goals of a profit of 11 800 Rz and 850 adult customers.

A profit of 11 800 Rz can be achieved by any action (X, Y) satisfying $50X + 40Y = 11\,800$. The feasible actions are those lying on the segment UV in Figure 9.2.

Having achieved the profit goal, we now try to attain the adult customers goal of 850. Using the argument of graphical linear programming, consideration is now restricted to the feasible region consisting of the line segment UV and the optimum value of objective function (9.15) must be at one of the extremes. The coordinates of U can be calculated as (108, 160) producing a value for Z_2 of 835.2. For

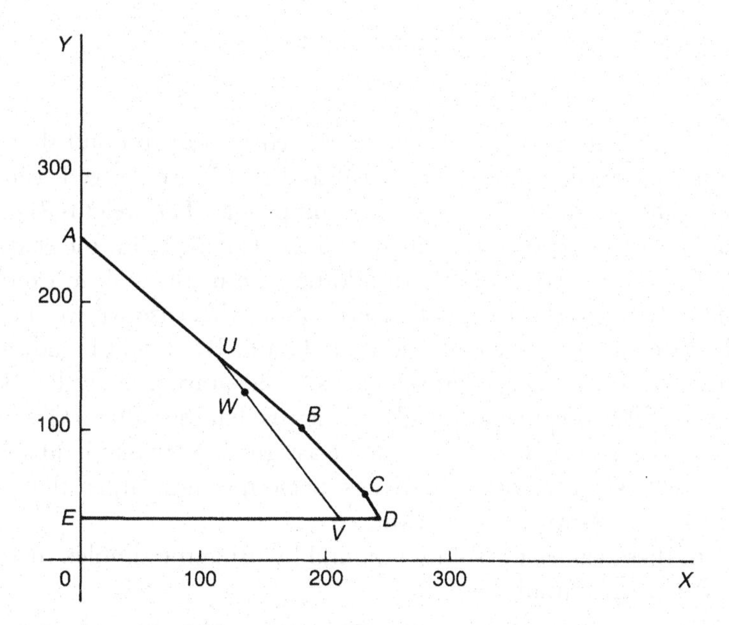

Fig. 9.2 Graphical illustration of profit goal of 11 800

V whose coordinates are (212, 30), the value of Z_2 is 616.8. Thus the best value of the second goal is 835.2, a shortfall of 14.8, and this is achieved at $X = 108$, $Y = 160$, the point U. Note that if we had been less ambitious and sought for a goal of 810 adult customers together with the profit goal of 11 800 Rz we would achieve this at the point $W(X = 120, Y = 145)$. Since W is an internal point both of the objectives can be improved by feasible actions. The goals in this case have been set too low.

Now suppose that use is to be made of the more orthodox goal programming model. The goals of 11 800 for Z_1 and 850 for Z_2 are to be used, so that the additional constraints corresponding to equation (9.13) are imposed:

$$50X + 40Y + d_1^- - d_1^+ = 11\,800 \tag{9.16}$$

$$2.4X + 3.6Y + d_2^- - d_2^+ = 850. \tag{9.17}$$

These augment the previous constraints

$$50X + 60Y \leqslant 15\,000 \tag{9.18}$$

$$Y \geqslant 30 \tag{9.19}$$

$$X + Y \leqslant 280 \tag{9.20}$$

$$2.5X + 1.5Y \leqslant 650. \tag{9.21}$$

The objective function to be minimized is

$$Z = W_1^+ d_1^+ + W_1^- d_1^- + W_2^+ d_2^+ + W_2^- d_2^-. \tag{9.22}$$

In selecting values for the weights we need to bear in mind that the two original goals from which (9.16) and (9.17) are derived were in totally different units, that is money and people. The weights W_1^-, W_2^- should therefore express the relative utilities of falling below the goals by 1 Rz (Z_1) and 1 person (Z_2). In addition we might wish to exercise the facility to exceed the goals. Certainly in this case we would not wish to penalize any excesses, so that W_1^+ and W_2^+ should be no more than zero, and in fact negative values could appropriately be used. Suppose, therefore, that a considered view is that the values $W_1^+ = -1$, $W_1^- = 5$, $W_2^+ = -10$, $W_2^- = 20$ are to be assigned. Note that in absolute values the weights with negative superscripts are larger than the weights with positive superscripts.

With these values the solution, found by using the simplex method of linear programming, is

$$X = 180, \quad Y = 100, \quad d_1^+ = 1200, \quad d_1^- = 0, \quad d_2^+ = 0, \quad d_2^- = 58,$$

and Z, the value of the objective function, is -40. The solution corresponds to the point B in Figure 9.2. Because of the weights, the optimal action has actually fallen short of the second goal by 58 units (persons) and at the same time exceeded the first goal by 1200 Rz.

Alternatively, suppose that the decision-maker argues that he is not particularly interested in exceeding his goals, but would like to put the emphasis on achieving levels as close as possible to them. One interpretation of this policy would be to put zero weights on the excesses. Retaining the other weights, this problem is as before, but with an objective function:

$$\text{minimize } 5d_1^- + 20d_2^-.$$

The d_1^+, d_2^+ are retained in the constraints (9.16) and (9.17) but dropped from the objective function (9.22). The resultant linear programming problem now has the solution $X = 108$, $Y = 160$, $d_1^- = 0$, $d_1^- = 0$, $d_2^- = 14.8$, $d_2^+ = 0$ with a minimizing value of 296 for the objective function. This is the point U in Figure 9.2. The first goal has been achieved, and the second has been maximized conditional on this. The solution is in fact identical with that when priorities were allocated with Z_1 coming before Z_2.

Both these models have produced solutions on the efficient frontier.

They are non-inferior actions as defined in Section 9.2. This will not necessarily happen if the goals are defined too conservatively. For example, if we use again the objective function $5d_1^- + 20d_2^-$ but have a goal for Z_2 of 810, one solution will be $X = 120$, $Y = 145$, $d_1^- = 0$, $d_1^+ = 0$, $d_2^- = 0$, $d_2^+ = 0$, the point W in Figure 9.3. Any action in the triangle UWS is not only feasible but is at least as good as the set goals. These are multiple solutions to the linear programming problem, but in practice the decision-maker would surely take an action which in terms of the objective functions was an improvement on $X = 120$, $Y = 145$. The argument suggests using an action somewhere on the segment US, which in turn is a segment of the region of non-inferior activities. With the alternative objective of rewarding excesses over the set goals, i.e. with $W_1^+ = -1$, $W_1^- = 5$, $W_2^+ = -10$ and $W_2^- = 20$, the solution appears once more as $X = 180$, $Y = 100$, the point B in Figure 9.3. The more plausible form of objective function has avoided the dilemma of choosing from several apparent optima and in this case the different levels of goal have produced the same solution. Whilst this is likely to happen when the changes in goal levels are relatively small it will not always be the case. The parallel with sensitivity analysis on the constraint constants discussed in Section 8.6 should be apparent.

Enough has been said about the difficulties of selecting goal levels to justify the alternative approach where the goal levels are taken as the maximum attainable value of each single objective. Some plausible departure below these goal levels is then to be accepted. For example, the maximum profit for the Delluge Park proprietor is 13 500 Rz and the maximum number of adult customers is 900. We might, therefore, take these as goals and maximize a weighted sum of the proportional departures. We are only considering d_1^- and d_2^-, and in this case the weights would be $W_1^- = W_1/13\,500$ and $W_2^- = W_2/900$, where W_1 and W_2 are constants whose values express the relative priorities of the two objectives.

Alternatively, we could argue that since any action will produce values of the objective function which are less than the goals we should try to minimize the maximum proportional departure. The simplest way of formulating the problem is to minimize each proportional departure in turn subject to the constraints that this particular departure is greater than each of the others. As stated, each minimization is a linear programming problem, and can be solved by the simplex procedure. The solution to the original problem is found by selecting the

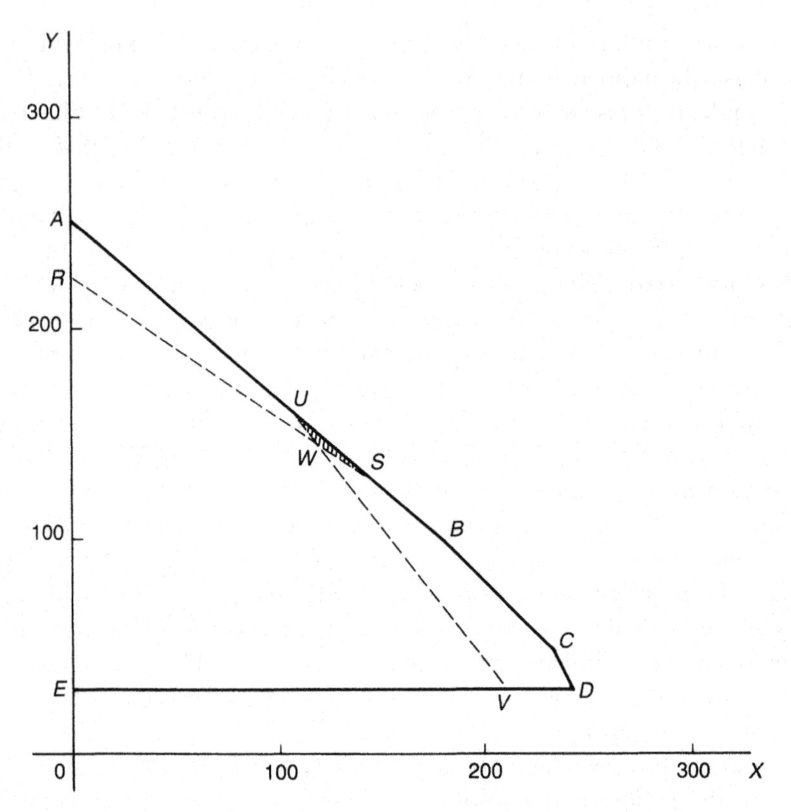

Fig. 9.3 Goal programming leading to an inferior action

smallest solution of the above set of problems. In general we shall be considering only a limited number of objectives, and since this is the number of individual minimizations to be carried out, the task will be of manageable proportions. For example, in the Delluge Park case the two problems would have the common constraints:

$$\begin{aligned} 50X + 60Y &\leqslant 15\,000 \\ Y &\geqslant 30 \\ X + Y &\leqslant 280 \\ 2.5X + 1.5Y &\leqslant 650 \\ 50X + 40Y + d_1^- &= 13\,500 \qquad (9.23) \\ 2.4X + 3.6Y + d_2^- &= 900 \qquad (9.24) \end{aligned}$$

Note that in equations (9.23) and (9.24), since the right-hand sides are the individual maxima, there is no possibility of excesses and therefore d_1^+, d_2^+ must be zero. The two linear programming problems supplement the constraints with

Problem (A) $d_1^-/13\,500 \geqslant d_2^-/900$, minimize $d_1^-/13\,500$ (or equivalently minimize d_1^-)

Problem (B) $d_2^-/900 \geqslant d_1^-/13\,500$, minimize $d_2^-/900$ (or equivalently minimize d_2^-).

Both problems have the same solution $X = 1500/11$, $Y = 1500/11$, and this is therefore the answer sought. The solution actually corresponds to the case where the proportional departures are equal. Note also that in this case the solution consists of fractional numbers of places. Investigation of the integer solutions surrounding this point establishes that the integer solution which best satisfies the given criterion is $X = 136$, $Y = 136$. This single search practice for integer solutions is really only feasible when there are at most three variables.

9.6 General comments

The methods in this chapter describe two approaches to an impossible problem. It is in general impossible to optimize more than one criterion, and the models used here should be taken as providers of insight rather than mechanical forms of analysis. They require the decision-maker to think objectively about the compromises he has to make between his objectives. The danger is that a fairly sophisticated form of analysis may give a spurious respectability to an approach whose basic assumptions have been insufficiently argued.

A generalization of the underlying principles of goal programming called *indifference mapping* has been devised by Rivett. This is an adaptation of the idea of structural mapping originated by D. G. Kendall, and it relies on a listing of all the available decision outcomes between members of which there is a measure of distance. This measure can be extremely crude, perhaps to the extent of stating that the decision-maker is indifferent between two outcomes or he is not. From these 'distances' a 'map' can be drawn which locates the outcomes in a way that is consistent with the measures. It usually follows that the outcomes which can be identified as most preferable and conversely as less preferable will lie at the extremes of the map. Thus, although the technique does not solve the problem uniquely, it can direct attention to a limited number of good alternatives. For further information reference may be made to the two articles by Rivett (1977a, b) listed in the following section.

9.7 Further reading

Bunn, D. W., *Analysis for Optimal Decisions*, Wiley, New York (1982), Chapter 5.

Daellenbach, H. G. and George, J. A., *Introduction to Operations Research Techniques*, Allyn and Bacon, Boston, Mass. (1978), Chapter 2.

Hsiao, J. C. and Cleaver, D. S., *Management Science*, Houghton Mifflin, Boston (1982), Chapter 7.

Moskowitz, H. and Wright, G. P., *Operations Research Techniques for Management*, Prentice-Hall, Englewood Cliffs, N.J. (1979), Chapter 14.

Rivett, B. H. P., 'Policy Selection by Structural Mapping', *Proceedings of the Royal Society*, London, A, **354**, 407–423 (1977a).

Rivett, B. H. P., 'Multidimensional Scaling for Multiobjective Policies', *Omega*, **5** (4), 367–379 (1977b).

9.8 Problems and practical exercises

1 Take the original situation of the hotel owner in Exercise 1 of Chapter 8. Suppose that he also owns a nearby golf course and that he sees as one of his objectives the maximization of the number of hotel guests using the golf course. Not only is this profitable but it creates an attraction for golf aficionados to stay in his hotel. Past experience shows that 80 per cent of the occupants of single rooms make daily use of his golf course. Usage by occupants of double rooms is a little lower at 60 per cent.

Find the mixture of single and double rooms which maximizes the usage of his golf course by residents. Show that this differs from the mixture found in the question as previously stated. Characterize the set of non-inferior strategies for these two objectives. List all integer solutions.

Suppose that the hotel owner now sets as goals the maximum of each objective considered independently and proposes to back off by a minimum percentage of each goal. What is the resultant strategy in terms of single and double bedrooms? (This must be in integers.)

2 Refer to Exercise 2, Chapter 8. In addition to the objective of maximizing the income from each trip, the operators also want to sail as many trips as possible, as they fear that rival operators may otherwise be attracted to this service. They would therefore like to improve the turn-round time. Loading a car takes on average 1 minute, but loading a car plus caravan takes 3 minutes.

Minimizing the loading time under the given conditions leads to the absurd result of running an empty ferry. Show how by accepting a lower goal for the profit, improvements can be made in loading time and specify the actual relationship between these goals.

3 Now refer to Exercise 3 of Chapter 8. In this situation it is argued that the sports hall is a service to the community and it should therefore be available to the maximum number of people. A 'quiet' evening normally caters for 80 people, but on a 'noisy' evening, because the sports are more vigorous, there is only room for 50 people. What range of non-inferior activities is defined by the objectives of profit maximization and attendance maximization?

The Local Authority which runs the sports hall accepts that for every reduction in the profit by £1 there should be an increase in the annual attendance of 10. What now is the optimum mix of 'quiet' and 'noisy' evenings?

10
The method of branch-and-bound

OBJECTIVES

This chapter introduces the technique of branch-and-bound and readers should have an understanding of its general uses and in particular its approach to

(a) assignment problems;
(b) travelling salesman problems;
(c) integer programming problems.

10.1 The underlying principle

The method of branch-and-bound is used to find the optimal action when there is a large number of alternatives. A complete enumeration of the outcomes would be theoretically possible but in practice very time consuming. Most of the applications are in what would otherwise be termed mathematical programming, and some can be solved by specific tailor-made techniques. We shall see, for example, that the assignment problem of Section 8.13 can be solved by branch-and-bound, thereby avoiding the degeneracy difficulties inherent in the stepping-stone method.

The method depends on a very obvious principle. Take, for example, a minimization problem. Suppose that the possible actions can be divided into sets, and it can be shown that all outcomes in one set are greater than a particular one in another set. In such a case all the actions in the first set can be eliminated from further consideration. The sets remaining are then refined until they consist of only one action, at which point optimization is a matter of direct comparison. The method only works effectively where large numbers of actions can be eliminated at a relatively early stage. It is particularly effective when actions can be developed sequentially, so that if at a common intermediate stage all the actions in one set are worse than some complete action, there is no need to progress further with this set.

The sets are developed by a branching process which at any node splits the actions in a branch into two sub-sets, forming the branches from this node. The similarity between this aspect of the solution process and decision trees will become apparent as the examples are described. Three illustrations will be taken from different applications of mathematical programming. For ease of handling the illustrations will be of modest proportions. It may, therefore, be possible for the reader to deduce or guess the solution without this technique, but this will certainly not be the case with larger practical problems.

10.2 Application to assignment problems

British Electronic Rotary Calculators (BERC) plc is a manufacturer of large computers. To service the machines they have installed in the United Kingdom they have established teams at centres in Aberystwyth, Birmingham, Cambridge, Dover and Edinburgh (Northern Ireland has its own team). Each service team can cope with one maintenance or repair call on any day, the assignment of teams to calls being decided from BERC's central office. On a particular day calls have been received from Leicester, Manchester, Nottingham, Oxford and Portsmouth. Given Table 10.1, which shows the distances (in miles) between service centres and calls, how should the service teams be assigned to minimize the total distance travelled?

This is the standard assignment problem of Section 8.13. In this case, since we are minimizing, the minimum distance to be travelled from each centre can be subtracted from all the distances in the cor-

Table 10.1

Service centre			*Call location*		
	Leicester	*Manchester*	*Nottingham*	*Oxford*	*Portsmouth*
Aberystwyth	151	131	156	157	218
Birmingham	41	81	49	63	141
Cambridge	69	155	84	80	126
Dover	171	257	196	129	131
Edinburgh	284	213	259	353	431

responding row, and the same procedure adopted for the reduced distances in the columns. The equivalent problem now has at least one zero distance in every row and column (Table 10.2). There is now a possibility of finding by inspection an assignment using only zero distance routes, which must therefore be the optimum. Unfortunately in this case no solution with a total cost of zero exists.

Table 10.2

Reduced distances					
Service centre	*Call location*				
	Leicester	*Manchester*	*Nottingham*	*Oxford*	*Portsmouth*
Aberystwyth	20	0	17	26	85
Birmingham	0	40	0	22	98
Cambridge	0	86	7	11	55
Dover	42	128	59	0	0
Edinburgh	71	0	38	140	216

The sources or service centres are now taken in turn, building up the total assignments sequentially and at the same time recording the accumulated distances travelled. Although with care a better ordering of the sources could be devised (e.g. by size of difference between the best (zero) distance and the next best), in this case we will work through them in the order listed above.

The best location for the service team at Aberystwyth is Manchester, producing a distance of zero. Failing this, any other assignment would involve a distance of at least 17 miles, when Aberystwyth is assigned to Nottingham. The situation is represented in the form of a tree, the branch being the choice between including Aberystwyth–Manchester (AM) or not including it ($\overline{AM}$). The bar notation is used for 'not including'. Although the distance appropriate to this branch refers to the Aberystwyth–Nottingham pairing, the choice is not between Manchester and Nottingham but between Manchester and not Manchester. At each node the branches must exhaust all the possibilities present. The distance of 17 is a minimum distance for the branch, not an actual one as is the case with the zero distance between Aberystwyth and Manchester. It is convenient to distinguish between the two cases, and therefore on the diagram the number 17 is enclosed in a circle. The initial choice is shown in Figure 10.1.

We then scan the figures at the ends of branches and find the smallest.

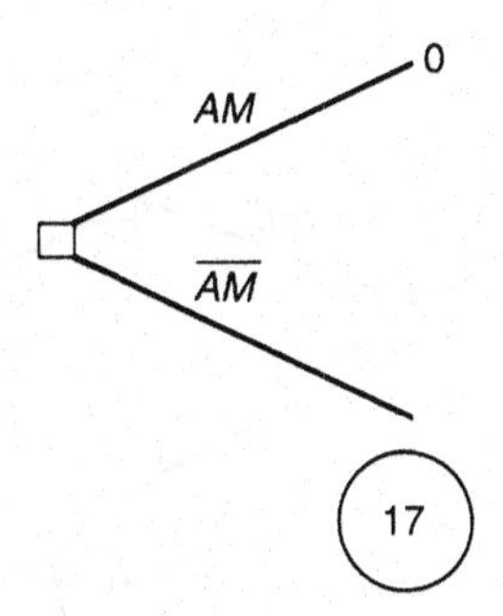

Fig. 10.1 The first branch

This is the node for the next branch. Having assigned Aberystwyth to Manchester, Birmingham is next considered. Birmingham must not be assigned to Manchester, but is given the minimum distance call location of those remaining. In this case there are two zero distance locations at Leicester and Nottingham. We arbitrarily choose one of these, say Leicester, so that the next branching is the choice between *BL* and $\overline{BL}$. The augmented tree is now shown in Figure 10.2.

The same argument adds the Cambridge–Nottingham pairing to the *AM*, *BL* branch. Note that the zero-distance Cambridge–Leicester pairing is ruled out, since Leicester has already *on this branch* been assigned to Birmingham. Also the best alternative to Cambridge–Nottingham (7 miles) is Cambridge–Oxford (11 miles). Figure 10.3 shows the current position.

At this stage the minimum distance branch is *AM*, $\overline{BL}$ and this is now pursued. The best assignment available for Birmingham is Nottingham, a distance of zero, the best alternative being Oxford, a

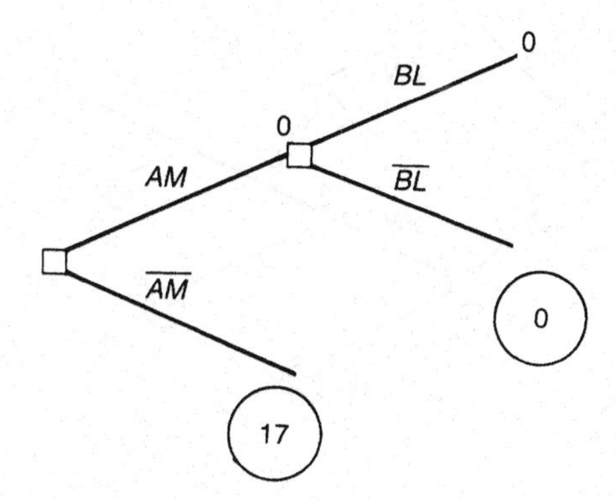

Fig. 10.2 The first two branches

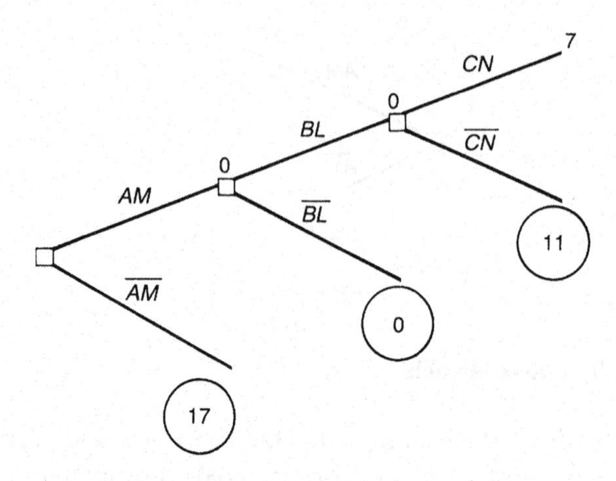

Fig. 10.3 The third branch

distance of 22 miles. From the zero-distance node, Cambridge can be assigned to Leicester, so that the accumulated distance is still zero, and moreover proceeding to the next node Dover can be assigned to Oxford, a further zero-distance pairing. This only leaves Portsmouth for Edinburgh, a distance of 216 miles. This is a good example of how short-sighted the approach can be. It only looks one step ahead and can end up with clearly inefficient actions. The branches at this stage are summarized in Figure 10.4.

At some of the nodes towards the ends of the branches no alternative choices remain. Note also that there is now one complete assignment, resulting in a distance of 216 miles. For any partial assignment

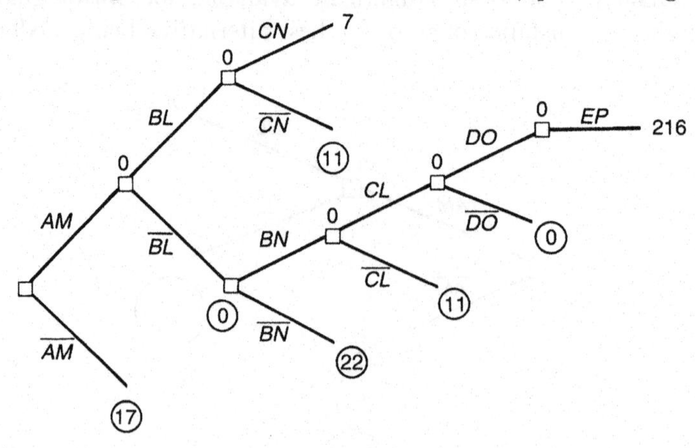

Fig. 10.4 The first complete branch

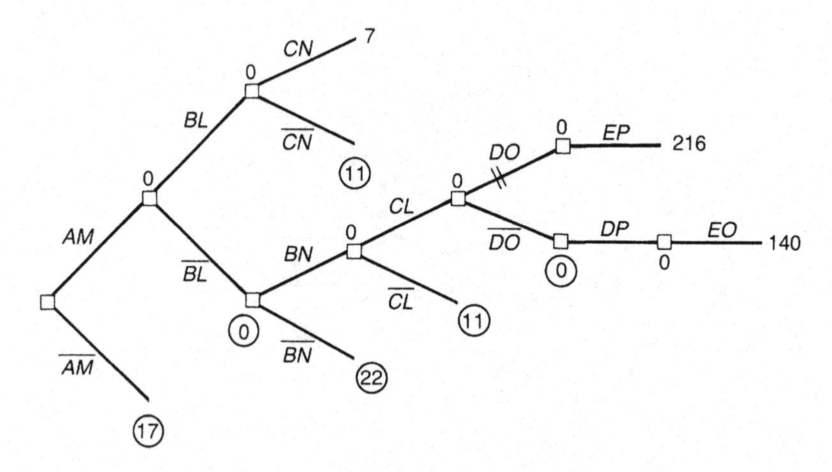

Fig. 10.5 First elimination of a branch

with an accumulated total greater than this amount, the branch may be terminated without further consideration. This hopefully is where the savings will occur, since had we originally simply decided to list all the possible assignments, there would have been factorial 5 or 120 of them. By the systematic approach we should only have to examine a small fraction of them. The current 'minimum' of 216 can, of course, be superseded as further branches are extended. In fact the next branch with the smallest distance is the *AM*, $\overline{BL}$, *BN*, *CL*, $\overline{DO}$ branch with a zero distance and this must be extended by *DP*, *EO* giving a total distance of 140 miles. The exclusion of the previous minimum is indicated in the diagram by the lines across the appropriate part of the branch as shown in Figure 10.5.

The method now returns to the 7 mile distance at the end of the *AM*, *BL*, *CN* branch, and proceeds as before. The remaining analysis is left as an exercise for the reader, but the final tree should be that shown in Figure 10.6.

The solution is therefore given by the branch $\overline{AM}$, $\overline{AN}$, $\overline{AL}$, *AO*, $\overline{BL}$, *BN*, *CL*, *DP*, *EM* or in terms of the assignment

Aberystwyth	→	Oxford
Birmingham	→	Nottingham
Cambridge	→	Leicester
Dover	→	Portsmouth
Edinburgh	→	Manchester

The total distance is 619 miles.

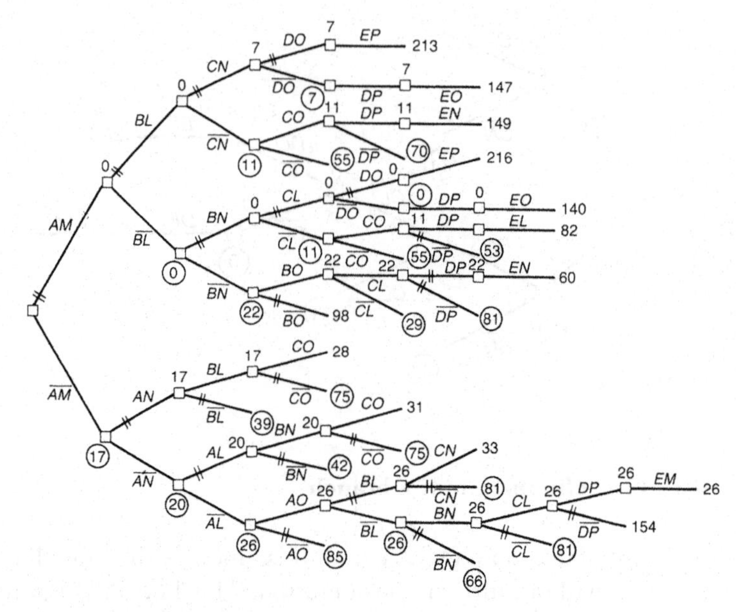

Fig. 10.6 The final tree

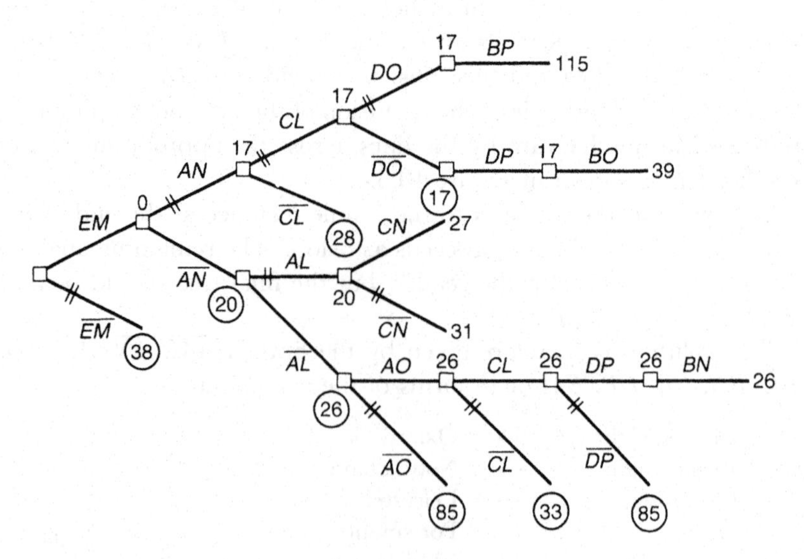

Fig. 10.7 Tree using efficient ordering

Had we applied a more efficient order to the sources, the size of the tree would have been reduced. Earlier in this section, a rule which ranks them by the differences between the best and the next-best (or the next-best-but-one in the case of ties) was mentioned. This produces an ordering Edinburgh, Aberystwyth, Cambridge, Dover, Birmingham and the simpler tree shown in Figure 10.7.

10.3 Application to travelling salesman problems

The travelling salesman problem is the basic problem faced by any sales representative, delivery truck, health visitor, travelling library, etc., where somebody has to visit a number of known locations and they wish to minimize the distance travelled. There are more elaborate versions of the problem where account is taken of the times of the visits, avoiding for example arriving in a main shopping street at the peak traffic time or a beer delivery during the publican's opening hours. Attention here is confined to the simple distance minimization problem. The problem can be stated as a programming problem with variables X_{ij} equal to 1 if location j is visited immediately after location i, and zero otherwise. Any problem of practical size would not be a feasible proposition as far as the computation is concerned. The method of branch-and-bound can be applied, although a problem with, say, ten locations would still be quite a formidable proposition. Note that a complete listing of all possible routes round ten locations would have factorial ten members (assuming that the reverse direction routes are counted; otherwise the number is halved). This would amount to approximately three-and-a-half million, which is scarcely a practical method of solution.

To illustrate the use of the branch-and-bound method in solving this problem, imagine that you are a salesman visiting the five customers A, B, C, D and E, one of whom is located at your own base. (It is clear that the actual starting point is immaterial in terms of minimizing the total distance if it is assumed that the salesman returns to his base at the end of the trip.) The distances (in kilometres) between the customers are given in Table 10.3. Note that here the reverse distances are the same as the forward ones. This will not necessarily always be the case, when obstructions like one-way streets occur.

Since each customer is joined by the route to another customer, the

Table 10.3

	A	*B*	*C*	*D*	*E*
A	—	3	16	20	19
B	3	—	14	18	20
C	16	14	—	5	10
D	20	18	5	—	7
E	19	20	10	7	—

distances in each row can be reduced by subtracting out the minimum value. The resulting problem is equivalent to the original. Also the same argument can be applied to the columns of the reduced-distance problem, producing a matrix of distance with at least one zero in each row and column. If a solution to this equivalent problem can be found where only zero-distance routes are used, then clearly it must be the optimum. The reduced distances are shown in Table 10.4. By inspection, no zero total distance route exists.

Table 10.4

	A	*B*	*C*	*D*	*E*
A	—	0	13	17	14
B	0	—	11	15	15
C	11	9	—	0	3
D	15	13	0	—	0
E	12	13	3	0	—

The branch-and-bound method selects any customer as a starting point. Take customer *A*. The nearest customer to *A* is *B*, a distance of zero kilometres. If the link *AB* is not in the solution, the link from *A* will be at least 13 km. The hope is that we may be able to find a complete circuit, starting with *AB*, which has a total distance less than 13 km, in which case there is no point in examining further the branch where *AB* is not included.

The next step is to look at the link from customer *B*. Customer *A* has already been included, and the choice is therefore between *C*, *D* and *E*. Of these the nearest is *C* and the choice at the branch is therefore that link *BC* should be included or that it should not. In the former case the total distance so far is 11 km, and in the latter it is at least 15 km (which is more than the minimum when *AB* is not included). The branching so far is shown in Figure 10.8.

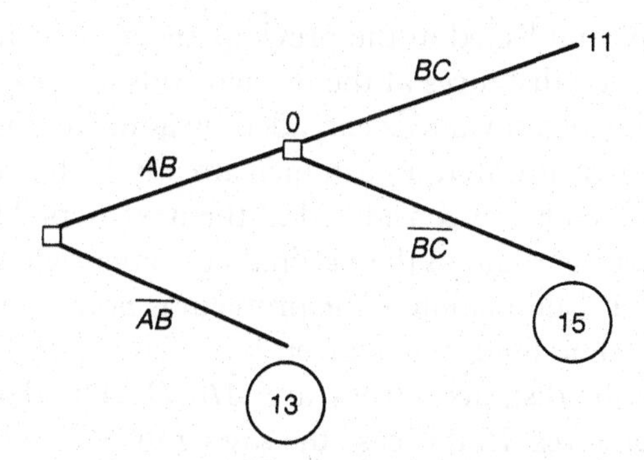

Fig. 10.8 First branch – travelling salesman problem

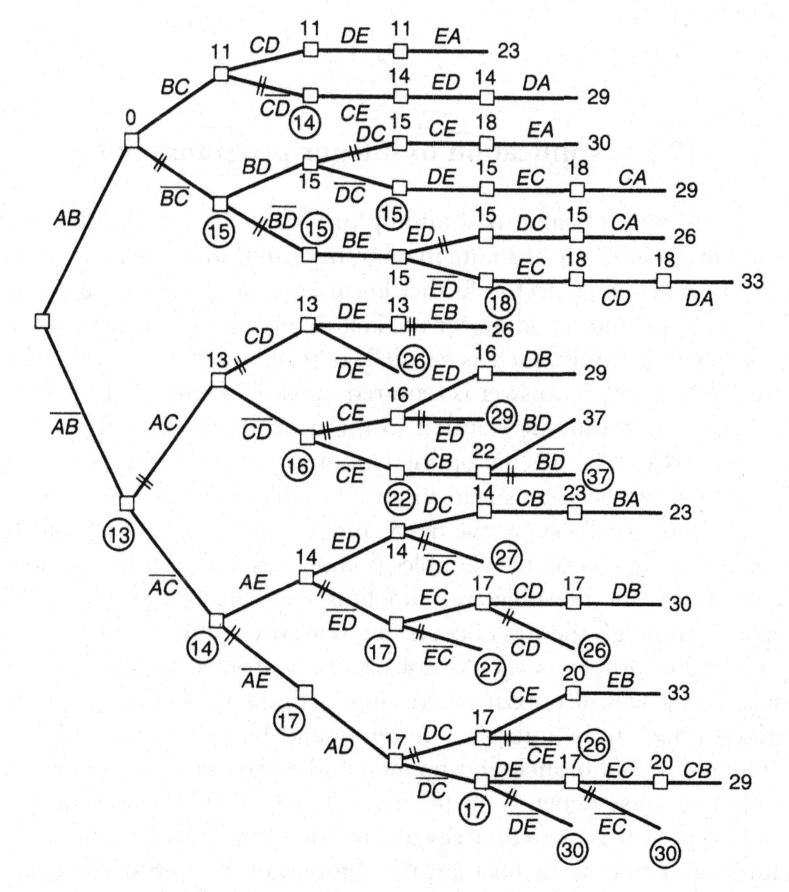

Fig. 10.9 Travelling salesman problem tree

The pattern established in the previous section is now applied. At each step the total distances at the branch ends are scanned, and the branch with the lowest value is extended by a further link. Care must be exercised to ensure that a customer already included in previous links of the branch is not considered at the next step. At the last step the link returns the route to the original customer, which in this case was customer A. Completion of the problem is now left as an exercise for the reader. The final tree is given for reference in Figure 10.9.

The minimum distance routes are $ABCDEA$ and its reverse, a total of 48 kilometres. In this case the saving in time and effort over a complete enumeration of all routes is very marginal, and indeed the nature of the travelling salesman problem is such that it is unlikely that any rigorous optimizing procedure will produce a really substantial benefit.

10.4 Application to integer programming

In dealing with linear programming in Chapter 8 a technique was used which found the absolute maximum or minimum of some linear objective function subject to further linear constraints on the variables. The technique did not allow for the fact that on occasions we shall be dealing with variables such as numbers of people, vehicles, production units, etc., where an answer is required in whole numbers. Rounding the linear programming solution to the nearest whole numbers can give an answer which is not optimal amongst all feasible integer solutions; it can on occasions lead to an infeasible solution.

The problem of solving the linear model when the solution has to be integer values for all the variables is known as integer programming. The linearity of the objective function and constraints is usually implied, although there is of course no reason why more complicated optimization problems should not also be restricted to integer solutions. Cases will also exist where only a designated number of the variables have to be integers, the remainder being unrestricted.

The method of branch-and-bound can be used to solve the integer problem. Its use depends on the very obvious fact that in solving a problem in which the variables are restricted to integer values, the solution must be no better than the solution of the unrestricted problem, simply because any solution restricted to integers is also a solution of the unrestricted problem. If, therefore, a sub-region of the feasible actions can be found such that its best unrestricted solution is inferior to an integer solution outside the sub-region, there is no point in further investigation of the sub-region. The skill lies in the careful definition of the sub-regions. The method will be demonstrated by means of an example.

Cobblestone Engineering plc have a plant making three of its products – Asterisks (A), Brackets (B) and Colons (C). The products all go through the milling and grinding processes, the times taken (in minutes) being shown in Table 10.5. Both processes are available for

Table 10.5

	Milling	*Grinding*
Asterisks	35	55
Brackets	25	85
Colons	90	45

the full eight hours of a shift, but for bonus and quality control purposes it is essential that each shift completes the production of any of the products within its working time. It can be assumed that the operations are sufficiently flexible that no time is lost unnecessarily by a process waiting for a product. Cobblestone make profits of £6, £10 and £8 respectively on the three products. How many of each product should be made on a shift to maximize the total profit?

The problem can be readily transformed into its mathematical model. Denoting the numbers of each product by the corresponding letters, the milling process limitation requires that

$$35A + 25B + 90C \leqslant 480,$$

or simplifying,

$$7A + 5B + 18C \leqslant 96. \tag{10.1}$$

Similarly the grinding limitation amounts to

$$55A + 85B + 45C \leqslant 480,$$

or

$$11A + 17B + 9C \leqslant 96. \tag{10.2}$$

The objective of maximizing profit can be expressed as

$$\text{maximize } 6A + 10B + 8C. \tag{10.3}$$

The variables A, B and C are non-negative and also restricted to integer values.

The first step in solving the problem is to solve it as an unrestricted linear programme. (In this section the unrestricted linear programme will be used on several occasions. The working of the solution will not be offered.) The unrestricted solution is

$$A = 0, \quad B = 3.31, \quad C = 4.41, \quad \text{profit} = 68.38.$$

No integer solution can therefore provide a profit greater than 68.38. Clearly the optimal integer solution should be somewhere in the vicinity of this solution, and it therefore forms the basis of the branching of possible solutions. Take first the variable B, since this contributes the largest amount per unit to the objective function. The value found is intermediate between the integer values 3 and 4, and we shall, therefore, solve two further unrestricted problems with $B \leqslant 3$ and $B \geqslant 4$ respectively. Branching in this way covers all possible integer solutions. The solutions now are

$$\begin{array}{lllll} B \leqslant 3; & A = 0.6, & B = 3, & C = 4.27, & \text{profit} = 67.73 \\ B \geqslant 4; & A = 0, & B = 4, & C = 3.11, & \text{profit} = 64.89. \end{array}$$

Thus the former set ($B \leqslant 3$) looks the more promising, having a larger profit. Indeed if an integer solution with profit greater than 64.89 can be found, there will be no need for further investigation of the latter branch. The first stage of the tree is shown in Figure 10.10.

Investigation of the branch $B \leqslant 3$ amounts initially to refining the criterion into $B = 3$ and $B \leqslant 2$. We already have the optimal solution for $B = 3$, but a further linear programme with the additional constraint $B \leqslant 2$ produces the solution

$$A = 2.53, \quad B = 2, \quad C = 3.79, \quad \text{profit} = 65.54.$$

The highest profit has, of course, occurred when $B = 3$. Attention is now turned to the other variables, neither of which currently takes an integer value. The variable C has the higher coefficient in the objective function, and applying the same argument as we did for the B variable, the unrestricted problem is solved for $B = 3$, $C \leqslant 4$ and $B = 3$, $C \geqslant 5$. In fact the latter case has imposed restrictions which are incompatible with the original constraints, so that the problem is infeasible and this branch need be explored no further. The alternative, with $C \leqslant 4$, produced the solution

$$A = 0.82, \quad (B = 3), \quad C = 4, \quad \text{profit} = 60.55.$$

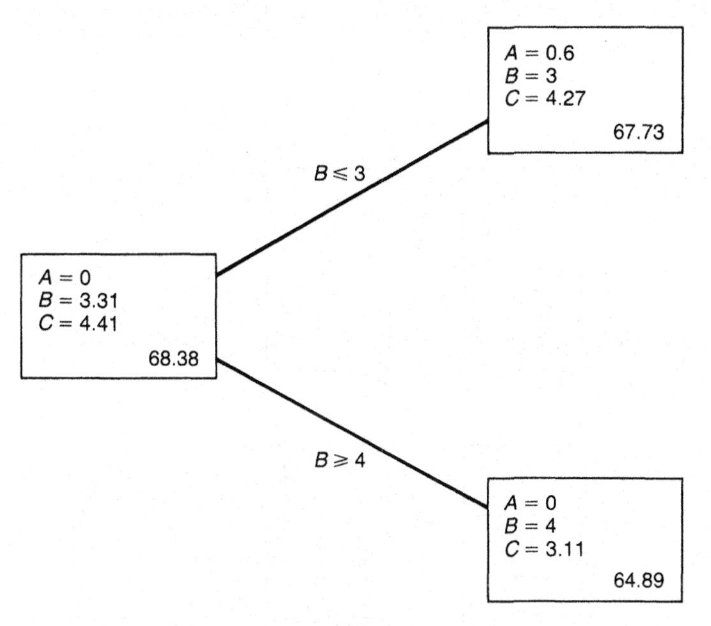

Fig. 10.10 Integer programming, the first branch

The augmented tree is shown in Figure 10.11.

The highest profit branch end is now that of $B \leqslant 2$, a profit of 65.54. This is now further divided by $B = 2$, $B \leqslant 1$, and the corresponding optimization carried out. The problem where $B = 2$ is that already solved and it therefore retains its position as the branch to be investigated further. Taking the division between $C \leqslant 3$ and $C \geqslant 4$ as the next branch we have solutions

$$C \leqslant 3; \quad A = 3.18, \quad (B = 2), \quad C = 3, \quad \text{profit} = 63.09$$
$$C \geqslant 4; \quad A = 2, \quad (B = 2), \quad C = 4, \quad \text{profit} = 64.$$

The latter is a totally integer solution. By chance it has occurred one step before the same approach would have been applied to the variable A.

The implication now is that any branch with a profit less than 64 need be investigated no further, irrespective of the integer or non-integer state of the variables. The complete branch-and-bound tree is shown in Figure 10.12.

In this case there are multiple optimum integer solutions, namely

$$A = 0, \quad B = 4, \quad C = 3; \qquad A = 2, \quad B = 2, \quad C = 4.$$

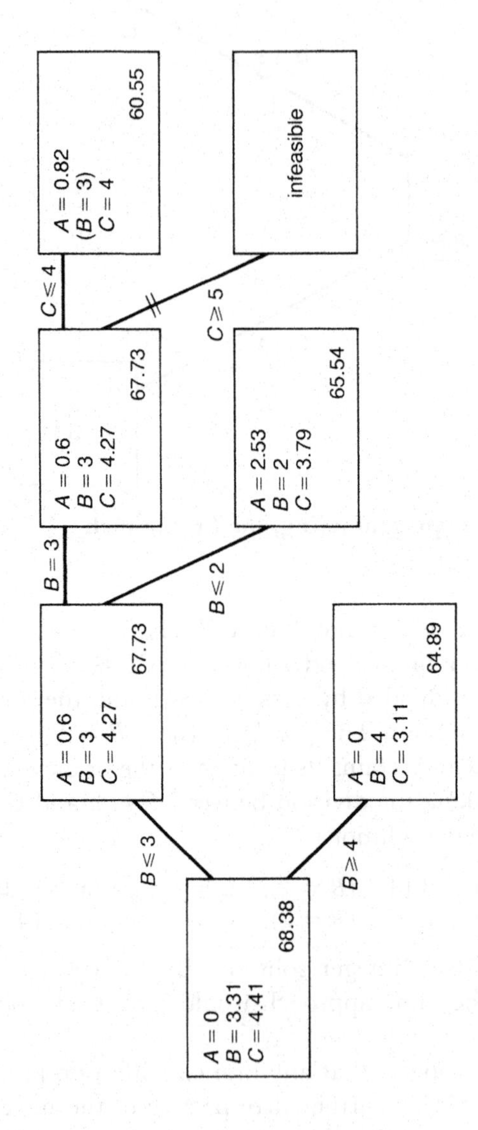

Fig. 10.11 Integer programming, second stage

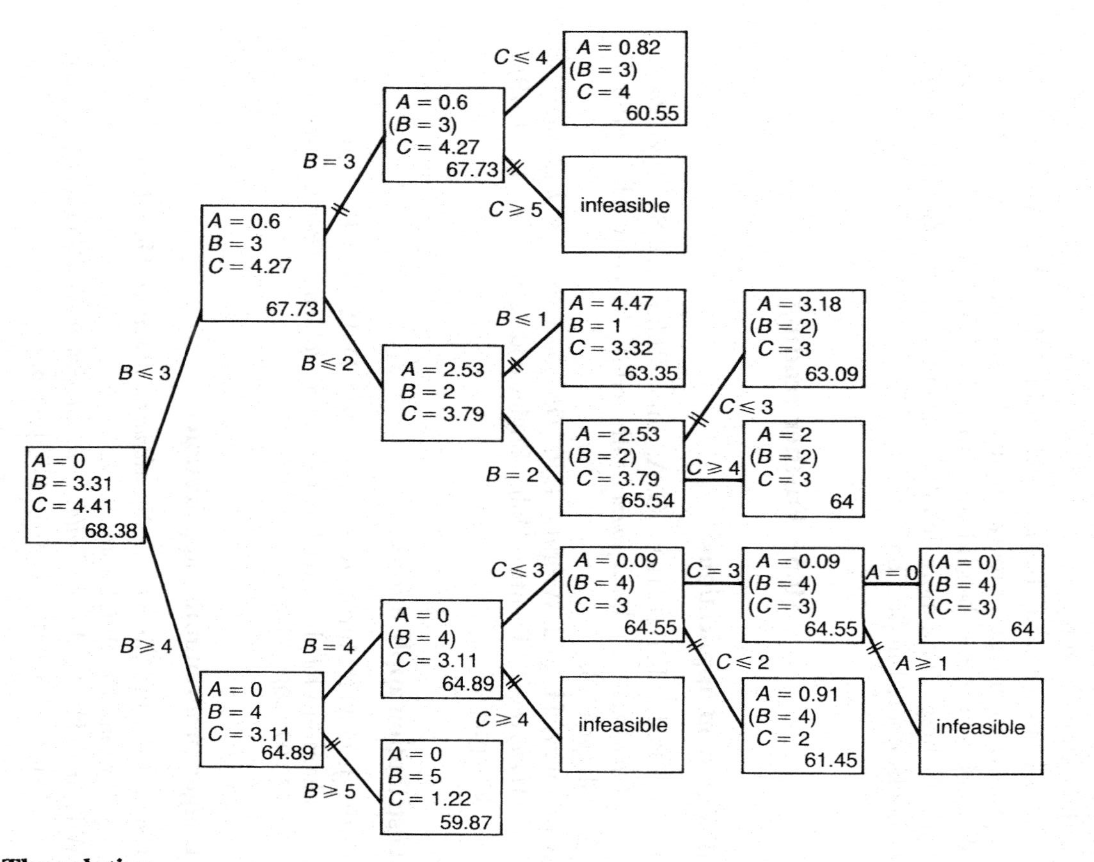

Fig. 10.12 **The solution**

Both produce profits of 64. Note that if we had taken the simple approach of finding the unrestricted optimum and taking all combinations of values which are the nearest integers rounding both up and down, that is $A = 0,1$; $B = 3,4$; $C = 4,5$, we would have missed both of the optimal solutions.

This has necessarily been a brief outline of the branch-and-bound approach to integer programming. It is claimed to be efficient, although more could be said about the selection of the variable chosen for branching. For a more detailed account reference should be made to books quoted in Section 10.5.

10.5 Further reading

Assignment applications

Lewin, R. I., Kirkpatrick, C. A. and Rubin, D. S., *Quantitative Approaches to Management* (5th edn) McGraw-Hill (1982), Chapter 13.

Moskowitz, H. and Wright, G. P., *Operations Research Techniques for Management*, Prentice-Hall, Englewood Cliffs, N. J. (1979), Chapter 13.

Distribution applications

Little, J. D. C., Murty, K. G., Sweeney, D. W. and Karel, C., 'An algorithm for the travelling salesman problem', *Operations Research*, **11**, 979–989 (1963).

Integer programming applications

Hsiao, J. C. and Cleaver, D. S., *Management Science*, Houghton Mifflin, Boston (1982), Chapter 6.

Williams, H. P., *Model Building in Mathematical Programming*, John Wiley, Chichester (1978), Chapter 8.

10.6 Problems and practical exercises

Use the method of branch-and-bound to solve the following problems.

1 Five firms have submitted bids to build five telephone exchanges. Each firm can only take one contract. How would you assign firms to contracts to minimize the total cost, given the bids shown in Table 10.6?

Table 10.6

Bids (£000 000)			*Contract*		
Firm	*V*	*W*	*X*	*Y*	*Z*
A	45	45	45	45	45
B	53	50	48	55	40
C	61	60	57	59	42
D	65	60	58	63	48
E	58	59	55	60	43

2 Starting from location *A*, a salesman has to visit customers at locations *B*, *C* and *D* before returning to *A*. Given the distances (miles) between the locations shown in Table 10.7, how does he plan the trip to minimize the total distance?

Table 10.7

	A	*B*	*C*	*D*
A		10	12	8
B			14	7
C				9

3 A hotel is considering a reorganization of its facilities. The basic choice is between accommodating paying guests or conference delegates. The sleeping accommodation is identical, consisting of space for 150 people. In the dining room 1.9 square metres is allowed for a paying guest, but because fewer tables would be required, the allowance for conference delegates is 1.5 square metres. The planned size for the dining room is 260 square metres (all residents use the dining room at one sitting). The lounge accommodation (which can be conveniently split between paying guests and conference delegates) consists of 616 square metres. For comfort a paying guest should have 3.2 square metres, but a conference delegate needs 4.4 square metres.

The contribution from paying guests is £18 per day and from conference delegates £15 per day. What is the optimum split between the two types of residents?

11
Dynamic programming

OBJECTIVES

This chapter introduces the reader to the principle on which the technique of dynamic programming is based. Having understood this the reader is then given examples of the use of dynamic programming in a simple deterministic production planning situation followed by applications of the stochastic model to a problem of deciding on the timing of currency conversion and to a standard inventory problem. With this information the reader should be able to apply the technique to larger problems involving computer programs, having gained insights into the nature of the model, the type of input data required and the form of the solution.

11.1 Bellman's principle of optimality

Dynamic programming is a technique which can be of value in situations where a sequence of decisions has to be taken. Each decision in the sequence is interspersed with a decision outside the control of the decision-maker. This is, of course, the basic situation for the decision tree which has already been discussed in Chapter 6. There will, however, be instances where essentially the same decision is being taken repeatedly and also essentially the same chance events are occurring between the decisions. For example, a retailer who reviews his stock on a regular weekly basis and then decides how much, if any, to re-order is faced with the same problem; it differs only in as much as he takes into account the amount of stock he holds at the time. The chance events between these decisions can be summarized as the demand during the week. Thus although a decision tree could be drawn up showing each possible re-order level for the retailer followed by each possible weekly demand level, continued over several weeks, it would be a very cumbersome diagram to handle. Dynamic programming uses the repetitiveness of the decisions to save computational effort. As a word of caution note that even in situations where the chance elements are repeated exactly, it does not follow that each cycle can be analysed in isolation. If, for our retailer, there is a significant delivery charge, it may pay him to place orders which are likely to cover him for several further weeks, rather than order every week.

The technique depends on a deceptively simple but remarkably fruitful principle. It is generally referred to as *Bellman's Principle of Optimality*, first stated in Bellman (1957). It reads:

> An optimal policy has the property that, whatever the initial state and initial decision are, the remaining decisions must constitute an optimal policy with regard to the state resulting from the first decision.

At first reading the definition appears almost a tautology. In this chapter the power of this principle will be demonstrated when it is used in the important context of production planning. Before this, simply as an illustration of the technique, the standard illustration of finding the minimum distance from A to B will be used.

11.2 The road journey problem

A motorist is planning a journey between Leicester and Boston (Lincolnshire), and wishes to find the minimum distance route using only main roads. The network of roads together with the distances (in miles) between neighbouring towns is shown in Figure 11.1.

The dynamic programming man now argues as follows. He starts at his destination Boston and notes that to reach Boston he must go through Sleaford, Osbournby or Spalding. If he is at Spalding his optimal decision is to drive directly to Boston, a distance of 16 miles. He therefore associates the optimal value of 16 with Spalding. Similarly values of 21 and 17 are associated with Osbournby and Sleaford respectively. (He has to take care that a lateral journey, say from Osbournby to Boston via Sleaford, is not shorter.) Next he looks at the position he would be in if his journey took him through Stamford. He rules out the possibilities of returning to Leicester or crossing back to Melton Mowbray or Grantham, since these are clearly longer than would be possible using either of the two remaining routes. If he takes the route to Spalding and then acts in an optimal manner he reaches

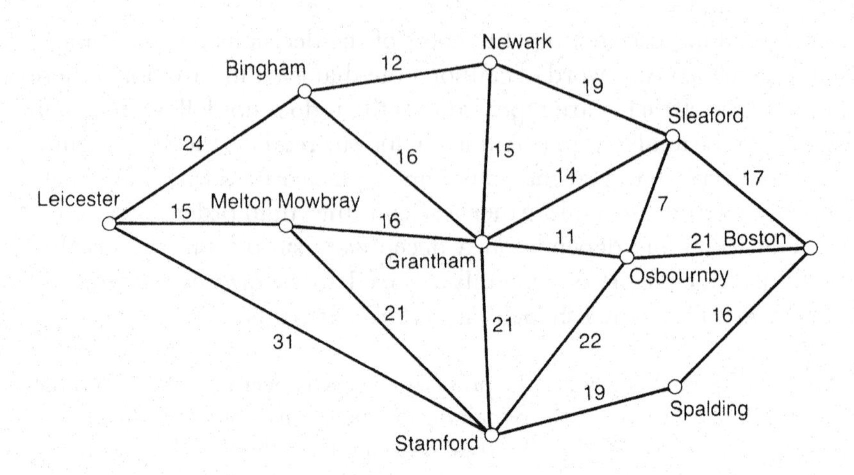

Fig. 11.1 Available routes, Leicester to Boston

Boston in 19 + 16 = 35 miles. Alternatively, he could take the road to Osbournby and reach Boston in 22 + 21 = 43. Using the principle of optimality, his optimal distance is the smaller of these, 35, and this value is now associated with Stamford. Similar arguments for Grantham and Newark produce associated optimal distances of 31 and 36 respectively. On each occasion, a town being a junction of routes, there is a decision to be taken and the choice is governed by the relative values of the sum of the distance to the next town and the optimal distance from that town. In this way optimal distances can be found working back from the final destination at Boston to the origin at Leicester. The final route is shown in Figure 11.2, where arrows indicate the roads to be taken and the circled numbers by each town give the optimal distance to Boston from that town. The optimal route is Leicester, Melton Mowbray, Grantham, Sleaford and Boston, a total distance of 62 miles.

This is not intended as a serious application of dynamic programming. Another drawback is that it does not explicitly define the decisions in a sequence since the derived optimal route involves four decision stages, whereas an alternative via Stamford and Spalding has only three, of which one has no alternatives. The aim here has been to illustrate the approach, particularly the backwards ordering of the decisions and the combination of the immediate consequences of the decision with the optimal value thereafter.

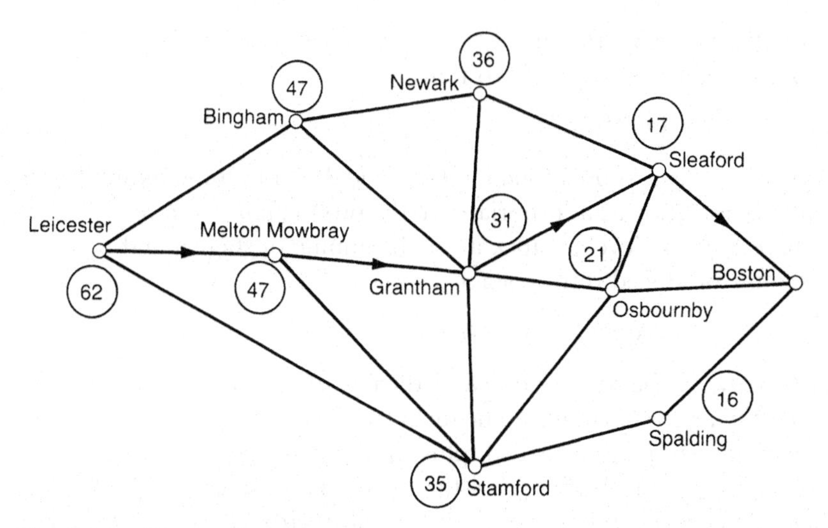

Fig. 11.2 The optimal route

11.3 The deterministic model

In dynamic programming we speak of a *policy*, this being a sequence of decisions taken at a number of stages. The number of stages may be infinite or we may work towards a finite horizon with given end conditions.

Consider first the case of the finite horizon with N stages. At each stage the system can be described completely by a *state variable* S_n. This may, for example, be the level of stock present at the time of re-ordering, in situations where this is the only variable to influence the decision. The action to be taken at stage n is the *decision* variable, denoted by X_n. Finally, there is the objective function. This is defined for any stage and is the value of the function appropriate to that stage and all subsequent stages. In the deterministic model where all the subsequent outcomes are known, the value of the objective function is an expression in all the decision variables still to be taken, together with the value of the current state variable. In the deterministic case, of course, once these variables have been determined, all the remaining state variables are also known.

Finally, suppose also that $C_n(X_n)$ is the value of the objective between stages n and $n + 1$ when action X_n is taken. Bellman's principle of optimality now permits a statement of the problem in terms of its optimal policy. Choose X_n so that

$$f_n(S_n) = \text{Opt}\ \{C_n(X_n) + f_{n+1}(S_{n+1})\} \tag{11.1}$$

where S_{n+1} is a known function of S_n and X_n and Opt is minimization or maximization as appropriate. In the production planning example, where S_n is the level of stock at the beginning of stage n and X_n is the amount produced during stage n,

$$S_{n+1} = \max\ \{0, S_n + X_n - D_n\},$$

where D_n is the known demand during the stage. Maximizing the quantity against zero has to be included to avoid negative stock levels where demand exceeds stock plus production.

The procedure begins from the most remote stage (as the form of equation (11.1) implies) and works backwards to the present. The variable S_1 is the known initial state, and $f_1(s_1)$ is therefore the total objective which is to be optimized. Where successful the technique not only does this but also specifies the values of the decision variables to be taken at each stage. The main requirements for the procedure are the data for each stage and the assumption that the objective function can be separated additively into parts appropriate for each stage.

11.4 Illustration of the deterministic model

Stringybark Engineering plc is a manufacturer of large electrical transformers. They have orders for the next four months in the following quantities:

January	February	March	April
3	2	6	2

At present there are none in stock. In any month there is sufficient capacity to make up to 7 transformers. Because of the uncertainties in the market Stringybark do not wish to plan for any surplus stock at the end of April.

Any transformer not sold in its month of manufacture incurs a penalty cost of £200 per month. This is to cover storage costs and interest on the investment in work in progress. Also if the firm decides to produce in any month there is a set-up cost of £500 irrespective of the number of transformers produced. This is because they have to hire specialist consultants by the month and also with a monthly planning system they cannot use the production space for any other purpose.

Finally, there is a price rise for their raw materials which will come into effect for the March and April production. This will add £100 to the cost of manufacture of any transformer made in these months. Note that since all transformers made will be sold, there is no need to include the selling price and the standard production costs in the model. Moreover, the short time of four months covering the duration of the system means that it is not necessary to discount costs.

The problem could be modelled as a decision tree with decision nodes corresponding to each month. Even allowing for the curtailment of the options at each month because of production requirements (e.g. always at least 3 transformers in January and never more than 2 in April) the tree becomes rather cumbersome. An idea of the size of the tree can be detected from Figure 11.3, where only a fraction of the branches has been included.

The state variables S_1, S_2, S_3, S_4 and S_5 are the numbers of transformers in stock at the beginning of each month. The variable S_5 denotes the final stock at the end of April, and the problem requires that $S_5 = 0$. We are also given that $S_1 = 0$.

The decision variables X_1, X_2, X_3 and X_4 are the numbers to be produced in each month. The only requirement is that they should be integers between 0 and 7.

At first glance it would seem that the price rise of £100, being less than the monthly stock holding cost, does not in itself justify making in January or February for sales in March and April. On the other hand, it is possible to make sufficient transformers in January and February to meet the demand for all four months, thereby saving on the later set-up costs. Optimization is a matter of balancing these costs, and dynamic programming is basically a systematic approach to the problem.

The procedure starts at the final month, April. By the end of April there should be no transformers in stock, and since the demand in that month is for 2 transformers, any policy which is a contender for optimality will have no more than 2 transformers in stock at the beginning of the month. We have, therefore, three possibilities:

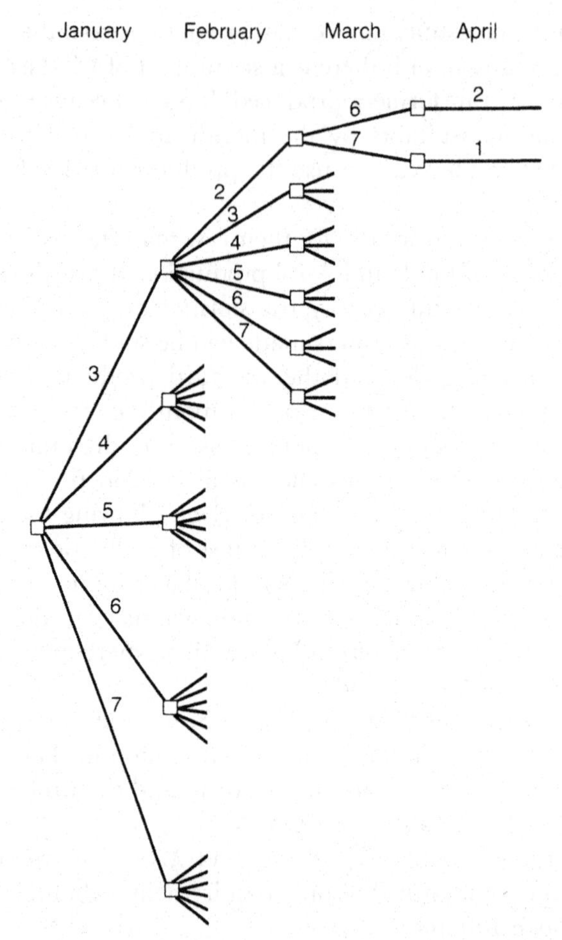

Fig. 11.3 Decision tree (part) for a dynamic programming problem

(a) $S_4 = 0$, in which case $X_4 = 2$
(b) $S_4 = 1$, in which case $X_4 = 1$
(c) $S_4 = 2$, in which case $X_4 = 0$.

The costs appropriate to this month consist of no more than three components – a set-up cost (£500), a price increase penalty (£100/unit made), and a stock holding cost (£200/unit in stock at the beginning of the month). These costs are, respectively,

(a) $500 + 200 + 0 = 700$
(b) $500 + 100 + 200 = 800$
(c) $0 + 0 + 400 = 400$.

(From now on all costs will be in pounds sterling and we shall omit the £ sign.)

The question of optimizing over this stage does not arise, since there is only one decision associated with any value of the state variable S_4.

Also $f_5(S_5) = 0$. Hence in terms of equation (11.1)

$$f_4(0) = 700, \quad f_4(1) = 800, \quad f_4(2) = 400.$$

The procedure is summarized in Table 11.1.

Table 11.1

			Costs			
S_4	X_4	S_5	*Set-up*	*Price rise*	*Inventory*	*Total*
0	2	0	500	200	0	700*
1	1	0	500	100	200	800*
2	0	0	0	0	400	400*

The asterisk on the total value is used to denote that this is the lowest cost for the corresponding S_4 value. Again, since there is only one decision for each S_4 value, this is unnecessary at this stage and is included only for consistency with later stages, where the optimization is not trivial..

Now consider March, month 3. The total demand for March and April is 8 transformers. Also since the total demand for the first two months is 5 and the total capacity is 14, a stock level of 8 transformers at the beginning of March is both feasible and plausible. At the other extreme a stock level of zero could occur and also the demand for March could be met by that month's production. Thus all values of S_3 between 0 and 8 have to be considered as candidates for inclusion within the optimal strategy.

If S_3 has the value 8, there is enough stock for the remaining demand, and thus we must have X_3 equal to zero, resulting in a final stock level, S_4 equal to 2. The costs in this stage are zero for the set-up and price increase but 1600 for the inventory carried over. Thus

$$f_3(8) = 1600 + f_4(2) = 1600 + 400 = 2000.$$

With intermediate values of S_3 the calculations are a little more involved, since it is possible for S_4 to take 2 or all 3 of its range of values. For-example if $S_3 = 4$, we must produce at least 2 to meet the current month's demand, but no more than 4 so that the total demand is not exceeded. The three possibilities are

(a) $X_3 = 2$ and hence $S_4 = 0$
(b) $X_3 = 3$ and hence $S_4 = 1$
(c) $X_3 = 4$ and hence $S_4 = 2$.

The corresponding costs for this stage are

(a) $C_3(2) = 500 + 200 + 800 = 1500$
(b) $C_3(3) = 500 + 300 + 800 = 1600$
(c) $C_4(4) = 500 + 400 + 800 = 1700$.

The optimization at this stage is over the current costs plus the optimum costs for subsequent stages, that is over $C_3(X_3) + f_4(S_4)$. These are $1500 + 700 = 2200$, $1600 + 800 = 2400$, $1700 + 400 = 2100$. Thus the minimum is 2100 and occurs when $X_3 = 4$ and subsequently $X_4 = 0$. The implication of this is that, should the decision-maker find himself at the beginning of March with 4 transformers in stock, his best strategy is to manufacture 4 more transformers in March, but none in April. This is Bellman's principle of optimality in action. We do not yet know whether or not an optimal policy starting from the beginning of January will put us in the position of having 4 transformers in stock at the beginning of March, but by similar calculations to this we can be prepared with optimal subsequent decisions for any stock level at this epoch.

The full calculations for this stage are again summarized in Table 11.2.

A similar argument now follows for the production in February. With a maximum production of 7 transformers in January and a demand for three with no initial stock, the stock level at the beginning of February can be no more than 4. We have, therefore, to consider values of S_2 ranging from 4 to zero. The resultant costs for this stage are shown in Table 11.3.

Finally we turn our attention to the first month of the policy. The initial state variable $S_1 = 0$. The decision variable X_1 can take any value which covers the demand of 3 in the first month and is within the capacity limit of 7. Again the computations can be expressed in a simple table, linking up with the optimal strategies for subsequent state variables.

Hence the optimal decision for January is to manufacture 5 transformers producing a stock of 2 transformers at the end of the month.

Table 11.2

			Costs				
S_3	X_3	S_4	*Set-up*	*Price rise*	*Inventory*	$f_4(S_4)$	*Total*
8	0	2	0	0	1600	400	2000*
7	0	1	0	0	1400	800	2200*
	1	2	500	100	1400	400	2400
6	0	0	0	0	1200	700	1900*
	1	1	500	100	1200	800	2600
	2	2	500	200	1200	400	2300
5	1	0	500	100	1000	700	2300
	2	1	500	200	1000	800	2500
	3	2	500	300	1000	400	2200*
4	2	0	500	200	800	700	2200
	3	1	500	300	800	800	2400
	4	2	500	400	800	400	2100*
3	3	0	500	300	600	700	2100
	4	1	500	400	600	800	2300
	5	2	500	500	600	400	2000*
2	4	0	500	400	400	700	2000
	5	1	500	500	400	800	2200
	6	2	500	600	400	400	1900*
1	5	0	500	500	200	700	1900
	6	1	500	600	200	800	2100
	7	2	500	700	200	400	1800*
0	6	0	500	600	0	700	1800*
	7	1	500	700	0	800	2000

From the previous table the optimal decision for February consequent on the January action is to make no transformers leaving a stock of zero at the end of this month. For the March decision reference is made to the corresponding table, showing that 6 transformers should be made, again leaving no transformers in stock. Finally, the analysis for April showed that in this situation, 2 transformers were to be made. In terms of the variables of the model the optimal policy is $X_1 = 5$ ($S_2 = 2$), $X_2 = 0$ ($S_3 = 0$), $X_3 = 6$ ($S_4 = 0$), $X_4 = 2$. The total of the set-up, price rise and inventory costs is 2700.

Table 11.3

S_2	X_2	S_3	Costs				
			Set-up	*Price rise*	*Inventory*	$f_3(S_3)$	*Total*
4	0	2	0	0	800	1900	2700*
	1	3	500	0	800	2000	3300
	2	4	500	0	800	2100	3400
	3	5	500	0	800	2200	3500
	4	6	500	0	800	1900	3200
	5	7	500	0	800	2200	3500
	6	8	500	0	800	2000	3300
3	0	1	0	0	600	1800	2400*
	1	2	500	0	600	1900	3000
	2	3	500	0	600	2000	3100
	3	4	500	0	600	2100	3200
	4	5	500	0	600	2200	3300
	5	6	500	0	600	1900	3000
	6	7	500	0	600	2200	3300
	7	8	500	0	600	2000	3100
2	0	0	0	0	400	1800	2200*
	1	1	500	0	400	1800	2700
	2	2	500	0	400	1900	2800
	3	3	500	0	400	2000	2900
	4	4	500	0	400	2100	3000
	5	5	500	0	400	1900	2800
	6	6	500	0	400	2200	3100
	7	7	500	0	400	2000	2900
1	1	0	500	0	200	1800	2500*
	2	1	500	0	200	1800	2500*
	3	2	500	0	200	1900	2600
	4	3	500	0	200	2000	2700
	5	4	500	0	200	2100	2800
	6	5	500	0	200	1900	2600
	7	6	500	0	200	2200	2900
0	2	0	500	0	0	1800	2300*
	3	1	500	0	0	1800	2300
	4	2	500	0	0	1900	2400
	5	3	500	0	0	2000	2500
	6	4	500	0	0	2100	2600
	7	5	500	0	0	1900	2400

Table 11.4

			Costs				
S_1	X_1	S_2	*Set-up*	*Price rise*	*Inventory*	$f_2(S_2)$	*Total*
0	3	0	500	0	0	2300	2800
	4	1	500	0	0	2500	3000
	5	2	500	0	0	2200	2700*
	6	3	500	0	0	2400	2900
	7	4	500	0	0	2700	3200

With larger problems, the computations can readily be carried out by computer, although the problem does not have to be particularly extensive before it will tax the storage capacity of the most modern system. Readers may refer to Chapter 4 of Norman (1975) for a fuller discussion.

The time effect of money can be catered for by a discount factor applied as a multiple to the $f_{n+1}(S_{n+1})$ term in equation (11.1). It is only really appropriate where each decision stage covers at least a year.

The analysis in this section has assumed a finite planning horizon of four months. Although the plant is intended to run for many years yet, at the time of planning no firm orders for sales beyond April have been received. To use this method, and indeed to carry out any form of objective planning, some assumption has to be made about the stock level at the end of the planning period. The simplest approach is to argue that the system should be in a position where it will be able to cope with anything but an extreme demand in the subsequent months. Within this proviso, the final stock level should be minimized. In this case, if we are prepared to believe that a demand for six units is the most that can reasonably be expected, with a production capacity of seven units there will be no embarrassment if the stock level is run down to zero. There is a further safeguard in that an unusually high demand in May is likely to make itself known in time for a revision of the March or April production figures.

11.5 The stochastic model

So far in this chapter we have assumed that all the costs and demands are known. Often this is not the case. Monthly demands for a product, for example, could be fully determined without error for the current month, but looking twelve months ahead they would be nothing better than forecasts with all their inherent chance variations. At intermediate months there will be some firm orders with the possibility of additional orders before the month is reached. If, therefore, we are planning production, we shall have to take account of the unpredictable nature of our data.

The stochastic model requires the same fundamental assumptions as the deterministic. At each stage there will be an explicitly known state variable S_n. The decision variable is again denoted by X_n, but whereas in the deterministic model S_n and X_n led to a unique state variable S_{n+1} at the next stage, in this case there is a probability distribution, dependent on S_n and X_n, over the next state variable. For example, in the production planning situation, if S_n is the stock at the beginning of the month, X_n is the amount produced and therefore

$$S_{n+1} = \max\{0, S_n + X_n - Y_n\} \tag{11.2}$$

where Y_n is the demand for the product during the month. The demand, Y_n, can be assumed to be a random variable, which with given values of S_n and X_n produces through (11.2) a probability distribution for S_{n+1}.

The cost of the stage, $C_n(X_n)$, is usually assumed to be known without error. The equivalent form of the fundamental equation (11.1) replaces the term $f_{n+1}(S_{n+1})$, which would otherwise be a random variable, by its expected value. This, it will be recalled, is the weighted average of all possible values of $f_{n+1}(S_{n+1})$, where the weights are the corresponding probabilities. We write this as $E[f_{n+1}(S_{n+1})]$. The problem now becomes one of choosing X_n so that

$$f_n(S_n) = \text{Opt}\{C_n(X_n) + E[f_{n+1}(S_{n+1})]\} \tag{11.3}$$

11.6 Illustration of the stochastic model

Walkback Travel Services plc have a large sum of money which they must convert into Paratanian dollars. The transaction must take place within the next six days. The Paratanian economy is rather volatile and consequently the exchange rate varies in a random fashion. The research division at Walkback estimate that in the immediate future the following probabilities apply to the exchange rate values (dollars per pound sterling)

dollars	20	21	22	23	24
probability	0.1	0.1	0.4	0.3	0.1

The entire transaction must be carried out on one day, at which time of course the exchange rate for that date will be known. What policy should Walkback follow to maximize its expected return from the exchange?

The state and decision variables need careful definition. Obviously a decision on purchasing the dollars will have to be made only if it has not been made positively on any of the previous dates. Thus we can define the decision variables

$X_n = 0$ if the purchase is not made on day n
$X_n = 1$ if the purchase is made on day n.

Both of these are conditional on the purchase not having been made previously. The state variable S_n is the exchange rate for that day, and the return from stage (day) n is the exchange rate if a purchase is made or zero if it is not. This can be expressed as $X_nS_n = C_n(X_n)$. Also the expected return from stage $n + 1$ onwards is zero if the purchase is made in stage n, but $E[f_{n+1}(S_{n+1})]$ if the purchase is not made. Again this can be expressed succinctly as $(1 - X_n)E[f_{n+1}(S_{n+1})]$. Hence the object of the policy is to find X_n, which satisfies

$$f_n(S_n) = \max \{X_nS_n + (1 - X_n)E[f_{n+1}(S_{n+1})]\}. \tag{11.4}$$

This is not precisely in the form of (11.3), but it is nevertheless obtained directly using Bellman's principle of optimality.

Returning to the illustration, we again work backwards from the final stage. If, by day 6, a purchase has not been made, Walkback have no option but to accept the exchange rate of that day. Thus $X_6 = 1$ and $f_6(S_6) = S_6$ from (11.4). Furthermore,

$$\begin{aligned} E[f_6(S_6)] &= (0.1)(20) + (0.1)(21) + (0.4)(22) + (0.3)(23) + (0.1)(24) \\ &= 22.2. \end{aligned}$$

Now turn to day 5. From equation (11.4),

$f_5(S_5) = \max \{X_5S_5 + (1 - X_5)\ 22.2\}$,
the maximization being taken over $X_5 = 0$ and $X_5 = 1$.

The maximization process can be expressed as in Table 11.5.

Table 11.5

S_5	$X_5S_5 + (1 - X_5)22.2$ $X_5 = 0$	 $X_5 = 1$
20	22.2*	20
21	22.2*	21
22	22.2*	22
23	22.2	23*
24	22.2	24*

Maximum values are indicated by asterisks. The policy is therefore if $S_5 = 20$, 21 or 22, $X_5 = 0$ (do not purchase); if $S_5 = 23$ or 24, $X_5 = 1$ (purchase).

The expected value of this policy is found by weighting the above maximum returns by the probabilities of the corresponding S_5 variables, namely

$$E[f_5(S_5)] = (0.1)(22.2) + (0.1)(22.2) + (0.4)(22.2) + (0.3)23 + (0.1)(24)$$
$$= 22.62.$$

On day 4, a similar argument follows from the definition of

$$f_4(S_4) = \max \{X_4S_4 + (1 - X_4)\ 22.62\},$$

and therefore if $S_4 = 20$, 21 or 22, $X_4 = 0$ and $f_4(S_4) = 22.62$ (do not purchase); but if $S_4 = 23$ or 24, $X_4 = 1$ and $f_4(S_4) = 23$, 24 respectively (purchase).

$$E[f_4(S_4)] = (0.1)(22.62) + (0.1)(22.62) + (0.4)(22.62) + (0.3)23 + (0.1)(24)$$
$$= 22.87.$$

For days 3, 2 and 1 the results together with those previously found may be summarized as in Table 11.6.

The strategy therefore amounts to purchase on the first two days only if the exchange rate is at its maximum value of 24. For the next three days, if a purchase has not already been made, the transaction is carried out if the exchange rate is 23 or 24. If by the final day no purchase had been made, the firm must then accept the exchange rate on that date. With this policy the expected rate of exchange, assuming that many similar decisions are to be made is 23.21. Note

Table 11.6

Day	Decision variable X_2 when S_n takes the value					$E[f_n S_n]$
n	20	21	22	23	24	
6	1	1	1	1	1	22.2
5	0	0	0	1	1	22.62
4	0	0	0	1	1	22.87
3	0	0	0	1	1	23.02
2	0	0	0	0	1	23.12
1	0	0	0	0	1	23.21

that this is a significant improvement on the expected rate of 22.20 if Walkback simply accept the exchange rate on some arbitrarily (randomly) selected date for the transaction.

11.7 Application to inventory problems

Many basic inventory/production problems are of a type which might appear to lend itself to solution by dynamic programming. In the simplest form there is a single product which is produced in batches for a finished goods stock. Regular decisions have to be taken about setting up a production batch and about the size of the batch. There is a set-up cost and also a cost resulting from keeping goods in the finished goods stock. In addition there could be a penalty cost for being unable to meet a demand. With a known demand the problem would be very similar to that illustrated in Section 11.4. In many practical situations the demand is unknown, the amount of uncertainty increasing as we look further into the future. If, however, it is possible to express the uncertainty in the form of probability distributions for the demands, then in theory the basic equation (11.3) can be applied to this problem. The state variable S_n is the inventory level at a time n (the units being months or weeks) and the decision variable X_n is the amount to be produced in this period. If Y_n is the demand during the period, then

$$S_{n+1} = \max\,[0,\, S_n + X_n - Y_n],$$

a random variable at the time of period n. Thus $E[f_{n+1}(S_{n+1})]$ can be evaluated, knowing the distribution of Y_n, and hence the optimum strategy can be determined working backwards from a finite horizon of N periods. In practice the approach is feasible only if N is relatively small (say 5 or 6) and if the range of variation of the demand as a discrete variable is similarly limited.

Consider for example a situation where a company is manufacturing a substantial unit (e.g. a machine tool). The demand for these units in any month has never been less than 5 nor more than 8. Next month they have orders for 6 units and reckon that there is a 50–50 chance of an order for a further unit. In the following month their forecasts of the demand become less precise, but they are prepared to work to the probabilities given in Table 11.7.

Table 11.7

	Months			
Demand (units)	*June*	*July*	*August*	*September*
5	0	0.1	0.2	0.2
6	0.5	0.4	0.3	0.3
7	0.5	0.4	0.3	0.3
8	0	0.1	0.2	0.2

Further relevant information is

Cost of stock £20/unit/month
Cost of shortage £50/unit
Set-up cost for production in any month £200
Maximum production in any month 15 units
Initial stock 4 units.

Since all units made will be sold, the cost of manufacture is not included in the analysis. This assumes that manufacturing costs will not change appreciably over the period under study. There is, of course, a carrying cost when units are not sold in the month when they are available. This is expressed by the cost of £20 per unit per month, an amount charged at the end of each month proportional to the number of units unsold. The shortage cost (£50 per unit) is the penalty for not meeting a demand. Shortages are not carried over to the following month. It is as if the units deficient were bought out, resulting in a loss to the firm of £50 per unit. Finally, although the initial stock is 4 units, the plan at this stage is to have no stock at the end of September. The aim will be achieved by minimizing the overall cost.

The analysis starts by examining the best strategies for all values of S_4, the stock at the beginning of September. Each value is matched with all possible values of X_4 the number of units to be manufactured in September, and the expected cost evaluated, so that the minimum of these will determine the optimum X_4 for each S_4. For example if $S_4 = 7$, Table 11.8 may be produced, resulting in

$$E[f_4(7)] = (0.2)(40) + (0.3)(20) + (0.2)(50) = 24.$$

Table 11.8

	Set-up cost	*Demand* Y_4	*Shortage*	*Cost of shortage*	*Final stock*	*Cost of stock*	*Total cost*	*Probability*
$X_4 = 0$	0	5	0	0	2	40	40	0.2
	0	6	0	0	1	20	20	0.3
	0	7	0	0	0	0	0	0.3
	0	8	1	50	0	0	50	0.2

Alternatively, taking $S_4 = 7$ but now $X_4 = 3$, we obtain Table 11.9, giving

$$E[f_4(7)] = (0.2)(300) + (0.3)(280) + (0.3)(260) + (0.2)(240) = 270.$$

A full evaluation of all X_4 possibilities produces a minimum when $X_4 = 0$. In a similar way all minimum cost strategies can be produced for this stage, resulting in Table 11.10

In summary the policy for September is: If the opening stock is 2 units or less, manufacture so that the total is 7 units; if the opening stock is 3 units or more do not manufacture any units. A little thought

Table 11.9

	Set-up cost	*Demand* Y_4	*Shortage*	*Cost of shortage*	*Final stock*	*Cost of stock*	*Total cost*	*Probability*
$X_4 = 3$	200	5	0	0	5	100	300	0.2
	200	6	0	0	4	80	280	0.3
	200	7	0	0	3	60	260	0.3
	200	8	0	0	2	40	240	0.2

Table 11.10

S_4	0	1	2	3	4	5	6	7	$\geqslant 8$
Optimum X_4	7	6	5	0	0	0	0	0	0
$E[f_4(S_4)]$	224	224	224	175	125	75	39	24	$20S_4 - 130$

will show the logic of this policy and the way it is influenced by the decision to aim for zero stock at the end of September.

Next consider August, period 3. The analysis is similar to that of period 4, except that the cost for period 4 is used for the optimization, where it can be assumed that the optimum period 4 strategy is used. Suppose for example the case $S_3 = 2$ is taken. Try $X_3 = 0$, which gives Table 11.11, from which

$$\begin{aligned} E[f_3(27)] &= (0.2)(374) + (0.3)(424) + (0.3)(474) + (0.2)(524) \\ &= 449. \end{aligned}$$

Table 11.11

Set-up cost	*Demand Y_3*	*Shortage*	*Cost of shortage*	*Final stock*	*Cost of stock*	*Cost of period 4*	*Total cost*	*Probability*
0	5	3	150	0	0	224	374	0.2
0	6	4	200	0	0	224	424	0.3
0	7	5	250	0	0	224	474	0.3
0	8	6	300	0	0	224	524	0.2

Alternatively, trying $X_3 = 8$, we obtain Table 11.12, and

$$\begin{aligned} E[f_3(2)] &= (0.2)(375) + (0.3)(405) + (0.3)(435) + (0.2)(464) \\ &= 419.8. \end{aligned}$$

The optimum value of $E[f_3(2)]$ in this case occurs when $X_3 = 11$ and the actual value is then 369.9. Table 11.13 can now

Table 11.12

Set-up cost	*Demand Y_3*	*Shortage*	*Cost of shortage*	*Final stock*	*Cost of stock*	*Cost of period 4*	*Total cost*	*Probability*
200	5	0	0	5	100	75	375	0.2
200	6	0	0	4	80	125	405	0.3
200	7	0	0	3	60	175	435	0.3
200	8	0	0	2	40	224	464	0.2

be determined. Again the policy can be summarized: If the opening stock is 3 or less, manufacture so that the total is 13 units; if the opening stock is 4 or more units, do not manufacture any units.

Table 11.13

S_3	0	1	2	3	4	5	6	7	8
Optimum X_3	13	12	11	10	0	0	0	0	0
$E[f_3(S_3)]$	369.9	369.9	369.9	369.9	349	299	263	248	244.2
S_3	9	10	11	12	13	14	15	>16	
Optimum X_3	0	0	0	0	0	0	0	0	
$E[f_3(S_3)]$	239.5	219.8	192.8	174	169.9	184	212.8	$40S_3 - 390$	

Next the costs in period 2 (July) are analysed, producing Table 11.14. Note that because of the limit of 15 units on the monthly production, and the initial stock of 4 units, the opening stock for period 2 could never be more than 13.

Table 11.14

S_2	0	1	2	3	4	5	6
Optimum X_2	7	6	5	0	0	0	0
$E[f_2(S_2)]$	586.9	586.9	586.9	544.9	494.9	444.9	401.9
S_2	7	8	9	10	11	12	13
Optimum X_2	0	0	0	0	0	0	0
$E[f_2(S_2)]$	386.9	399.9	417.8	424.4	412.5	394.5	388.7

Finally it is known that S_1 equals 4. For this period it is therefore only necessary to consider $S_1 = 4$ matched with all possible X_1 values. The optimum occurs when $X_1 = 0$ and the cost $E[f_1(4)]$ is 711.9. In this case the optimum policy is as follows:

June: $X_1 = 0$, i.e. do not manufacture. There will be penalties for shortages and the final stock value S_2 will be zero.

July: $S_2 = 0$. Therefore $X_2 = 7$, i.e. manufacture 7 units. There will be at most a shortage of 1 unit, and the final stock value will be 0, 1 or 2.

August: $S_3 = 0$, 1 or 2. Manufacture to a total of 13 units. There will be no shortages, and the final stock will be 5, 6, 7 or 8 units.

September: $S_4 = 5$, 6, 7 or 8. In each circumstance $X_4 = 0$, i.e. do not manufacture. There will be at most a shortage or a surplus of 3 units.

This particular case has been dealt with in some detail to demonstrate the arithmetic complexities. It is, in practical terms, a very simple case, and clearly the technique lends itself to programming on a computer. This would present the additional benefit of regular updating as more information in the form of modified demand probabilities is accrued.

11.8 Further reading

Bellman, R. E., *Dynamic Programming*, Princeton University Press, Princeton, N. J. (1957).

Daellenbach, H. G. and George, J. A., *Introduction to Operations Research Techniques*, Allyn and Bacon (1987), Chapter 7.

French, S., Hartley, R., Thomas, L. C. and White, D. J., *Operational Research Techniques*, Edward Arnold, London (1986), Chapter 12.

Hsiao, J. C. and Cleaver, D. S., *Management Science*, Houghton Mifflin, Boston (1982), Chapter 9.

Mitchell, G. H., *Operational Research*, English Universities, London (1972), Chapter 3.

Norman, J. M., *Elementary Dynamic Programming*, Edward Arnold, London (1975).

White, D. J., *Dynamic Programming*, Oliver and Boyd, Edinburgh (1969).

11.9 Problems and practical exercises

1 Suppose that in the illustrative example of Section 11.4 Stringybark Engineering decided that the demand in May could be greater than their production limit of 7 transformers. Because of this they wish to plan for a stock of 3 transformers at the end of April. What now should be their production plan?

2 In the illustration of Section 11.6 Walkback Travel Services had to make their money transaction within the next six days. Suppose now that they had an extra day allowed. Show that there is now a very simple extension to their previous policy.

3 Wally Pratt wishes to sell his shares in Imperial Drainpipes. They currently stand around the 100p mark and Wally is persuaded that over the next seven days – the period in which he must make his sale – the shares will vary between 95p and 110p with the following probabilities:

Price (p)	95	100	105	110
Probability	0.5	0.2	0.2	0.1

The price on each day is assumed to be independent of the prices on other days. What policy should Wally adopt in deciding when to sell his shares?

4 A firm has an advertising budget of £50 000 which it can spend on any or all of the media of television, newspapers or roadside posters. The firm allocates its budget to the media in units of £10 000 and estimates the returns (in extra sales units) shown in Table 11.15.

Table 11.15

Allocation (£)	0	10 000	20 000	30 000	40 000	50 000
Television	0	4	8	15	25	38
Newspapers	0	15	20	25	28	30
Roadside posters	0	10	20	24	27	28

Show how this problem can be formulated as a dynamic programming problem and solve it.

12
Simulation

OBJECTIVES

From this chapter the reader should have an understanding of the power of the technique of simulation and should then be able to

(a) generate a simple series of pseudo random numbers;
(b) convert the problem of sampling from a distribution (either empirical or theoretical) to one of associating values to corresponding random numbers;
(c) carry out a simple simulation;
(d) understand the need for computer simulation and hence for computer packages.

Finally the reader is introduced to the associated topics of System Dynamics, Decision Support Systems, Artificial Intelligence and Expert Systems.

12.1 What is simulation?

As its name implies, simulation is simply a method of analysing a problem situation by creating a model of the situation which can then be manipulated by trial-and-error methods. Any model is an approximation to the real situation and certainly the closer the approximation becomes, the more difficult and time-consuming the model will be to analyse. The skill of the management scientist lies in the way he balances the reality of his model with the effort it will take to find an answer to his problem. To illustrate this point, suppose that you were asked to find the surface area of the floor of the room in which you are now sitting. The chances are that the room is rectangular. You would then simply measure the lengths of two appropriate walls and multiply these two figures together to obtain your answer. In so doing you have modelled the floor by a rectangle and used a very simple geometrical property, namely that the area of a rectangle is the product of its length and breadth. The collection of data appeared to be straightforward and the level of mathematics used was basic. The solution would for example be perfectly adequate if you were estimating the amount of paint needed to stain the floor.

Suppose that you are proposing to cover the floor with a rather expensive carpet. You may then wish to take note of the recesses by the doors and the intrusion into the floor of the fireplace. Data collection and the calculations resulting from the geometry of the model are now not quite as simple as before, but are well within the range of school mathematics. The answer should be closer to reality and you pay a modest price, in terms of your own effort, for it. Having taken this first step, there is in theory no reason why you should not refine the model further. For example the corners of the room may not be exact right-angles, the boundaries of the floor may not be straight lines and the floor itself may sag, converting the model from two to three dimensions. With modifications like these data collection becomes a major undertaking and the derivation of the solution could require mathematics at quite a sophisticated level. Clearly in this instance the modelling is taken to absurd extremes, but in many practical situations the point where refinement of the model stops may not be so obvious. For the mathematician it is sometimes tempting to go a little further than is justified so that a challenging problem can be tackled. Advances in computer technology are liable to encourage this tendency.

Simulation is a technique for solving models. In the context of the above illustration it permits the use of models which require a method of solution not covered by analytical techniques. A good analogy for the floor problem is to draw accurately on graph paper the mathematical representation used and then to count up the number of squares included. It is a simple technique used on a sophisticated model.

12.2 The simulation approach

The complicating factor in most mathematical models arises from chance or stochastic elements. For example, customers arriving at a building society counter do not appear at regular intervals, nor do they require transactions of a standard duration. There is a chance element in both the arrival pattern and the service time. If the inter-arrival times for consecutive customers were measured, it would be feasible to draw up a histogram and from the histogram make some plausible assumption about the distribution of the inter-arrival time. With luck, it might be possible to fit a standard distribution (e.g. beta, gamma, negative exponential) to the data. A similar procedure for the service times might, with further luck, produce an appropriate distribution for the random variable. There is now a technique known as 'queueing theory' which might be able to assist with the analysis of the model. The name queueing theory arises from the nature of the situation where customers require a service whose provision is limited, and consequently congestion is likely to occur. For an introduction to the theory see the references Harper and Lim (1982) and Stainton (1977) at the end of the chapter. More advanced treatment is given in Lee (1966) or in Cox and Smith (1961).

Unfortunately queueing theory produces only limited results for a very limited class of inter-arrival and service distributions. In designing their counter system the building society may, for example, be interested in how long customers have to wait, or in the probability that a customer has to wait more than fifteen minutes for service. Alternatively they may wish to know the proportion of time that the servers behind the counter spend actually serving customers. They may also wish to modify the system. What would be the effect of introducing an extra server on the customer waiting time and on the server idle time? They could also study the effect of changing the system from one where customers on arrival joined one of the separate queues before each server to a system of a single queue called upon by each server as he or she becomes free. It is highly unlikely that queueing theory would give anything but a very rough approximation to the answers to these questions. Instead we might resort to simulation.

In the simple congestion situation described above there are two chance inputs, the inter-arrival times and the service times. The simulation technique takes samples randomly from each of the corresponding distributions and operates the system as if these were the actual arrival and service values. It can go on doing this for as long as is felt necessary. At the same time, appropriate measures of the system, such as queue size or number of idle servers, can be recorded.

The strength of the simulation technique is its flexibility. The user is no longer restricted to standard distributions for the inputs. The actual data, possibly smoothed to remove clearly chance irregularities, can be used. Mechanically this is equivalent to drawing the input from a lottery where the tickets consist of the actual observations on the initial data, with the proviso that each drawn ticket is immediately returned to the lottery. Variants on the arrival pattern can be introduced; customers may, for example, balk at joining the queue when too many people are already waiting. Accommodation can be made for customers arriving in groups. The arrival rate may vary over the day. The service mechanism can also be modified. Servers may speed up when the queue becomes longer; they may require breaks at specified times irrespective of the queue size. There could also be some further interacting congestion, perhaps resulting from the servers using some central information processor to determine the credit standing of customers. In practice some very complicated models with many interacting demands and services have been used. Whole sections of an organization have been simulated – the limits are only the size of the computer and the time taken to work and test the program. Much of the work has been facilitated by devising special computer simulation languages (see Section 12.5).

12.3 The mechanics of simulation

Individual components of a simulation require the generation of data from a distribution. Central to the problem of generation is the concept of the random number (introduced in Section 2.3.5). This is a random variable taking integer values between 0 and 9, each with probability 0.1. Note that this means that if a random number between 0 and 99 is required, it can be generated by taking two random numbers (0 to 9) using one for the tens digit, the other for the units. Similarly a range covering any number of digits can be generated. The effect of generating a random number between 0 and 99 is equivalent to drawing from a lottery with tickets numbered 0 to 99. If a lottery with say only 87 tickets were required, every time a random number between 87 and 99 turned up, it would be discarded and a further number generated.

Suppose therefore that 87 inter-arrival times have been recorded and summarized in Figure 12.1. The simulation will be based on these data. Using random numbers between 0 and 99, to make the generations equivalent, 9 numbers must be associated with a time of 0, 17 with a time of 1 etc, leaving 13 to be discarded. Strictly it does

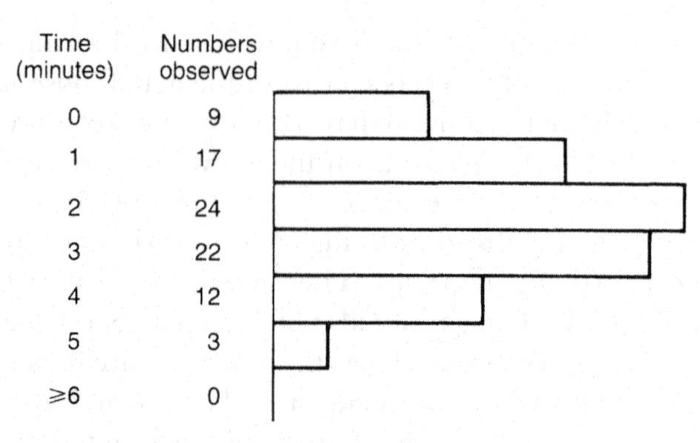

Fig. 12.1 Histogram of inter-arrival times

not matter which random numbers are used for each association, although for convenience consecutive increasing sets of numbers are used as shown in Table 12.1. It is customary to put an initial digit of 0 before single-digit numbers.

Table 12.1 Allocation of random numbers to the inter-arrival distribution

Time (minutes)	*Numbers observed*	*Associated random numbers*
0	9	00–08
1	17	09–25
2	24	26–49
3	22	50–71
4	12	72–83
5	3	84–86
Discard		87–99

Thus, for example, if the sequence of random numbers 49, 85, 88, 19, 37, 37, 04, 79, 64, 11, 83, 67 were generated, this would translate into the sequence of inter-arrival times 2, 5, discard, 1, 2, 2, 0, 4, 3, 1, 4, 3. Note that even in this small sample the proportions of each number approximate to the proportions in the original 87. Note also that in the set of random numbers the number 37 occurred twice in succession – such peculiarities do happen, and it would be a mistake to reject a set of numbers simply because they did not 'look random'.

In effect the procedure above means that the generation of random observations from any distribution can be reduced to the generation of random numbers. This equivalent problem remains, and since any practical simulation will be carried out by computer, a means of generating random numbers by the computer has to be devised. It would, of course, be possible to input the numbers to the computer, and this was the practice some years ago. Clearly it is time consuming and therefore expensive. An alternative, known as pseudo-random numbers, is now employed. These are numbers which when written down in sequence have most of the properties of random numbers, yet paradoxically they are generated by an explicit non-random procedure.

One popular method of generating pseudo-random numbers is the congruential method. For this two positive integers k and m are required. The first random number is then taken arbitrarily, from which the second is found as the remainder when k times the first is divided by m. In mathematical terms this is written as

$$x_1 = kx_0 \quad (\text{modulo } m). \tag{12.1}$$

Subsequent random numbers are generated sequentially using equation

$$x_{j+1} = kx_j \quad (\text{modulo } m). \tag{12.2}$$

The disadvantage with this method is that as soon as any x_j value repeats (and there can be no more than $m - 1$ different values) the values following repeat in the same order. The sequence cycles. The effect of this can be reduced by making k and m large. Also if k and m are mutually prime, zero values will not occur.

For example $k = 43$, $m = 101$ will yield numbers between 0 and 100. The number zero will never occur (unless we are unwise enough to start with it, in which case it will repeat endlessly) and thus if 100 is taken as the number 00 we have in effect two digit numbers between 0 and 99. Taking 84 as the initial x_0 value, the sequence then runs as in Table 12.2. Following this the sequence then repeats itself exactly.

Table 12.2

84	77	79	64	25	65	68	96	88	47
01	43	31	20	52	14	97	30	78	21
95	45	16	82	92	17	24	22	37	76
36	33	05	13	54	00	58	70	81	49
87	04	71	23	80	06	56	85	19	09

There is thus a cycle of 50 numbers. The complementary cycle of 50 can be tapped by starting the same procedure with any number not in the original cycle.

Larger values of k and m would usually increase the length of the cycle. The 'randomness' can be improved by taking only a limited number of the digits produced. In this case we might, for example, discard the first digit.

A similar procedure is the mid-square method, where each number generates its successor as the middle digits of its square. For example $63^2 = 3969$, $96^2 = 9216$, $21^2 = 0441$, $44^2 = 1936$ etc. is a typical sequence. The method also suffers from the disadvantage of cycling, and it can come to a stop if small numbers are used, e.g. $27^2 = 0729$, $72^2 = 5184$, $18^2 = 0324$, $32^2 = 1024$, $02^2 = 0004$, $00^2 = 0000$ etc.

The modification to the procedure when data from an assumed statistical distribution are to be used rather than from a set of observations is minimal. For example, it might be assumed that a service time has a normal distribution with mean 5 minutes and standard deviation 1 minute. Suppose that times are to be recorded at 15 second intervals. This means for example that any value within $\pm$ 7.5 seconds of 5 minutes will be recorded as 5 minutes. The probability of a value of 5 minutes is therefore the area under the normal curve between 4.875 and 5.125, which can be calculated from tables of the normal distribution (see Appendix 2). The geometry of the calculations is shown in Figure 12.2.

The probabilities are then used to determine the corresponding proportional sets of random numbers as shown in Table 12.3.

With most of the standard distributions there is a special method

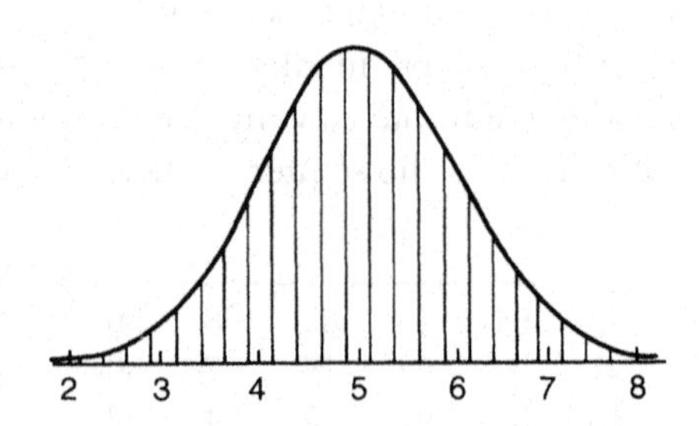

Fig. 12.2 Normal distribution split into discrete intervals

Table 12.3

Value (minutes)	*Probability*	*Random numbers*
2	0.0020	0000–0019
$2\frac{1}{4}$	0.0023	0020–0042
$2\frac{1}{2}$	0.0045	0043–0087
$2\frac{3}{4}$	0.0080	0088–0167
3	0.0136	0168–0303
$3\frac{1}{4}$	0.0217	0304–0520
$3\frac{1}{2}$	0.0324	0521–0844
$3\frac{3}{4}$	0.0458	0845–1302
4	0.0605	1303–1907
$4\frac{1}{4}$	0.0751	1908–2658
$4\frac{1}{2}$	0.0879	2659–3537
$4\frac{3}{4}$	0.0964	3538–4501
5	0.0996	4502–5497
$5\frac{1}{4}$	0.0964	5498–6461
$5\frac{1}{2}$	0.0879	6462–7340
$5\frac{3}{4}$	0.0751	7341–8091
6	0.0605	8092–8696
$6\frac{1}{4}$	0.0458	8697–9154
$6\frac{1}{2}$	0.0324	9155–9478
$6\frac{3}{4}$	0.0217	9479–9695
7	0.0136	9696–9831
$7\frac{1}{4}$	0.0080	9832–9911
$7\frac{1}{2}$	0.0045	9912–9956
$7\frac{3}{4}$	0.0023	9957–9979
8	0.0020	9980–9999

which can be adopted and which avoids calculations similar to the above. The normal distribution itself can be approximated closely by the sum of, say, six random numbers. This is because of a result known as the *central limit theorem*, which broadly states that the sum of independent observations of any random variable is approximately normally distributed, the approximation improving as the number of observations in the sum increases. With an appropriate adjustment to the sum, a normal distribution with any mean and standard deviation can be generated in this way. For a fuller account of these and other methods of generation reference should be made to Tocher (1963) or to Chapter 15 of Daellenbach and George (1978).

12.4 The physiotherapist's clinic – an illustration of simulation

Rhianon Adam is a physiotherapist who operates a clinic where she sees patients by appointment. She starts at 9.00 a.m. and works without a scheduled break throughout the morning, hoping to finish around midday. Patients are booked in at 15-minute intervals, but of course they do not arrive exactly on time. Also each consultation does not last for exactly the fifteen minutes allowed. If the patient arrives early and the physiotherapist is free, the patient is seen immediately. Rhianon has told her receptionist to turn away any patient who arrives more than five minutes late, irrespective of whether or not she is free. This is in the belief that the resultant delay would cause a backlog of patients who would not be cleared until well after midday. If the first patient arrives early, he or she is not seen until 9.00 a.m., and the morning session concludes when all scheduled patients, other than those who have been turned away, have been seen.

Rhianon is aware that operating in this way there are times when she is sitting waiting for the next patient. A certain amount of idle time is quite useful but she does not want to overdo it. She also finds that patients become irritated if they have to wait too long, particularly if they find that the patient scheduled to be seen before them is still waiting. Finally she finds that she is quite regularly working until well after midday. She wonders now what would be the effects of changing a number of her operating parameters. She could make appointments at 20-minute intervals (any finer tuned interval would not be practical). She could make the last appointment earlier than she does at present. She could change her 'turn away' lateness criterion to say 7 or 8 minutes; she could also only operate this policy in the latter half of the morning. She could make appointments at 15-minute intervals in the first two hours, but at 20-minute intervals in the last hour so that she would have a better chance of working off any backlog of patients before midday.

All of these possibilities could be tested by trying them out in practice. Clearly there are a lot of systems when all combinations of alternative strategies are tried, and the strength of the mathematical modelling approach is that the alternatives can be tried quickly and cheaply without any risk of disrupting the physiotherapist's service when a particularly inefficient combination is used.

Consider first the original set-up. There are two random elements in the system. Firstly there is the difference between a customer's appointment time and his actual arrival time. This could be a positive or a negative quantity. Secondly there is the consultation time or time which the patient spends with the physiotherapist. This is basically a queueing situation, but the complexities of even the simplest version are such that a realistic analytical solution is impossible. Moreover the simulation approach is straightforward and is clearly the one to use.

Suppose therefore that data were recorded on arrival times and consultation times for 16 morning sessions which were felt to be typical. The data, converted to relative frequencies and hence to appropriately corresponding random numbers, are given in Tables 12.4 and 12.5.

Using the first block of pseudo-random numbers generated by the congruential method in Section 12.3 one morning's operations can now be simulated. Note that the random numbers have to be in sets of three, and that this can be achieved by simply reading across the

Table 12.4

Consultation times			
Time taken (minutes)	*Frequency*	*Relative frequency*	*Random numbers*
10	2	0.011	001–011
11	3	0.016	012–027
12	3	0.016	028–043
13	5	0.026	044–069
14	11	0.059	070–128
15	18	0.096	129–224
16	23	0.122	225–346
17	27	0.144	347–490
18	27	0.144	491–634
19	21	0.112	635–746
20	17	0.090	747–836
21	12	0.064	837–900
22	9	0.048	901–948
23	5	0.026	949–974
24	4	0.021	975–995
25	1	0.005	996–000
	Total 188	1.000	

Table 12.5

		Arrival times	
Early/late Correction (minutes)	*Frequency*	*Relative frequency*	*Random numbers*
−5	12	0.063	001–063
−4	26	0.135	064–198
−3	46	0.240	199–438
−2	34	0.177	439–615
−1	29	0.151	616–766
0	15	0.078	767–844
1	10	0.052	845–896
2	7	0.036	897–932
3	5	0.026	933–958
4	3	0.016	959–974
5	1	0.005	975–979
Turned away	4	0.021	980–000
	Total 192	1.000	

rows. With a manual operation of the simulation the results can be set out as in Table 12.6.

Extracting the principal relevant data from Table 12.6, we obtain

Session overrun	+ 31 minutes
Physiotherapist idle	1 minute
Max. number waiting	2
Total waiting time	202 minutes

This is only the evidence from the simulated operations of one morning. To compare strategies the results from several runs would be needed, from which we might for example compare averages or alternatively extreme values. Note that it would be inappropriate simply to run the session on well beyond the three hours of the morning. Repeat sessions each starting with an appointment at 9.00 a.m. and lasting three hours are needed. To do this manually would be tedious in the extreme, but fortunately the simple arithmetic operations required in a simulation can be carried out very efficiently by an electronic computer and this is the normal practice. More will be said about the uses of computers in Section 12.5.

To conclude the treatment of the illustration, suppose that a 20-minute appointment interval is to be considered and compared with

Table 12.6

	Actual arrival			Duration of consultation					
Appointment	*Random no.*	*Correction*	*Time*	*Random no.*	*Time*	*Time of consultation*	*Number waiting at end of consultation*	*Patient waiting time*	*Physiotherapist idle time*
09.00	847	+1	9.01	779	20	9.01– 9.21	1	0	1
09.15	642	−1	9.14	565	18	9.21– 9.39	1	7	0
09.30	689	−1	9.29	688	19	9.39– 9.58	2	10	0
09.45	470	−2	9.43	143	15	9.58–10.13	2	15	0
10.00	312	−3	9.57	052	13	10.13–10.26	1	16	0
10.15	149	−4	10.11	730	19	10.26–10.45	2	15	0
10.30	782	0	10.30	195	15	10.45–11.00	1	15	0
10.45	451	−2	10.43	682	19	11.00–11.19	2	17	0
11.00	921	+2	11.02	724	19	11.19–11.38	2	17	0
11.15	223	−3	11.12	776	20	11.38–11.58	2	26	0
11.30	363	−3	11.27	305	16	11.58–12.14	1	31	0
11.45	135	−4	11.41	400	17	12.14–12.31	0	33	0

the original. The same form of simulation can be carried out, and moreover to make the comparisons more direct, the same random numbers may be used, a facility made more feasible by the generation method for pseudo-random numbers. The actual run is shown in Table 12.7.

This summarizes into

Session overrun	1 minute
Physiotherapist idle	24 minutes
Max. number waiting	1
Total waiting time	2 minutes

Comparing the two single runs it is apparent that patient waiting time has been exchanged for physiotherapist idle time. Further runs would give a more reliable measure of this phenomenon. The analysis also indicates the possible benefits from switching from a 15-minute appointment interval to a 20-minute interval during the last hour.

A further point is that it has been implied that the same input data would be used for all alternative strategies to be simulated. It could be argued, for example, that when the appointment interval is increased to 20 minutes the physiotherapist will tend to take a little longer time with each patient. If this is so, a fairly simple scaling adjustment to the distribution of consultation times can be effected.

12.5 Computer simulation

It should be clear by now that, if the technique of simulation is to be used to analyse a model, almost without exception the magnitude of the task necessitates some form of automatic computation. The case is enhanced by the fact that the logic of a simulation can be set down in an explicit way requiring only basic arithmetic operations. To assist the user further a number of languages and packages have been developed. These basically relieve the user of the need to program parts of the system which appear regularly in this type of exercise. These include up-dating the time mechanism and statistical routines such as taking averages, drawing up histograms, etc. The word *package* is used when these subroutines are offered in one of the standard general-purpose languages such as FORTRAN or Algol. The programmer simply calls on these facilities as he needs them when he is developing his total model. An extension of this concept is the *simulation language*,

Table 12.7

	Actual arrival			Duration of consultation			Number waiting	Patient	
Appointment	*Random no.*	*Correction*	*Time*	*Random no.*	*Time*	*Time of consultation*	*at end of consultation*	*waiting time*	*Physiotherapist idle time*
09.00	847	+1	9.01	779	20	9.01– 9.21	1	0	1
09.20	642	−1	9.19	565	18	9.21– 9.39	1	2	0
09.40	689	−1	9.39	688	19	9.39– 9.58	1	0	0
10.00	470	−2	9.58	143	15	9.58–10.13	0	0	0
10.20	312	−3	10.17	052	13	10.17–10.30	0	0	4
10.40	149	−4	10.36	730	19	10.36–10.55	0	0	6
11.00	782	0	11.00	195	15	11.00–11.15	0	0	5
11.20	451	−2	11.18	682	19	11.18–11.37	0	0	3
11.40	921	+2	11.42	724	19	11.42–12.01	0	0	5

where steps in the program logic itself are replaced by a reduced number of statements, thereby simplifying the task of programming at the expense of some effort in learning the language and also at the expense of some loss of flexibility required by the language. Crookes (1982) claimed that in 1981 there were 131 different languages and packages, and forecast that the number would increase. Subsequent events have justified the prediction. There is therefore a considerable problem for the user in deciding which package to use or which language to use. To some extent he will be strongly influenced by the computer he is using and by the general-purpose language with which he is familiar. He will also need to consider the approach used.

The first distinction in the approaches is the way time is considered. Ideally time is continuous and events within the simulation such as arrivals and departures should be recorded as they occur in the modelled system. This may be possible on analogue computers, which are generally designed for a specific purpose and are based on some physical analogy. The digital computer has to take time in discrete intervals up-dating the system according to programmed relationships for the behaviour of the system during each interval. By making these intervals small, a close approximation to continuous time can be achieved, although at a cost of slowing down the whole procedure. Continuous-time simulations are not used to any appreciable extent at present.

The alternative to continuous is, of course, discrete simulations. Here the program is driven by examining the system at distinct epochs, where the behaviour in the intervals between consecutive epochs can be readily spelt out in forms which are repetitive in their logic. In a queueing situation this may for example be arrival epochs of customers. This can best be seen in a flow diagram, which is a useful intermediate stage between the model and the actual computer program. The flow diagram demonstrates the fundamental logic without the details of the programming mechanism. In the case of the illustrative example of Section 12.4, the flow diagram could be as shown in Figure 12.3.

For simplicity the procedure for determining the queue size at an arrival epoch has been omitted, but the reader should check through this flow diagram to see that it does in fact represent the logic of the system. The flow diagram depicts one run of the simulation, but it would be a simple task to extend it to require a specified number of runs.

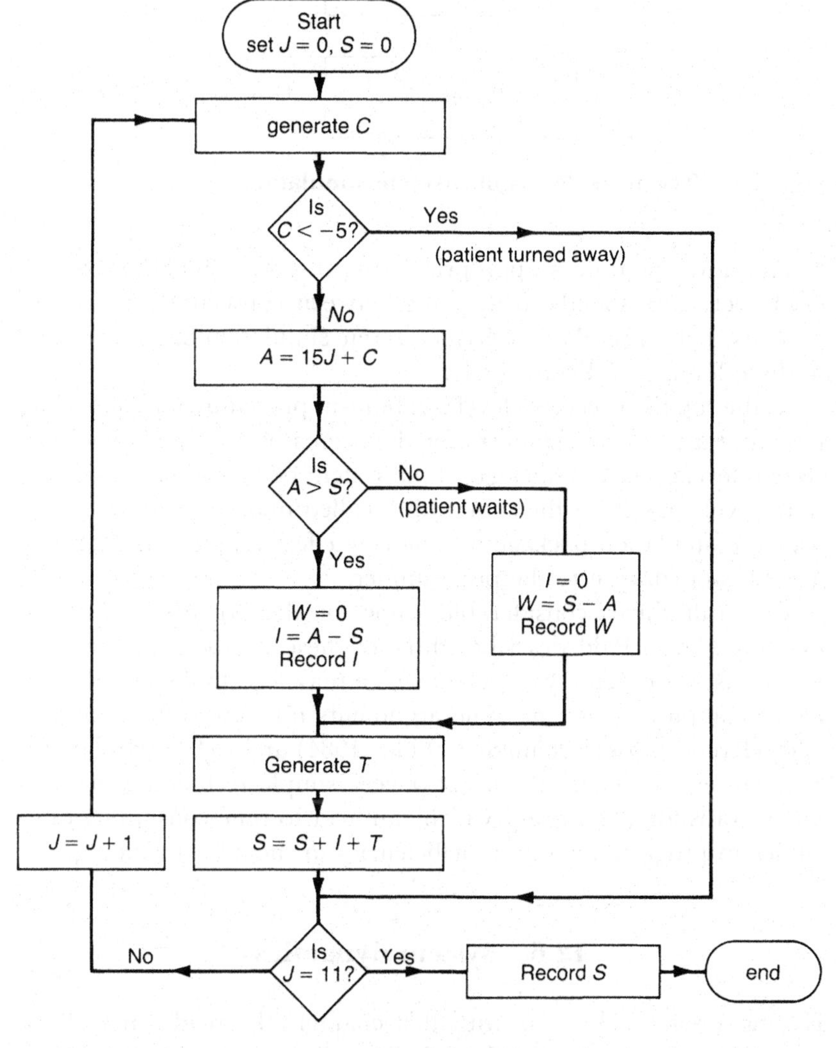

Fig. 12.3 Flow diagram for physiotherapist's clinic simulation

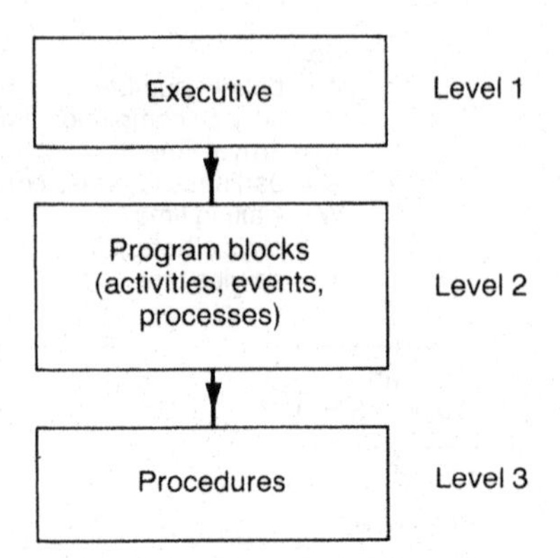

Fig. 12.4 The three levels of discrete simulation

The flow diagram is appropriate to a discrete simulation, where each cycle corresponds to the time between consecutive arrivals of patients. Following Pidd (1984) a discrete simulation has a hierarchy of three levels (see Figure 12.4).

At the top the executive level is a built-in procedure for controlling time and a trigger mechanism for the events of the simulation. The second level describes the logic of the system being simulated and the third level consists of the subroutines called upon from level 2. The use of a simulation package or language effectively means that only level 2 is of concern to the programmer.

This still leaves considerable scope for the organization of the executive level. Pidd describes three of these – process interaction, event bases and activity – each of which may lead to the use of a different computer language. The arguments now are rather complex and reference should be made to Pidd (1984) and to Crookes (1982). Simulation, although in concept a very simple technique, presents many traps for the unwary which can lead to time-consuming difficulties in programming and inefficiencies in the actual running.

12.6 System dynamics

Having progressed so far with this chapter, the reader may have formed the impression that the only limitations to the size of problem which simulation can tackle are the capacity of the computer and the mental capability of the programmer. Certainly very large models have been developed for corporate planning in large organizations. Sometimes for feasibility the system being modelled is decomposed into interacting but still relatively large subsystems. These may for example be separate factories or production centres within a company; sales regions or some other convenient geographical division of, say, a bank's activities. In effect the modeller is using a 'bottom-up' approach, amalgamating a mass of simple relationships and interactions into what he hopes will be meaningful measures of the decisions open to management.

Much of this objectivity is illusory. Most large systems have what may be termed a 'soft' mechanism present, where, although some clear reaction to a decision or external impulse may be detected, precise quantification is infeasible – despite the laws of physics. For example, in weather forecasting, there is no shortage of data for the meteorologists. The model can be specified with great accuracy and moreover the laws which gases obey are also known precisely. There is – give or take the odd sun-spot – a very well-defined model. Weather forecasters in general do a remarkably efficient job in predicting temperatures a day or two ahead and on such matters as whether or not it will rain. They draw the line at saying precisely how much rain will fall and exactly where and when, although it might be thought that with a sufficiently comprehensive model this would be possible. In practice they introduce a soft systems element where from knowledge of current regional barometric pressures and conditions generally they can make reasonable predictions for the immediate future. The same concept is used in industrial systems, where for example it can reasonably be argued that an increased investment in skill training will improve productivity, both in quantity and quality. No law of physics or even of economics establishes this, but the effect is there and it should be distinguished from the 'hard' systems elements such as the effects of price increases on contribution per unit (not on sales volume, which would be a soft relationship), or the effects on company profits of a change in taxation laws.

Arguments along these lines are partly responsible for the development of the topic of system dynamics. The early work is largely attributed to J. W. Forrester (1961) of the Massachusetts Institute of Technology, and although the original application was in management, Forrester subsequently extended the concept to urban and global modelling. In principle system dynamics describes a system by a set of physical flows of men, materials, cash, etc. and also a set of information flows. Together these form feedback loops, where for example the knowledge that an intermediate stock is being depleted will cause an increase in the throughput of the immediately preceding production stage. This in turn will then deplete the stocks supplying it and hence an interlocking system of loops or influence diagrams is formed. Part of a typical diagram is shown in Figure 12.5.

Arrows denote the direction of the influence, the full line indicating a physical flow, the dotted line an information flow. In addition the + or − sign indicates a positive or negative relationship between the changes in rates and the changes in level. Thus in Figure 12.5, an increase in the production rate of stage A causes an increase in the intermediate stocks, but the information that these stocks have increased should cause a reduction in the production rate of stage A. The systems dynamics diagram is then translated into a set of mathematical equations which in turn are solved, usually by computer.

Recent work by the system dynamics research group in the University of Bradford Management Centre has been reported by Wolstenholme (1982) and by Wolstenholme and Coyle (1983). They point out the emphasis that their group has developed on the system description/qualitative analysis aspect rather than the more traditional quantitative analysis. They argue that this is a closer representation of management's approach to what is usually an ill-structured problem and by spending more time on exploring the problem context through something like an influence diagram, a greater understanding will be achieved. Because of this it follows that better decisions will be taken. They do not discount the benefits of optimization through simulation, but argue for a balance between the two phases of analysis, as can be detected in their definition of system dynamics:

> A rigorous method of system description, which facilitates feedback analysis, usually via a continuous simulation model, of the effects of alternative system structure and control policies on system behaviour.

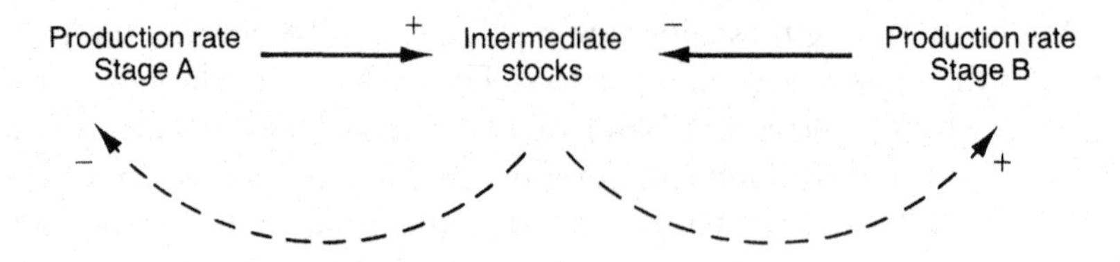

Fig. 12.5 Influence diagram, system dynamics

12.7 Decision support systems

The simplest way to think of decision support systems is that it is the logical stage in the evolution of management's use of computer technology. This began with the development of the computer itself, which presented a means of processing data quickly, accurately and relatively cheaply. At first management used the computer to replace slower methods of data processing, and later to extend the scope of similar operations. The computer was a machine to be called on when the need was felt. In the next phase of its evolution the computer was used to provide information for planning and control in well-defined areas using objective methods. The term Management Information Systems (MIS) fairly describes this phase. Since then there has been perceived a need for a more subjective approach to decision-making, where, although information is needed and furthermore a form of data processing is needed, the decision-maker requires a less structured approach to the way the data are used. This is the basis for Decision Support Systems (DSS). The development of interactive software has given much more flexibility to the way in which the decision-maker uses the data and the applied models, and this is really the essence of DSS.

An interesting application of DSS has been described in a paper by Thomas and Samson (1986). The authors relate the concept of DSS to decision analysis, showing how modules representing different parts of the system can be varied using plausible data inputs. From this, different scenarios are built up against which improved strategies can be tested. The key factor is that by means of a user-friendly DSS, the managers themselves can build up the scenarios based on their own experience and creativity. Thomas and Samson illustrate the practice with the case of an insurance company, where the modules are concerned with underwriting, investment and reinsurance activities, all of which interact on retained earnings.

12.8 Artificial intelligence and expert systems

A common theme of the preceding two sections has been the need for a different methodology in cases where the problems are both complex and ill structured. In system dynamics the approach used was to model the situation using rather more loosely defined relationships (feedback loops, physical flows, information flows) but then to translate these into a model suitable for computer analysis. Decision support systems argues broadly that the human element in the decision process should be retained, but that access to data should be made as comprehensive and straightforward as possible. Although these approaches are not strictly alternatives for all large problems, there is some application overlap, and indeed a third possible approach should also be considered. This is based on the idea that it should be possible to tackle complex problems by imitating the processes used by humans. These are generally subjective and heuristic or 'trial-and-error'. The obvious advantage of doing this is that the limit on the size of problem which the human mind can confidently tackle is thereby extended. Although there is some confusion over precise definition the general concept described above is known as *artificial intelligence* (AI). The program acquires its intelligence to take good decisions by its own experience applied to a specified form of reasoning. Essentially it is computer based, but logic and network relationships make up the representation of human cognitive processes. A full description of recent developments in AI is well beyond the scope of this book, and interested readers are advised to consult the references at the end of the chapter.

The section of AI directly concerned with computers is generally known as an *expert system*. Since this is at the heart of AI some authors seem to use both terms for the same approach, particularly when it is dominated by the computer program. Kastner and Hong (1984) put forward the definition 'an expert system is a computer program that solves problems that heretofore required significant human expertise by using explicitly represented domain knowledge and computational decision procedures.' By 'domain' the authors mean the appropriate application area.

The two elements 'human expertise' and 'computational decision procedures' present their own difficulties. Persuading the expert to express his knowledge in a form which can be quantified and then programmed will, often for good reason, be problematic (see Bell, 1985). Secondly the decision procedures must ensure that all possible alternative decisions are being considered when the interrogation program is designed.

Expert systems is a relatively new topic and although there are a number of texts on the subject, published case studies of successful applications are relatively rare. A paper by O'Keefe *et al.* (1986) claims experience with applications in agriculture, banking, personnel management and experimentation with discrete simulation models. The banking application is a very relevant use in analysing company accounts with a view to extending a loan. This is equivalent to the credit scoring approach normally used by banks, where commercial customers are given credit ratings on the values of certain critical ratios, e.g. current assets/current liabilities, stocks/turnover, etc. Approached scientifically this leads to an exercise in discriminant analysis (see for example Hope (1968)) which essentially amounts to determining the weights for a linear function of the ratios and then a critical value discriminating the good risks from the bad risks. The expert systems approach is to take each ratio in turn and to classify the current value for the company as either favourable, unfavourable or (as an intermediate) satisfactory. It does this by rules based on expert opinions and in treating each ratio sequentially claims to follow the normal practices of the bank manager involved. The outcome may not be able to claim optimality in the sense that discriminant analysis does, but, in behaving more like an expert in creditworthiness assessment, it is likely to achieve a greater level of acceptance.

12.9 References and further reading

General

Daellenbach, H. G. and George, J. A., *Introduction to Operations Research Techniques*, Allyn and Bacon (1978), Chapter 15.

Emshoff, J. R. and Sisson, R. L., *Design and Use of Computer Simulation Models*, Macmillan, New York (1970).

Fishman, G. S., *Principles of Discrete Simulation*, John Wiley, New York (1978).

Jones, G. T., *Simulation and Business Decisions*, Penguin, Harmondsworth, Middlesex (1972).

Morris, C., *Quantitative Approaches in Business Studies*, Pitman, London (1983), Chapter 17.

Stratfold, P. J., *BASIC Business Simulation*, Butterworth, London (1987).

Tocher, K. D., *The Art of Simulation*, English Universities Press, London (1963).

Queueing theory

Cox, D. R. and Smith, W. L., *Queues*, Methuen, London (1961).

Harper, W. M. and Lim, H. C., *Operational Research*, M & E Handbooks, Pitman, London (1982), Chapter IX (also Chapter VIII for simulation generally).

Lee, A. M., *Applied Queueing Theory*, Macmillan, London (1966).

Stainton, R. S., *Operational Research and its Management Applications*, Macdonald & Evans, Plymouth (1977), Chapter VI (also Chapter VII for simulation generally).

Computer simulation

Crookes, J. G., 'Simulation in 1981', *European J. Operational Research*, **9** (1) (1982).

Pidd, M., 'Computer Simulation for Operational Research' in *Developments in Operational Research*, Eglese, R. W. and Rand, G. K. (eds) Pergamon, Oxford (1984).

System dynamics

Forrester, J. W., *Industrial Dynamics*, MIT Press, Cambridge, Mass. (1961).

Sharp, J. A. and Price, D. H. R., 'System Dynamics and Operational Research', *European J. Operational Res.*, **16** (1), 1–12 (1984).

Wolstenholme, E. F., 'System Dynamics in Perspective', *J. Opl. Res. Soc.*, **33**, 547–556 (1982).

Wolstenholme, E. F. and Coyle, R. G., 'The Development of System Dynamics as a Methodology for Systems Description and Quantitative Analysis', *J. Opl. Res. Soc.*, **34**, 569–581 (1983).

Decision support systems

Ginzberg, M. J., Reitman, W. and Stohr, E. A. (eds.) *Decision Support Systems*, North-Holland, Amsterdam (1982).

Sol, H. G., 'DSS: Buzzword or OR Challenge', *European J. Operational Res.*, **22** (1), 1–8 (1985).

Thomas, H. and Samson, D., 'Subjective Aspects of the Art of Decision Analysis, Exploring the Role of Decision Analysis in Structuring, Decision Support and Policy Dialogue', *J. Opl. Res. Soc.*, **37** (3), 249–265 (1986).

Artificial intelligence and expert systems

Bell, M. Z., 'Why Expert Systems Fail', *J. Opl. Res. Soc.*, **36** (7), 613–619 (1985).

Forsyth, R. (ed.) *Expert Systems – Principles and Case Studies*, Chapman-Hall, New York (1984).

Hayes-Roth, F., Waterman, D. A. and Lenat, D. B. (eds) *Building Expert Systems*, Addison-Wesley, New York (1983).

Hope, K., *Methods of Multivariate Analysis*, London University Press, London (1968).

Kastner J. K. and Hongs S. J., 'A Review of Expert Systems', *European J. Operational Research*, **18**, 285–292 (1984).

O'Keefe, R. M., 'Expert Systems and Operational Research – Mutual Benefits', *J. Opl. Res. Soc.*, **36** (2), 125–129 (1985).

O'Keefe, R. M., Belton, V. and Ball, T., 'Experiences with using Expert Systems in O.R.', *J. Opl. Res. Soc.*, **17** (7), 657–668 (1986).

Paul, R. J. and Donkidis, G. L., 'Further Developments in the Use of Artificial Intelligence Techniques which Formulate Simulation Problems', *J. Opl. Res. Soc.*, **37** (8), 787–810 (1986).

Phelps, R. I., 'Artificial Intelligence – an Overview of Similarities with OR', *J. Opl. Res. Soc.*, **37** (1), 13–20 (1986).

12.10 Problems and practical exercises

1 Use the congruential method to generate pseudo-random numbers with $k = 37$, $m = 103$. Start with the number 22. The purpose is to generate two-digit random numbers. How do you cope with generated numbers greater than 99? When does the method cycle?

2 A firm supplies a special form of industrial boiler. Monthly demand for this product is variable, and recent experience suggests that the demand can be characterized as a probability distribution:

Monthly demand	0	1	2	3	4
Probability	0.1	0.2	0.3	0.2	0.2

The firm makes the boilers in batches of 12. Stock is reviewed at the end of each month, and a decision taken then to produce a batch of 12 would mean that this further stock would be available one month later.

Identify the characteristics of the system which would influence the choice of a production policy and show how a simulation could be used to find a good policy. (Use random numbers from Appendix 3.)

3 Consider the example of the physiotherapist's clinic from section 12.4. Run a simulation where the physiotherapist schedules patients at 15 minute intervals in the first two hours, but at 20 minute intervals in the last hour. What would be the relative merits of this and the previous policy modified so that the last appointment was made at 11.30 a.m.?

4 A bank operates an external cash dispenser which also provides information on the customer's cash balance, orders a new cheque book and accepts requests for bank statements. On average 40 per cent of customers require cash only, the other 60 per cent requiring cash and other services. If a customer requires cash only, the probability distribution of time spent at the dispenser can be approximated by

Time (seconds)	45	60
Probability	0.8	0.2

Other customers spend time whose distribution is:

Time (seconds)	75	90	105	120
Probability	0.1	0.4	0.3	0.2

Simulate the system (use random numbers from Appendix 3) determining the proportion of time the queue is of length 1, 2, 3, etc.

Further investigation showed that the inter-arrival of actual customers was distorted since customers tended to refuse to join the queue when too many people were waiting. Observations showed that the inter-arrival time (including those who balked in this way) was:

Time (seconds)	20	30	40	50	60
Probability	0.1	0.3	0.3	0.2	0.1

Also the probability that a customer joined when there were already n people waiting (including the one being served) was:

Number waiting n	0	1	2	3	4	5 or more
Probability of joining	1	1	0.9	0.7	0.3	0

Modify your simulations to take the balking into account. (Use random numbers from Appendix 3.)

5 Because of bad weather football league matches in Britain are frequently postponed. When this happens on a substantial scale, a panel of 'experts' sits and, for the purposes of football pools gambles, the panel pronounces a result (win, lose or draw) for each postponed match. These consequently tend to be predictable, reflecting the genuine merits of the teams in opposition.

Construct a simulation model to replace the team of experts, introducing an appropriate chance element into the way the results are decided. Note that for most teams there is a considerable advantage in playing at home. Also, by the time that the bad weather occurs, approximately half of the season's matches will have been played.

13
Capital investment appraisal

OBJECTIVES

In this chapter, three of the standard criteria for project appraisal – net present value, internal rate of return and pay-back period – are taken and analysed further in the context of the uncertainties in the input data. It is then shown how subjective probability assessments of cash flows are transformed into probability distributions of the criteria. The reader should be able to demonstrate the effects of probability distributions of the project returns on the investment appraisal criteria considered.

13.1 Introduction

The basic methods of evaluating projects in which a company might invest its capital have been discussed in the companion volume by F. Livesey (*Economics for Business Decisions*, Pitman, London (1983)). The full treatment will not be repeated; rather, having outlined these methods, further attention will be given to the problem of dealing with uncertainties in the data, and in particular to the way uncertainties influence investment decisions.

The usual model for a capital investment project is to assume that there are cash flows from and to the investing company at regular intervals. Normally the interval is taken to be a year and the pattern of cash flows is an initial substantial outlay followed by returns over a fixed number of years. There will be occasions when there is an outflow of cash at some year away from the initiation of the project – for example, if an expensive piece of equipment needs replacing part way through the project – and in such cases there are complications for all three methods of investment appraisal to be discussed. For simplicity discussion here is confined to the simple case of an initial outlay followed by a finite number of returns.

A key element of all capital investment appraisals is the fact that the same amount of money is worth more to the investor now than it would be at some future date. This is not simply because of inflation, although of course inflation reinforces the argument. The real reason is that money earns money. It is possible to put £100 in a bank deposit account, a building society, national savings, etc., with the very real prospect that in twelve months time the money will be available to you with a modest positive addition. The addition is modest because the money will have been invested by somebody else (e.g. the bank) in a portfolio of somewhat riskier ventures, and you, the investor, are therefore passing on the risk.

In evaluating a capital investment project, the company can therefore think in terms of the return achieved from the initial investment. Unfortunately unlike bank deposits where the interest is paid at a steady and stated rate (until of course the bank changes its rates), most investment projects have very variable returns. For example, a new product usually goes through the phases of growth, maturity and decline, with the return on the initial launch (advertising, tooling-up, etc.) correspondingly increasing, flattening out and then tapering off. As mentioned at the beginning of this section, the return is also subject to random variation. Encapsulating the pattern of cash outflows and inflows in a single figure is certainly possible, but there may be other considerations. For example, an investment of £100 which brings a return of £110 one year hence is, under certain criteria, exactly equivalent to the same investment bringing a return of £121 two years hence. Scaled up, the effects on a company's cash flow could be quite different.

An alternative approach which avoids the need to consider interest rates simply measures the time by which the company will have received back in total the initial outflow.

13.2 Net present value

The Net Present Value (NPV) discounts the returns from the project to the present point in time using what is known as the *cost of capital*, i, as the discount rate. In broad terms the cost of capital is the return that the firm believes that it should be receiving from its capital investments. As a minimum it is the cost to the firm of raising capital, but to this should be added a margin for profit. For a fuller discussion of this parameter, refer to Van Horne (1980) or Merrett and Sykes (1973).

Suppose for example that a carpet manufacturer is considering the introduction of a new style of carpet to complement his range. The style is expected to last four years and the initial outlay will be £150 000, and each year of the life of the product will produce a net inflow:

Year	1	2	3	4
Cash inflow (£000)	40	70	60	20

There is no salvage value of any of the initial outlay at the end of the four years.

If the firm uses a cost of capital of 10 per cent, $i = 0.1$, and therefore

$$\begin{aligned} \text{NPV} &= -150 + \frac{40}{1.1} + \frac{70}{(1.1)^2} + \frac{60}{(1.1)^3} + \frac{20}{(1.1)^4} \\ &= 2.95. \end{aligned}$$

Thus the NPV is positive, which indicates that this project is worth considering on a financial basis. The implication is that the return on the project is better than the cost of capital. The fact that it is only just better, reflected in the relatively low value of the NPV compared with the cash flows involved, should give rise to doubts about the project's viability; it would need only a small reduction in any of the forecast cash flows to make the NPV negative, in which case the project would, by this criterion, be rejected.

13.3 Internal rate of return

The internal rate of return (IRR) approach takes a similar view to the NPV approach, except that in this case the basic cash flow equation is solved for the rate of return, r, assuming that the NPV is zero. The answer is therefore the rate of return corresponding to the given cash flows. For the carpet launch example of Section 13.2, the equation to be solved is

$$150 = \frac{40}{1+r} + \frac{70}{(1+r)^2} + \frac{60}{(1+r)^3} + \frac{20}{(1+r)^4} \cdot$$

The presentation can be simplified by writing

$$x = \frac{1}{1 + r},$$

and hence

$$20x^4 + 60x^3 + 70x^2 + 40x - 150 = 0. \qquad (13.1)$$

Equation (13.1) is a quartic in x which can be solved by trial and error methods on a programmable calculator. The solution is $x = 0.90118$ which in turn leads to $r = 0.11$. The solution to (13.1) is unique, a result which follows from applying a rule known as Descartes' rule of signs, since there is only one change of sign in the coefficients of the equation. Had this not been the case – as it would if there had been a cash outflow during the life of the project – uniqueness of solution would not have followed, and there could be an embarrassment of having two or more rates of return. The fault is not with the mathematics, but rather with the model where we would be assuming that any intermediate accumulated positive balance would be invested at the IRR of this project. Usually this will not be possible, and something like the cost of capital should be used for intermediate investments.

Note that the IRR at 11 per cent is just above the cost of capital (10 per cent) used for the NPV. The closeness of these figures reflects the fact that the NPV is correspondingly close to zero.

The equivalence of IRR to an investment can be demonstrated as in Table 13.1.

Both the NPV and the IRR can be used to make comparisons between projects, when perhaps because of a shortage of investment cash or of suitable management manpower a firm has to select between a number of alternatives. The practice is to select the project with the

Table 13.1

Year	1	2	3	4
Initial investment	150	126.5	70.4	18.1
Interest at IRR (11%)	16.5	13.9	7.7	2.0
Total investment	166.5	140.4	78.1	20.1
Withdrawn (cash flow)	40	70	60	20
Final investment	126.5	70.4	18.1	0.1

The final figure 0.1 differs from zero owing to rounding errors.

greatest NPV (assuming it to be positive) or the highest IRR (assuming it to be greater than the cost of capital). Unfortunately the two approaches, although apparently similar in philosophy, do not necessarily rank the same projects in the same order, even when the initial outlays are identical. For example, compare projects *A* and *B*, each of duration two years, with the cash flows shown in Table 13.2.

Table 13.2

Project	*Initial outflow*	*Year 1 inflow*	*Year 2 inflow*
A	100	10	130
B	100	100	30

At a cost of capital of 10 per cent per annum, project *A* has the NPV of 16.5 whereas project *B* has NPV equal to 15.7. Clearly on this criterion project *A* is preferable. When the IRR values are calculated it turns out that project *A* has an IRR of 19.1 per cent whereas the IRR of project *B* is 24.2 per cent, reversing the previous decision. The dilemma can be resolved by examining the very different patterns of cash returns, where project *A* provides a larger total return, although substantially in the second year.

13.4 Pay-back period

A further measure of project viability is to determine the time when the initial outlay will have been balanced by cash inflows. It is a simple measure, easily understood, but it does have the disadvantage that it ignores everything that happens after the pay-back time. It would be foolish to take the criterion alone in deciding between projects, but it must be reassuring to know that the money will be returned within a known period of time. Conversely it must also be a cogent argument against projects which, although profitable, take some time before the benefits are realized.

A simple modification of the pay-back period criterion is to use cash flows discounted at the cost of capital rate, accumulating these to determine the point in time when the initial outlay is balanced by the discounted returns.

In the example of Section 13.2 the pay-back period is almost 3 years. Discounted at 10 per cent per annum, the pay-back period is then close to 4 years.

Reference may be made to standard texts on management accounting for a fuller discussion of these and other criteria, for example Chapter 8 of Sizer (1979).

13.5 Uncertainties in the model

In the previous sections it has been assumed that the cash flows have been known precisely. It is a very unusual manager who finds himself in this happy position. More often he will be dealing with one or more investment projects which, although not exactly of a one-off nature, have sufficient novelty to make an objective assessment impossible. For example a dry-cleaning firm may have opened scores of shop outlets in the past, but no future location will be identical with any of these; moreover economic circumstances will have changed. What the firm does possess is managers who have had experience of this type of decision, and who can be persuaded to justify their decisions with their estimates of the model parameters. With a little further persuasion, and perhaps a little education, they might present their estimates with the associated risks.

The 'easy' way out of the dilemma of uncertainty is to use single (point) estimates of the cash flows, and then work out whichever of the NPV, IRR or pay-back date is preferred. Then if, for example, the NPV is only marginally positive, the project is rejected because owing to uncertainties in the data, the actual NPV could easily be negative. Equivalently a cost of capital somewhat greater than that which would be acceptable might be used, again providing a buffer against uncertainty. Similar safety measures can be devised for the IRR and pay-back period criteria.

All of these approaches, although understandable, treat the uncertainty problem at the stage of the final measure, rather than at the input data stage where it occurs. Unnecessary subjectivity is introduced, and furthermore a bias can be brought into the cash flow estimates if it is known that the final measure is to be treated in this way. This is particularly true when there are political considerations where a view of the cash flows may be taken to ensure that an appropriate decision is taken. Projects such as developing an aircraft, building a major bridge or modernizing a port offer opportunities for the introduction of (perhaps involuntary) bias into the estimates.

Clearly in situations like this uncertainty is always present, and no technique completely removes the possibility of making a decision with a costly outcome. Any decision-maker can be unlucky at times. What is possible is a quantification of the risk and a full understanding of the implications of the assumptions in the model used. This will be illustrated by an example.

13.6 Uncertainties in the model – an example

Consider the problem defined in Section 13.2 where a carpet manufacturer is considering the launch of a new style, with estimated cash flows of:

Year	0	1	2	3	4
Cash flow (£000)	– 150	40	70	60	20

On the assumption that the cost of capital is 10 per cent, the NPV has been shown to be 2.95 and the IRR to be 11 per cent, both of which indicate a marginally acceptable project. To assess the accuracy of these measures, some assumptions on the variability of the cash flow values must be made. Clearly any situation must be treated on its merits, and the following are offered only on the grounds of plausibility.

Assumption 1. The negative cash flow of £150 000 in year 0 is known without error. This is effectively the cost of the launch (advertising, samples, etc.) for which a specified budget has been made.

Assumption 2. Each of the remaining cash flows is a random variable whose median is the value quoted in the above table. Furthermore it is assumed that the probability distribution of each cash flow is known. This is an important assumption which is critical to the subsequent analysis, and one which must be discussed further.

Several methods of assessing these distributions have been proposed (see for example the two books by Hertz and Thomas (1983, 1984) and the article by Samson and Thomas (1986) listed at the end of the chapter). The most popular method for continuous distributions would appear to be the Continuous Distribution Fractile (CDF) method. The method is basically one of quantifying the opinions of the most experienced expert available (the assessor). He is first asked to give the value which the unknown variable is equally likely to be greater or less than, or in other words the median. For the first year's cash flow this might be £40 000. He then considers only the possibility that the cash flow is greater than £40 000, and is asked to estimate the 'halfway mark' amongst these outcomes. A value of £60 000 for this estimate therefore means that if all he knows is that the cash flow is greater than £40 000, he reckons that it is equally likely to be greater or less than £60 000. Compounding these two benchmarks, he is equivalently saying that in his opinion there is a probability of $\frac{1}{4}$ that the cash flow is greater than £60 000. A similar argument for the lower half of the cash flows might produce a lower 'halfway mark' of £30 000. Note that there is no reason why the two values £30 000 and £60 000 should be equally spaced from the median value of £40 000. It is claimed that the mind can repeat this process of finding the halfway values within each of the four intervals defined already, producing values of the cash flow variable corresponding to cumulative probabilities at the levels $\frac{1}{8}$, $\frac{3}{8}$, $\frac{5}{8}$, $\frac{7}{8}$ to augment those at $\frac{1}{4}$, $\frac{1}{2}$, $\frac{3}{4}$ already found.

Finally the assessor might be asked to estimate extreme values for the cash flow – what are the worst and the best values that can be anticipated? This last exercise can be difficult as we all tend to have different interpretations of 'worst' and 'best', and care has to be taken to ensure that the assessor thinks of plausible extremes, and does not consider implausible events like the closing down of all competing carpet factories (best) or the total destruction by fire of his manufacturing plant (worst).

Suppose then that in this case the following values are given for the cash flow in year 1:

Cumulative probability	0	$\frac{1}{8}$	$\frac{1}{4}$	$\frac{3}{8}$	$\frac{1}{2}$	$\frac{5}{8}$	$\frac{3}{4}$	$\frac{7}{8}$	1
Cash flow (£000)	10	25	30	35	40	45	60	80	100

A plot of these values with a smooth curve through them drawn by eye is shown in Figure 13.1. Note that in practice the cash flow would be made up of several costs and returns. The essential random variable would be the annual sales.

The exercise is now repeated for each annual cash flow in turn. This can be a time-consuming and, particularly if the same assessor is involved, exhausting process. One possible way of overcoming the difficulty is to argue that the probability distribution for each year is simply a scaled version of the one already found for the first year. If this is acceptable all that need be done is the estimation of the median cash flow for each year which, when divided by the median cash flow

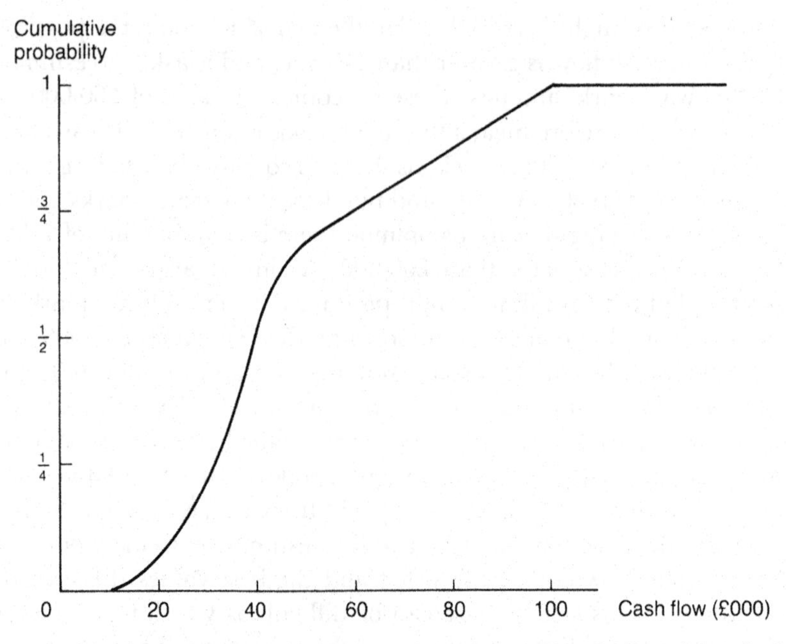

Fig. 13.1 Subjective probability distribution of cash flow in year 1

for the first year, provides the scaling factor. If this procedure is adopted for the given data we would have Table 13.3.

Generation of values for each distribution is then a matter of using one of the distributions (say Year 1) and applying to the generated variable a multiplying factor of 1.75, 1.5 or 0.5 for years 2, 3 or 4 respectively. The actual generation could be performed by reading off a table of transformations between random variables 00–99, where the values in the table are the random variable expressed as a fraction of 100 and the corresponding value for the Year 1 cash flow in the

Table 13.3

Cumulative probability	0	$\frac{1}{8}$	$\frac{1}{4}$	$\frac{3}{8}$	$\frac{1}{2}$	$\frac{5}{8}$	$\frac{3}{4}$	$\frac{7}{8}$	1
Cash flow (£000)									
Year 1	10	25	30	35	40	45	60	80	100
Year 2	17.5	43.8	52.5	61.2	70	78.8	105	140	175
Year 3	15	37.5	45	52.5	60	67.5	90	120	150
Year 4	5	12.5	15	17.5	20	22.5	30	40	50

cumulative probability graph (Figure 13.1). Thus a random variable of value 63 corresponds to a Year 1 cash flow of 45.6. If the same random number occurred for a Year 3 cash flow, the value would be $45.6 \times 1.5 = 68.4$.

Typically therefore for one run of the simulation we might have random numbers 63, 17, 02, 39 with corresponding cash flows of 45.6 (Year 1), 46.2 (Year 2), 20.4 (Year 3), 17.8 (Year 4). The NPV is then

$$\frac{45.6}{1.1} + \frac{46.2}{(1.1)^2} + \frac{20.4}{(1.1)^3} + \frac{17.8}{(1.1)^4} - 150 = 68.96 - 150 = -81.04.$$

This represents a substantial loss. To see whether this is an abnormal result, a full simulation should be carried out. The IRR for this set of input data is -0.07. Not only is the rate of return below the cost of capital (0.10), but it is negative, a result which might be expected since the total return is less than the initial outlay.

Some people might argue that the assumption in the above model is unrealistic where the actual cash flows in any year vary independently. What really happens is that if the project – in this case a new style of carpet – is a winner, the cash flows will be consistently greater than the median value, and conversely a flop will produce consistently lower cash flows throughout the project's life. In statistical language the generated random variables should be correlated and not independent as assumed so far. Dealing with correlation in the actual simulation is not difficult (see Hull (1977), for example); the problem is in matching the mathematical technique with the assessor's understanding of the concept.

One approach to the problem is to examine the two extremes of total independence and total dependence and to argue that the true situation lies somewhere between. It is hoped that for any measure of the viability of the project the extremes will give reasonably close values which will therefore be helpful in coming to a decision.

The total independence model has been considered already. For total dependence the assumption is made that each cash flow variable will be in the same position relative to its own distribution. This is achieved by the simple means of using the same random number for each variable X_1, X_2, X_3 and X_4 in the set. Suppose, for example, we have the random number 63. The corresponding cash flows are 45.6 (Year 1), 79.8 (Year 2), 68.4 (Year 3) and 22.8 (Year 4), resulting in an NPV of 24.4 and an IRR of 17.8 per cent.

Simulation runs under the assumptions of both the independent and the dependent models were carried out and the results plotted in the form of cumulative probability graphs for the NPV and IRR (Figures 13.2, 13.3). Both criteria are more variable for the dependent model as might be expected, since with this model the effect of an extreme random variable is repeated four times, whereas with the independent model there will be a tendency for extremes to be cushioned out by more moderate values for the other three random variables. A further phenomenon is the similarity in shape between the curves for

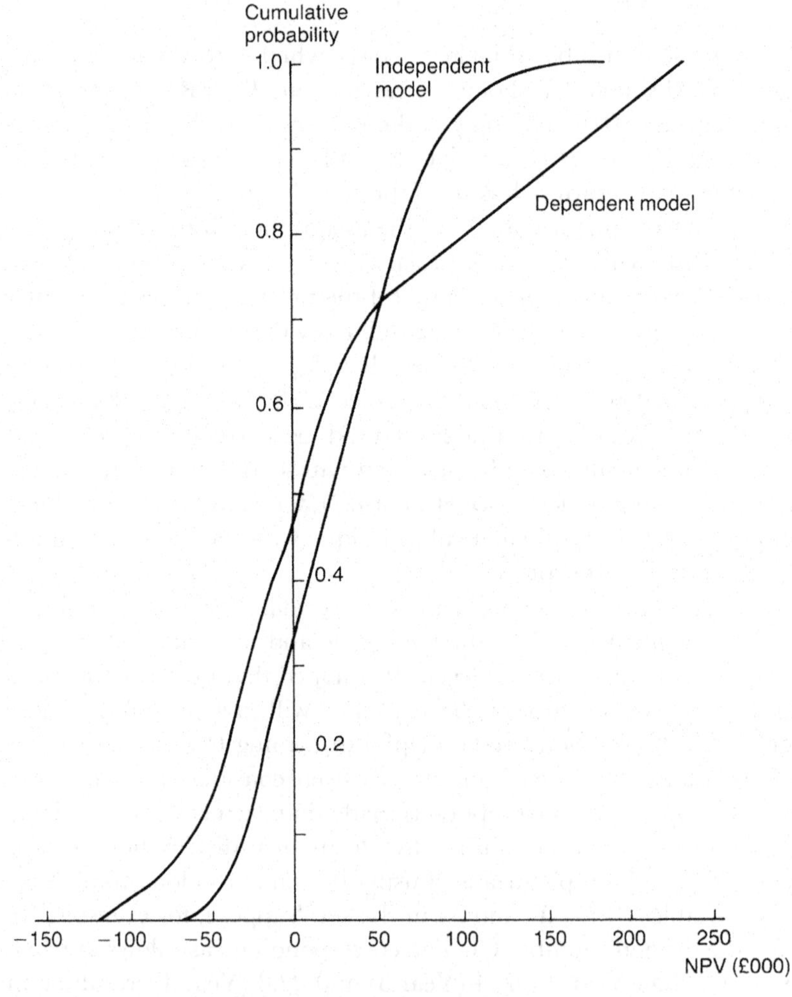

Fig. 13.2 Cumulative probability graphs for NPV

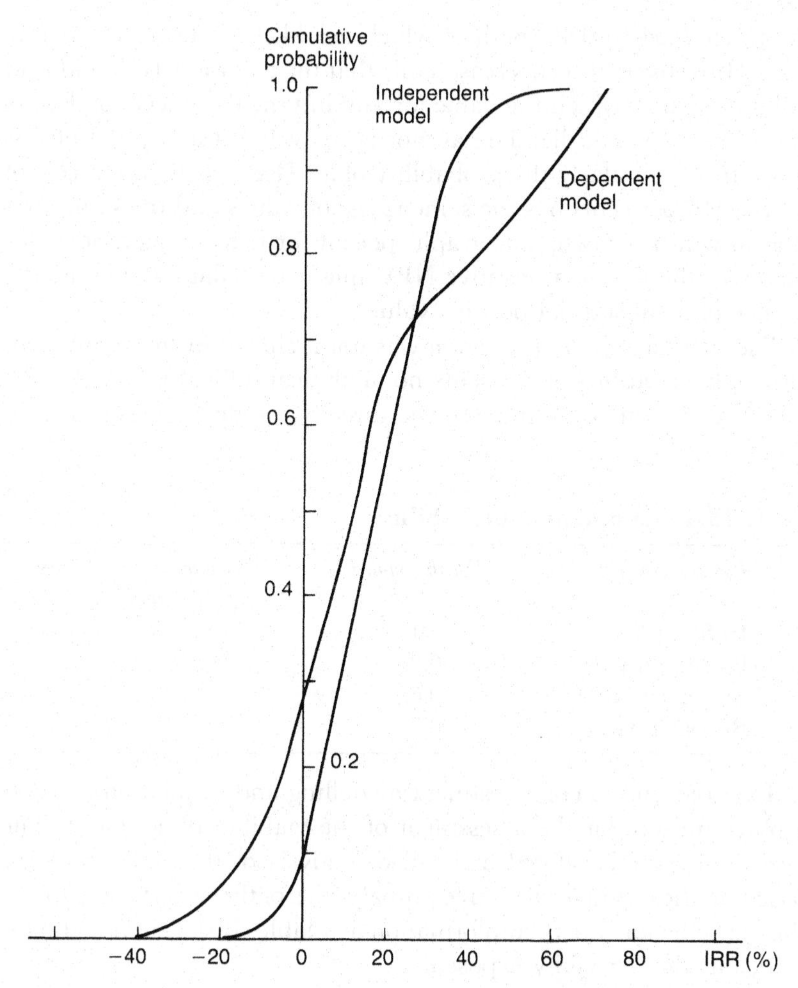

Fig. 13.3 Cumulative probability graphs for IRR

the dependent model and furthermore its similarity to the common cumulative curve for probabilities of the cash flows (Figure 13.1). This can be explained simply since for any random variable there is a unique cash flow for all four years and therefore there is a unique NPV and a unique IRR. Probability statements about cash flow, NPV and IRR are explicitly related, and it was in fact unnecessary to simulate for this model.

Using these graphs to assess the viability of the project the decision-maker might first observe that the probability of a negative NPV is in the range 0.34 to 0.48, both of which are likely to be unacceptably high. (Intuitively it seems reasonable that the two models should lead to limits within which the value for any intermediate level of dependence model would lie. The author is unaware of any proof of this conjecture.) Similarly the probability of an IRR less than the cost of capital (10 per cent) has the same range of values and the same conclusion will be drawn. The graphs permit further considerations: for example, the risk of a negative NPV might be balanced against the chance of a substantial positive value.

The pay-back period criterion was not included in the simulation, although the approach presents no analytical difficulties. As before, the dependent model can be solved directly, producing the results in Table 13.4.

Table 13.4 Cumulative probability

Pay back period	*Not discounted*	*Discounted at 10 per cent per annum*
In first year	0	0
By second year	0.28	0.22
By third year	0.62	0.42
By fourth year	0.71	0.51

The combination of investment modelling and simulation presents a powerful tool for the assessment of the viability of a project. The decision-maker is offered a detailed analysis of the risks involved, based on the original subjective probabilities of the expert assessor. In this sense he has the best information available. Risks are not eliminated, but they are fully exposed!

13.7 Further reading

Hertz, D. and Thomas, H., *Risk Analysis and its Applications*, John Wiley, Chichester (1983).

Hertz, D. and Thomas, H., *Practical Risk Analysis*, John Wiley, Chichester (1984).

Hull, J. C., 'Dealing with dependence in risk simulations', *Operational Research Quarterly* **28**(1), ii, 201–213 (1977).

Livesey, F., *Economics for Business Decisions*, Pitman, London (1983), Chapter 6.

Merrett, A. J. and Sykes, A., *The Finance and Analysis of Capital Projects*, Longman, London (1973).
Samson, D. A. and Thomas, H., 'Assessing probability distributions by the fractile method: evidence from managers', *Omega* **14**(5), 401–407 (1986).
Sizer, J., *An Insight into Management Accounting* (2nd edn) Penguin, Harmondsworth (1979), Chapter 8.
Van Horne, J. C., *Financial Management and Policy*, Prentice-Hall, Englewood Cliffs, N.J. (1980).

13.8 Problems and practical exercises

1 Calculate the NPV and IRR for a project lasting five years where the initial outlay is £200 000 and the returns over the life of the project are:

Year	1	2	3	4	5
Return (£)	50 000	80 000	100 000	100 000	60 000

Assume a cost of capital of 12 per cent.

2 A project has an initial outlay of £1000, a return in the first year of £2000, a further return in the second year of £1000 but an unavoidable outlay in the third and final year of £1000.

Calculate the NPV assuming a cost of capital of 10 per cent.

Under the same assumption, what would you regard as the IRR for this project?

3 A firm can invest in a project where the initial investment is £10 000. Returns occur only in the first two years. In Year 1 there is a probability 0.5 of a return of £8000 and a probability 0.5 of a return of £5000. In Year 2 there is a probability 0.7 of a return of £7000 and a probability 0.3 of a return of £5000. The outcomes in the two years are independent.

Determine the possible values of the NPV with their associated probabilities, assuming that the cost of capital is 10 per cent. What is the expected value of the NPV?

4 The firm in Exercise 3 has an alternative project for its investment of £10 000. This project also lasts for two years and in each year the return is £7000 with probability 0.4 and £6000 with probability 0.6, the outcomes again being independent. Assuming a cost of capital of 10 per cent calculate all possible NPVs with their associated probabilities, and the expected value of the NPV. Do you think that comparison of the expected NPVs is a reasonable criterion to use to choose between the projects?

5 A firm has the choice between two projects *A* and *B*. Both require an initial investment of £10 000 and the returns are as shown in Table 13.5. Calculate the IRR for each project.

Table 13.5

Year	1	2	3
Return project *A* (£)	4000	6000	6000
Return project *B* (£)	3000	5000	8500

Show also how the NPV varies for each project for different assumptions about the cost of capital. What would be your advice to the firm?

14

Risk analysis for large engineering projects*

OBJECTIVES

This chapter outlines the application of risk analysis to large engineering projects. It gives some of the initial reasons for undertaking risk analysis of a project, and it indicates some of the additional benefits that can be obtained in terms of lower risk exposure and better risk management. Risk analysis models and methods are described briefly, as are some of the other requirements for a successful risk analysis study.

14.1 Introduction to risk analysis

14.1.1 What is risk analysis?

Risk is defined here as the exposure to the possibility of delay, of economic or financial loss or gain, or of physical damage or injury, as a consequence of the uncertainty associated with pursuing a particular course of action.

Risk analysis can involve a number of approaches to dealing with the problems created by uncertainty, including the identification, evaluation, control and management of risk. *Risk engineering* is an integrated approach to all aspects of risk analysis, used as the basis of discussion here. Its aim is to identify and measure uncertainty as appropriate, and to develop the insight necessary to change associated risks through effective and efficient decisions. Risk engineering uses risk analysis for the purpose of better risk management.

* Chapter 14 is based upon a paper published by D. F. Cooper and C. B. Chapman in *Underground Space*, **9**, (1) (1985), and another published by C. B. Chapman and D. F. Cooper in *Further Developments in Operational Research*, Editors: G. K. Rand and R. W. Eglese, Copyright © Pergamon Press (1985). Published with permission. Pergamon, 1985.

Risk analysis of this risk engineering form has many potential applications in large projects. They may for example range over the economic evaluation of a hydroelectric energy-generation system, the optimization of an offshore oil development programme, the assessment of contract conditions for an underground development, or the analysis of seismic hazards for a nuclear power station. All major projects are unique. Some involve construction or operation in new or hazardous geographical or geological areas, some involve significant new or untried technology, and some involve risks with very large or potentially catastrophic consequences. Even where the individual risks may be known, their combined effects may be obvious, and the synergy inherent in large projects frequently leads to quite unexpected consequences.

Risk engineering provides a structured means of looking ahead in the life of a project. It differs from many conventional approaches, such as sensitivity analysis, in three key areas:

(a) It subdivides a project into a relatively small number of major elements, and then analyses each in detail in terms of uncertainty.
(b) It identifies the causes of time delays, cost changes and other impacts, and it evaluates responses to associated potential problems, prior to assessing net effects.
(c) It considers degrees of dependency between risks and between the project elements.

14.1.2 The need for risk analysis

Recent decades have been characterized by a vast proliferation of projects with a high level of risk. The real scale of projects and investment programmes has expanded dramatically, increasingly intractable areas have been developed, and economic instability in growth rates and prices has become endemic. With this uncertain and volatile environment, the need for risk analysis of potential projects and investments has increased. The need for analysis is particularly apparent when projects involve

large capital outlays;
unbalanced cash flows, requiring a large proportion of the total investment before any returns are obtained;

major geotechnical uncertainties;
significant new technology;
unusual legal, insurance or contractual arrangements;
important political, economic or financial parameters;
sensitive environmental or safety issues;
stringent regulatory or licensing requirements.

For many underground projects, the additional information needed to reduce risk and uncertainty to an acceptable level prior to commencing the development will not be available, and there may be large costs or delays in acquiring it. These factors increase the need for early assessment of the uncertainties and risks which affect the project before large sums of money are irrevocably committed.

14.1.3 The requirement for risk analysis

There are five somewhat different circumstances in which uncertainty may be a major factor and in which suitable forms of risk analysis may be appropriate:

(1) For pre-feasibility appraisal of a proposed project or investment, when a decision must be made, often on the basis of minimal information, to discard the project, to postpone it, or to proceed with more detailed feasibility studies.
(2) For deciding whether or not to undertake or become involved with a marginal project, when the rate of return calculated on the basis of the best estimates of capital requirements and cash flows is close to the opportunity cost of capital, or the net present value is close to zero.
(3) When a project or investment involves unusual risks or uncertainties, which may lead to a wide range of possible rates of return.
(4) For strategic decisions, when choosing between alternative projects or investments, for a project or investment concept which has already been justified at an earlier pre-feasibility or feasibility stage.
(5) For tactical decisions, when developing a detailed plan or optimising project specifications, for a project concept already given approval.

Within this broad framework, there may be formal requirements for risk analysis for many reasons, among which are:

economic viability assessment, for high-level strategic decision making within the organization;
financial feasibility assessment, for the bond or debt market when a finance package is being assembled;
insurance purposes, to assess premiums for unusual risks for which there may be little statistical or actuarial information;
accountability, for major project managers to demonstrate that they have fully assessed all the material risks, that the measures taken to control risk are appropriate, and that the economic reward for taking on the risk that remains is adequate;
contractual purposes, to assess alternative contractual and legal frameworks for the project, in the context of deciding who should bear what risks and determining an equitable allocation of risks and rewards between project owners, contractors and insurers;
regulatory purposes, for legislative, judicial or licensing agencies of government, or for public enquiries, to demonstrate accountability in a public or social context;
communication purposes, to provide information for project owners, contractors or joint venture partners, or to demonstrate capability and competence in an area.

14.1.4 The benefits of risk analysis

Risk analysis may be required initially for a limited range of purposes. However, the experience of many organizations suggests that it provides other benefits which may prove far more important in the long term. These benefits include

(a) Better and more definite perceptions of risks, their effects on the project, and their interactions.
(b) Better contingency planning and selection of responses to those risks which do occur, and more flexible assessment of the appropriate mix of ways of dealing with risk impacts.
(c) Feedback into the design and planning process in terms of ways of preventing or avoiding risks.
(d) Feed forward into the construction and operation of the project

in terms of ways of mitigating the impacts of those risks which do arise, in the form of response selection and contingency planning.

(e) Following from these aspects, an overall reduction in project risk exposure.

(f) Sensitivity testing of the assumptions in the project development scenario.

(g) Documentation and integration of corporate knowledge which usually remains the preserve of individual minds.

(h) Insight, knowledge and confidence for better decision making and improved risk management.

Of these benefits, it is the reduction in project risk exposure which provides corporate management with the bottom-line justification for undertaking risk analysis studies. At the project management level, better insight is the critical aspect, leading to better decision making and better risk management.

14.2 How is risk analysis performed?

Although risk analysis is usually thought of as quantitative, it need not be about measuring risk, and it need not use probabilities. Risk analysis is concerned with understanding what might happen and what should happen. As an aid in developing and communicating this understanding, structured verbal models can be extremely useful, especially when the nature of the risks and the associated responses may cause confusion or misunderstanding if an agreed definition is not provided.

Effective and efficient performance of risk analysis involves a number of contributing elements. These include models, methods and computer software, as well as less tangible skill-related elements such as methodology design, specialist expertise and study team management.

14.2.1 Risk analysis models

This section dominates the chapter in terms of pages, not because it covers the most important aspects, but because these aspects must be understood for comments on the other aspects to make sense.

Risk analysis is concerned with uncertainty and its consequences. Mathematically, risk analysis models manipulate probabilities and probability distributions, in order to assess the combined impact of risks on the project. The exact manner in which this is done depends on the purpose of the analysis.

There is no single all-purpose risk analysis model. Some models are very simple, while others may be very complex, embodying not only uncertainty about events or activities, for example, but responses to that uncertainty and the consequences of the responses. In general it is advisable to start with simple models, and to make them more complex only if doing so seems cost effective.

The next subsection considers a very simple initial model and computation procedure, concerned with the joint duration distribution for two sequential activities with independent individual duration distributions. Under the next main heading seven subsections consider generalization of this example within the project planning context in five ways: computation, specification, operators, statistical dependence, and model structures. A final subsection considers alternative contexts.

14.2.2 Initial model and computation procedure

An offshore pipeline project might have 'design' specified as the first of a sequence of activities, $i = 1 \ldots n$. Design may be associated with duration D_1 months and probability $\Pr(D_1)$ as indicated in Table 14.1.

The uncertainty represented by Table 14.1 might arise from uncertainty associated with the availability of design staff which is not worth more detailed analysis, giving rise to the common slightly asymmetric distribution shape of Table 14.1.

It may be convenient to interpret this uncertainty in one of two quite different ways. One is a discrete distribution illustrated by the

Table 14.1 Tabular form for the design distribution D_1 months

D_1	$\Pr(D_1)$
5	0.2
6	0.5
7	0.3

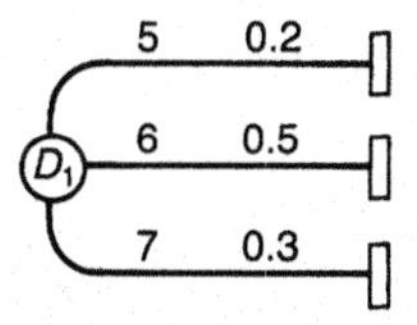

Fig. 14.1 Probability tree representation of the design distribution in Table 14.1

probability tree form of Figure 14.1. Alternative forms for the same interpretation which are worth noting but not particularly useful are conventional bar histogram representations of the density and cumulative probability distribution functions.

The other way to interpret this uncertainty is as a continuous distribution illustrated by the rectangular histogram probability density function form of Figure 14.2. An alternative form for the same interpretation which is often more convenient is the trapezoidal cumulative probability function form of Figure 14.3.

The equivalent histogram forms of Figures 14.2 and 14.3 ensure that expected values within the classes 4.5 to 5.5, 5.5 to 6.5 and 6.5 to 7.5 correspond to the classmarks 4, 5 and 6, providing consistency between the discrete and continuous distribution interpretations. However, smooth curves may be used as approximations to the shapes of Figures 14.2 and 14.3, or vice versa.

The design activity associated with Table 14.1 might be followed by a 'procurement' activity, associated with uncertainty defined by Table 14.2, assumed to be independent of that associated with Table 14.1.

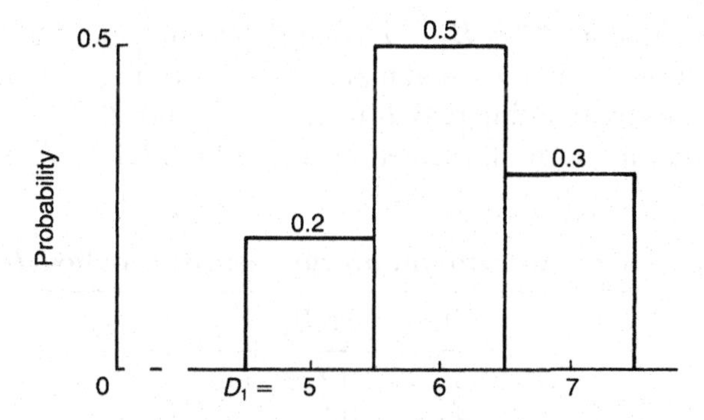

Fig. 14.2 Rectangular histogram probability density representation of Table 14.1

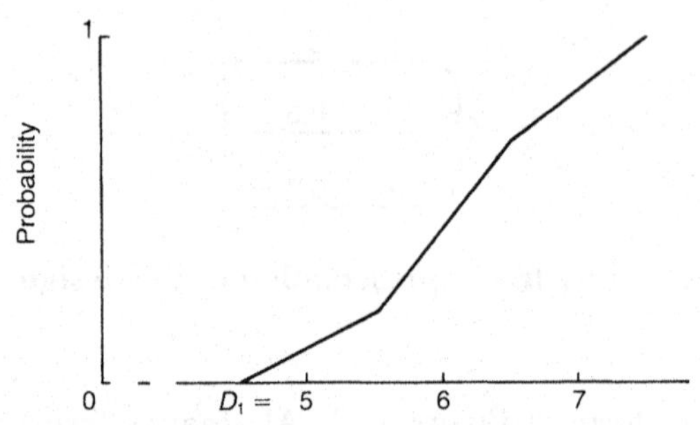

Fig. 14.3 Trapezoidal histogram cumulative distribution representation of Table 14.1

The uncertainty represented by Table 14.2 might be dominated by which of two possible suppliers the project will have to use for reasons of corporate policy beyond the control of the project team. Supplier A may be associated with a 3-month procurement process with virtual certainty, interpreting 3 months as 2.5 to 3.5 months. Supplier B may be associated with a 5-month procurement process with virtual certainty, interpreting 5 months as 4.5 to 5.5 months. The $\Pr(D_2) = 0.1$ associated with $D_2 = 4$ may reflect a very small chance of a tardy supplier A or a speedy supplier B.

Although it is important to interpret Table 14.2 as a continuous distribution when specifying probabilities, as was the case for Table 14.1, the discrete probability tree equivalent of Figure 14.1 is a useful basis for computation considerations. Adding this tree form to each of the branches of Figure 14.1 results in the two-level probability tree of Figure 14.4, D_1 defining 3 branches, each of which is associated with 3 D_2 branches.

Implied values of $D_1 = D_1 + D_2$ and $\Pr(D_1)$ may be obtained from the probability tree in the usual manner, as illustrated by Figure 14.4. However the common interval form of $\Pr(D_1)$ and $\Pr(D_2)$ allows the simplified tabular form illustrated by Table 14.3. Each possible com-

Table 14.2 Tabular form for the procurement distribution, D_2 months

D_2	$\Pr(D_2)$
3	0.2
4	0.1
5	0.7

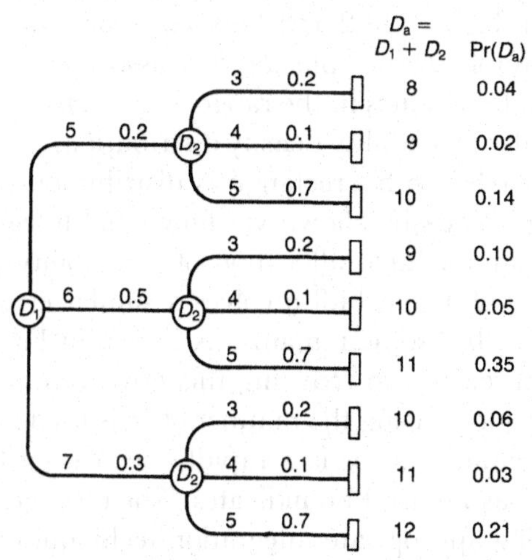

Fig. 14.4 Probability tree representation of the addition of independent durations

bination of D_1 and D_2 is considered, the joint probabilities producing the computation entries, and entries associated with the same D_a value are summed. In Markovian terms, if we do not need to remember D_1 or D_2 individually, we can choose to remember only their sum D_a.

14.3 Generalizations of the model

14.3.1 The computation process

If D_1 and D_2 are assumed to be discrete distributions of the form of Figure 14.1, the $\Pr(D_a)$ of Table 14.3 are free of computation error. However, if D_1 and D_2 are assumed to be continuous distributions of

Table 14.3 Calculation pattern for design plus procurement $D_a = D_1 + D_2$

D_a	*Computation*			$\Pr(D_a)$
8	0.2 × 0.2			0.04
9	0.2 × 0.1 +	0.5 × 0.2		0.12
10	0.2 × 0.7 +	0.5 × 0.1 +	0.3 × 0.2	0.25
11		0.5 × 0.7 +	0.3 × 0.1	0.38
12			0.3 × 0.7	0.21

the form used in Figures 14.2 and 14.3, an error is involved. It arises because, for example, $D_1 = 5$ plus $D_2 = 3$, associated with $D_a = 5 + 3 = 8$, should result in values in the range $4.5 + 2.5 = 7$ to $5.5 + 3.5 = 9$ with a triangular probability density distribution, rather than values in the range 7.5 to 8.5 with a rectangular distribution, as illustrated in Figure 14.5. This is easily shown via functional integration or finite difference techniques. Each allocation of probability in the computation procedure of Table 14.3 involves a similar error. Most of the error cancels out, but some remains, as shown in Figure 14.6.

A variety of approaches to reducing this error can be adopted. The simplest involves increasing the number of cells used for computation purposes: computation error is a rapidly decreasing function of the number of cells, as Figure 14.6 indicates. The most general approach involves adopting appropriate integration techniques with respect of each probability allocation. For example, the correct result in Figures 14.5 and 14.6 could be obtained by re-allocating the probability products in Table 14.3, with one-eighth to the cell below and one-eighth to

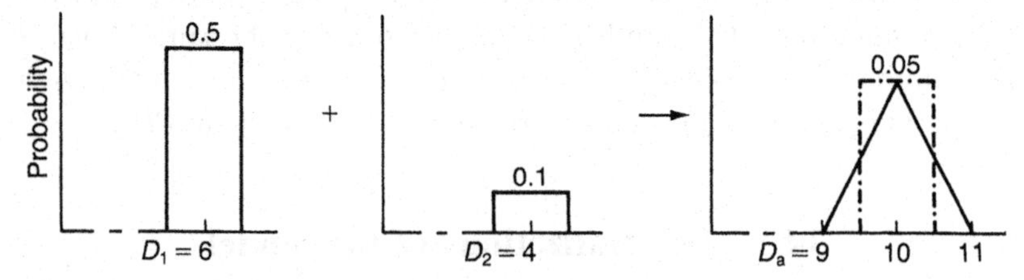

Fig. 14.5 Error associated with two cells

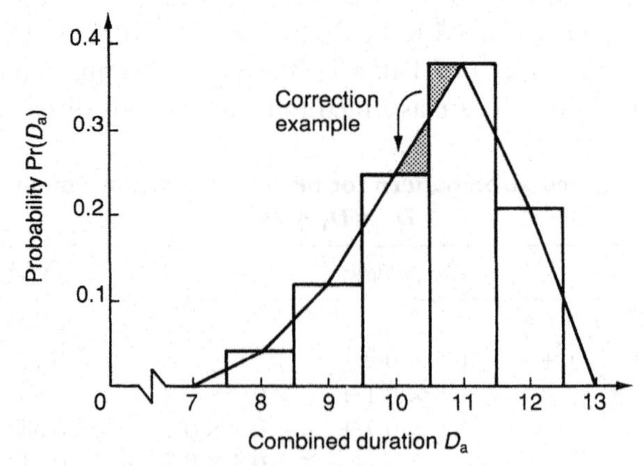

Fig. 14.6 Error associated with complete distribution

the cell above. This approach is complicated by the inconsistency associated with adding the results from earlier computations, which have trapezoidal rather than rectangular density forms. However, the bias induced by this inconsistency can be measured empirically and compensated for, within the integration process or via separate procedures associated with truncating undesirable distribution tails. In project planning, probability distribution tails involving probabilities less than 0.001 are of limited interest, and truncating them allows better representation of the rest of the distribution for the same level of computation effort.

Controlled interval (CI) procedures for the addition of independent distributions incorporate this general set of procedures, the common interval approach of Table 14.3 being a simple special case. In generalized CI computation, computation error is explicitly recognized, so it is controllable, and it can be reduced to zero at any significant level of measurement warranted in terms of a trade-off between computation effort and the value of computational precision.

14.3.2 The specification process

Error associated with the specification of probability distributions is often orders of magnitude more important than that induced by the crudest computation process.

In practice, if the common interval specification form of Tables 14.1 and 14.2 is used, more values are usually desirable, normally in the range 4 to 20, to represent the uncertainty in more detail. Any required level of precision of representation may be achieved, trading off specification effort and the benefits of that effort. Appropriate computer software can also generate a detailed specification, normally using 30 to 50 cells, from any convenient and appropriate distribution function and associated parameters. For example, Beta distributions can be used, defined by the minimum, maximum and most likely values (see Section 2.3.5).

The common interval aspect of the controlled interval form of Tables 14.1 and 14.2 could be relaxed. For example, given a generalized CI computation procedure, wider intervals could be used to represent the smaller probabilities, in order to reduce computer storage and computation requirements without sacrificing precision of representation of the distribution tails. The desirability of this generalization depends upon the value of precise representation of the tails. It has been deemed not worth the complications it causes in project planning.

Even the controlled aspect of unequal intervals could be relaxed in the sense that it could be ignored for specification purposes, but this may complicate computing and interpretation without significant benefits. Exceptions are truly discrete alternatives constrained to particular values when it may be necessary to abandon the continuous distribution interpretation of Figures 14.2 and 14.3, but even these exceptions can be considered within the overall controlled interval umbrella.

The generality of the CI approach from a specification perspective is important in terms of controlling specification error, allowing its reduction to a minimal level. However, this clearly does not guarantee freedom from specification error, unless a subjective probability perspective which simply ignores the existence of such error is adopted, an approach not recommended here. More specifically, the $\Pr(D_1)$ values of Figure 14.1 are associated with a first order probability model for D_1. In a second order model the $\Pr(D_1)$ may have a range of values with associated $\Pr(\Pr D_1))$, and so on, as indicated by Raiffa (1970). It may rarely prove worth using higher-order models explicitly. However, the optimistic bias implicit in assuming that the $\Pr(D_1)$ values are deterministic is always worth bearing in mind.

14.3.3 The operator set

In the context of a generalized CI addition procedure, further generalization to subtraction, multiplication, division, greatest, least, and other similar operators is straightforward, in principle if not in practice. It is given no more space here, apart from one comment related to the specification of Table 14.2. The bimodal nature of this specification makes obvious the two separate underlying cases. Often two or more separate underlying cases are not immediately obvious. Whether or not such separate cases result in multimodal distributions, they are often worth separate identification and estimation, to reduce specification error. Combining these separate cases using associated scenario probabilities requires another useful operator in practical computer software: 'weight distributions 1 . . . n by probabilities $p_1 \ldots p_n$'.

14.3.4 Statistical dependence

The procurement activity associated with Table 14.2 might be fol-

Table 14.4 Conditional specification of the delivery distribution, D_3 months

Procurement D_2	*Delivery* D_3	$\Pr(D_3)$
3	3	0.8
	4	0.2
4	3	0.7
	4	0.1
	5	0.1
	6	0.1
5	4	0.3
	5	0.4
	6	0.3

lowed by a 'delivery' activity associated with uncertainty in a conditional from defined by Table 14.4.

The uncertainty represented by Table 14.4 is explicitly dependent upon procurement duration. This might reflect the past delivery performance of the two different suppliers associated with Table 14.2.

A two-level probability tree defining $D_b = D_2 + D_3$ and $\Pr(D_b)$ is illustrated by Figure 14.7. A simplified tabular computation equivalent to Table 14.3 is provided by Table 14.5.

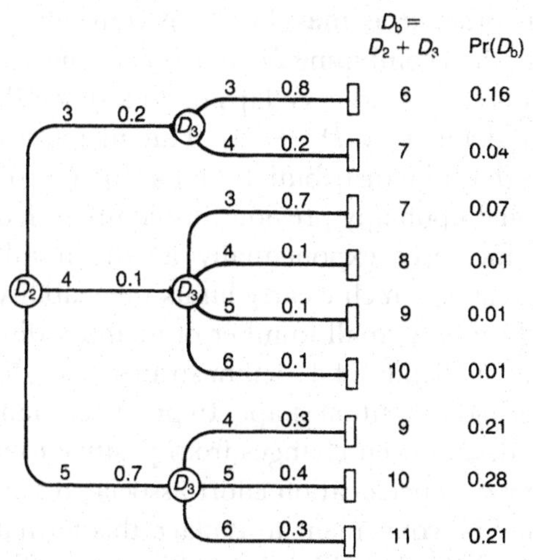

Fig. 14.7 Probability tree representation of the addition of dependent durations

Table 14.5 Calculation pattern for procurement plus delivery, $D_b = D_2 + D_3$

D_b	*Computation*			$Pr(D_b)$
6	0.2 × 0.8			0.16
7	0.2 × 0.2 +	0.1 × 0.7		0.11
8		0.1 × 0.1		0.01
9		0.1 × 0.1 +	0.7 × 0.3	0.22
10		0.1 × 0.1 +	0.7 × 0.4	0.29
11			0.7 × 0.3	0.21

Comparison of Figures 14.4 and 14.7 provides a useful demonstration of what independence means: second-level branch values and probabilities must be identical for all first-level branches.

Comparison of Tables 14.3 and 14.5 clearly indicates that dependence in the Table 14.4 form involves no significant additional computational effort, but it does involve additional specification effort. In practice, when larger numbers of cells might be associated with D_2 directly via a CI specification, or indirectly via a Beta distribution specification for example, it may be desirable to use computer software which interpolates $Pr(D_3|D_2)$ from a limited set of D_2 specifications, to reduce specification effort. This can increase computation effort significantly.

Having defined $Pr(D_b)$ as just discussed, $Pr(D_c)$ where $D_c = D_b + D_1$ could be computed using the independent addition operation of Table 14.3. In practice it may be more convenient to maintain a 'memory' of D_2 when combining D_1 and D_2, storing a matrix of $Pr(D_a)$ values by D_2 source instead of collapsing to a single $Pr(D_a)$ vector, so that D_c can be defined by $D_a + D_3$. This memory concept can be generalized, so that any distribution in a sequence to be operated on can be conditional upon any previous distribution. However, computational effort increases exponentially as the number of memory dimensions increases, which clearly limits the viability of the computational procedure to a small number of memory dimensions.

Examination of Table 14.4 demonstrates that dependence may involve changes in distribution shape. In practice, changes in skewness may be involved, and even changes from positive to negative dependence. However the specification effort associated with Table 14.4 is such that it may be convenient to replace this format by something simpler if simple uniform positive dependence is involved. This may be done using appropriate computer software by providing a percent dependence operator, which operates on unconditional specifications of the component distributions like Table 14.1 and 14.2 The software first combines the distributions assuming independence, corresponding to zero percent dependence. It combines them again assuming all percentile values of the component distributions coincide, corresponding to one hundred percent dependence. It then interpolates linearly at the specified percent dependence level. To familiarize a user with what percent dependence means, the results of the more general procedure of Tables 14.4 and 14.5 can be illustrated graphically in relation to associated zero and one hundred percent dependence bounds.

In practice it might be better to model 'procurement' plus 'delivery' for supplier A, do the same for supplier B separately, then collapse the associated tree. This avoids the dependence of Table 14.4. In general, if statistical dependence can be avoided by transforming it to a structured form it is worth doing so. However, it is not always possible to avoid dependence in this way. Experience suggests it is often present, despite best efforts to avoid it. If it is simply assumed away, very substantial errors result. Specification errors associated with unconditional distributions like those for D_1 and D_2 have negligible effects upon results compared to incorrect dependence specifications. Hence it is very important to face the statistical dependence issue with as flexible a model and as much determination as possible, after transforming statistical dependence into a structural form of dependence whenever possible.

14.3.5 Model structure: Markov processes through time

The separate treatment of suppliers A and B for the procurement and delivery sequence just discussed could be interpreted as a form of model structure generalization. Five further forms of model structure generalization are of interest in the present project planning context, the first of which is considered in this subsection.

Following the 'delivery' activity, 220 km of 32 inch steel pipe may require coating with six inches of concrete and anticorrosive materials. Estimation of an associated duration $\Pr(D_4)$ might be approached indirectly, via estimation of a coating rate R in km/month and associated probabilities $\Pr(R)$ indicated in Table 14.6.

The uncertainty represented by Table 14.6 might reflect productivity variations for a known coating yard operating on a single shift basis without any industrial disputes or other significant difficul-

Table 14.6 Coating rate distribution, R km/month

R	$\Pr(R)$
30	0.2
35	0.5
40	0.3

ties. It might be interpreted as a transition distribution for a simple Markov process (Howard, 1971) and used to create a sequence of corresponding state distributions, illustrated by Table 14.7. S_t is the state variable defining the stock in km of coated pipe available at the end of month t, $t = 0, 1, 2 \ldots$. The stock at the end of month t equates to the stock at the beginning of month $t + 1$, stock at the end of month $t = 0$ defining the initial conditions with no coated pipe. The $\Pr(S_t|t = 1)$ distribution is obtained by 'adding' $\Pr(R)$ to $\Pr(S_t|t = 0)$. $\Pr(R)$ is then 'added' to $\Pr(S_t|t = 1)$ to obtain $\Pr(S_t|t = 2)$, and so on, using the independent addition procedure of Table 14.3.

Table 14.7 Stock of coated pipe at the end of months 1, 2, 3 . . .

t	S_t	*Computation*			$\Pr(S_t)$
0	0				1
1	30	0.2 × 1			0.2
	35		0.5 × 1		0.5
	40			0.3 × 1	0.3
2	60	0.2 × 0.2			0.04
	65	0.2 × 0.5 +	0.5 × 0.2		0.20
	70	0.2 × 0.3 +	0.5 × 0.5 +	0.3 × 0.2	0.37
	75		0.5 × 0.3 +	0.3 × 0.5	0.30
	80			0.3 × 0.3	0.09
3	90	0.2 × 0.04			0.008
	95	0.2 × 0.20 +	0.5 × 0.40		0.060
	100	0.2 × 0.37 +	0.5 × 0.20 +	0.3 × 0.04	0.186
	105	0.2 × 0.30 +	0.5 × 0.37 +	0.3 × 0.20	0.305
	110	0.2 × 0.09 +	0.5 × 0.30 +	0.3 × 0.37	0.279
	115		0.5 × 0.09 +	0.3 × 0.30	0.135
	120			0.3 × 0.09	0.027
etc.					

In practice it would be convenient to generalize Table 14.6 to Table 14.8, so that when the required 220 km of pipe are coated, coating ceases, $S_t = 220$ becoming an absorbing state. The result is a semi-Markov process with a state dependent transition distribution. It would provide results identical to those of Table 14.7 until $t = 5$. After $t = 5$, $S_t = 220$ will limit the stock of coated pipe, with $\Pr(S_t = 220) = 1$ for $t = 8$ terminating the process. The $\Pr(S_t = 220)$ define the $\Pr(D_4)$.

Several important practical implications of this alternative approach to modelling an activity duration deserve note.

First, direct estimation of $\Pr(D_4)$ implies implicit intuitive use of Tables 14.7 and 14.8. Whenever a Markov process like this is a reasonably valid representation of reality, it is easier to estimate the equivalent of $\Pr(D_4)$ via $\Pr(R)$, and greater estimation precision can be expected.

Second the information provided is much richer, and very useful. Knowledge of the stock distribution for all time periods allows consideration of overlap between coating and the following pipelaying activity.

Third, the time period by time period structure is often useful as a basis for the further generalizations of model structure to be considered in the next two sections, although each can be employed without a Markov structure.

Table 14.8 Conditional coating rate distribution

S_t	R	$\Pr(R)$
< 220	30	0.2
	35	0.5
	40	0.3
220	0	1

14.3.6 Model structure generalization: responses

The $\Pr(D_4)$ distribution obtained above could be added to $\Pr(D_c)$ to obtain $\Pr(D_d)$, $D_d = D_1 + D_2 + D_3 + D_4$. In practice it may be more convenient to incorporate $\Pr(D_c)$ into the initial conditions of the Markov process. This would allow the consideration of a response to delays associated with any previous activity within the Markov pro-

Time 14.9 Coating rate with second shift option

t		S_t	R	$\Pr(R)$
< 14			30	0.2
or			35	0.5
$\geqslant 14$	and	$\geqslant 100$	40	0.3
$\geqslant 14$	and	< 100	60	0.1
			65	0.2
			70	0.3
			75	0.2
			80	0.2
$\geqslant 14$	and	220	0	1

cess. For example, a second shift might be used for coating, if by time period 14 measured from the planned start of the project less than 100 km of pipe are coated. This would require replacing the state-dependent transition distribution of Table 14.8 with state and time-dependent transition distributions illustrated by Table 14.9. A variety of decision rules for responses of this kind could be considered. As responses would be used in practice, incorporating them provides a much more realistic picture of duration uncertainty, modelling negative statistical dependence in a causal form, but this is simply a spinoff of a much richer and more useful planning process. If, for example, such modelling suggested that using a second shift was extremely likely, planning for it in advance and undertaking contracts with a second shift provided for might save large amounts of time and money.

14.3.7 Model structure generalization: sources of risk

When considering the 'pipelaying' activity following pipe coating, it may be useful to start with a focus on a single source of risk, weather variations. Table 14.10 portrays the kind of distributions which might be expected for L, the number of laydays in a month, defined in relation to a maximum wave height capacity for the assumed laybarge, using wave height data for the location of interest.

When weather effects and associated decision rules are understood, further sources of risk might be added, one at a time, like the laybarge

Table 14.10 Distributions of laydays available each month

L	*March*	*April*	*May*	*June*	*July*	*etc.*
0	0.3	0.2	0.1			
5	0.4	0.3	0.2	0.1		
10	0.2	0.3	0.3	0.1		
15	0.1	0.2	0.2	0.2	0.1	
20			0.1	0.3	0.2	
25			0.1	0.2	0.4	
30				0.1	0.3	

not being available when planned, the laybarge not working as fast as planned, and so on.

Considering sources of risk in this manner involves additional effort, but it provides a number of advantages.

First, different sources of information can be used for different sources of risk: wave data for weather, reliability and performance data for equipment, specialist engineering judgements for technical problems, procurement expertise for judgements involving markets, and so on.

Second, a portrait of the relative importance of each risk source is provided, illustrated by Figure 14.8 for an offshore platform jacket. Such pictures are extremely useful for risk analysts and other users of their analysis, especially if different sources of risk have been estimated with different degrees of confidence.

Third, it becomes clear which sources of risk have been considered and which have not. Generally a number of sources of risk will be identified which are not incorporated in the probabilistic analysis because they represent a change in scope and are best treated as conditions. For example, a management decision to change the route of a pipeline might be likely because of a new field discovery, but probabilistic treatment as part of the plan would not be helpful.

Fourth, it becomes possible to distinguish between responses that are specific to particular sources of risk and general responses that apply to large sets of risks. For example, specific responses to equipment failures are repairs or replacements, while additional equipment could be used to make up time lost because of earlier equipment failures, bad weather, late starts, and so on.

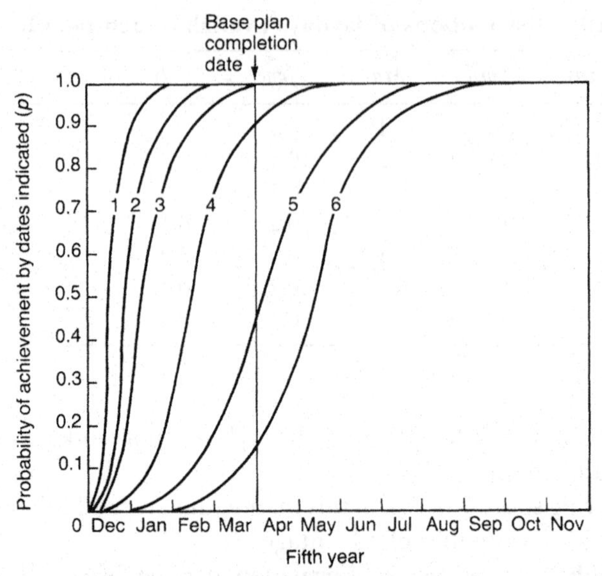

Fig. 14.8 Cumulative probability distribution with six risk service components. Components are: 1, yard not available/mobilization problems; 2, construction problems/adverse weather; 3, late delivery of subcontracted nodes; 4, material delivery delays; 5, industrial disputes; 6, delayed award of contract

14.3.8 Alternative contexts

Alternative contexts usually involve quite different terminology. For example, in a reliability context (Chapman, Cooper and Cammaert, 1984) 'sources of risk' become 'sources of outage', 'responses' may be limited to 'repair' and duration may become a period of outage translated into 'lost production'.

Alternative contexts usually involve somewhat different model structures. For example, an assessment of alternative ways to cross a river with a large gas pipeline (Chapman, Cooper, Debelius and Pecora, 1985) required simultaneous treatment of multiple-criterion measures of the consequences of damage scenarios, in terms of lost gas production, lost production of oil due to loss of an oil line on the same bridge, dollar cost of damage to the bridge and pipe lines, amount of spilled oil, and so on. A further special characteristic of the model structure was the probabilistic transformation of a wide set of initiating event levels (like earthquake ground acceleration) into a limited set of damage scenarios, akin to the process of Table 14.3 in some respects, but without a single direct measure of damage comparable to duration.

However, the modelling choices are much the same, a common probabilistic algebra is involved, and some very different circumstances can be catered for with any given model. For example, the pipe coating activity model associated with Table 14.9 is a production process and inventory model which could be linked to a probabilistic demand model using a structure like that of the pipe laying model associated with Table 14.10. This model could be used to study integrated pricing, advertising, production and inventory policies for a coal-mining operation, a timber development, a manufacturer of sailboats, or a producer of children's toys. Sources of variation in production and demand could be considered individually, at various levels of detail, or not at all. Responses to competitive action or industrial disputes or other sources of variation in demand or production could be linked to specific and general responses, or just the latter. Marketing children's toys and laying offshore pipelines involve very different considerations, but common model structures may prove useful.

Although very similar model structures may be involved in very different contexts, an important feature of the risk engineering model set is their generality and flexibility.

No matter how general and flexible a model set, there clearly are limitations. Solutions looking for problems are always a source of risk. The sailboat manufacturer just considered might be better with a dynamic programming model, or a linear programming model, or some other quite different model, if indeed he needs a model at all.

14.4 Risk analysis methods

Risk analysis methods must be designed to suit the model and the circumstances in which it is used. There is no single all-purpose method for risk analysis. Families of related methods have been developed, involving similar concepts and characteristics. They follow a systematic series of steps, but the steps are related to the specific risk analysis context.

When it was tentatively agreed that models like that associated with pipe laying in the last section would be used for offshore North Sea projects, a tentative method outline was developed, tested, modified

Activities associated with platform structure and topside facilities:

- conceptual design
- steel procurement and delivery
- procurement and delivery of long lead time items
- other material procurement and delivery
- detailed design
- fabrication, erection, onshore testing
- load out, tow to site, installation
- hook-up and commissioning

Activities associated with subsea pipelines:

- design and awarding contracts
- surveys for route selection
- steel procurement and delivery
- other material procurement and delivery
- pipe coating and delivery to onshore base
- pipe laying
- pipe trenching or buying
- pipeline tie-ins to platforms and/or shore
- pipeline tests

Fig. 14.9 Activities for an offshore project

as appropriate and documented during the period 1976–7. This method has since been used for about 20 major projects, with some further minor modifications. It is described in more detail elsewhere (Chapman, 1979). A brief outline will suffice here.

The first phase is qualitative. The project is divided into a very limited number of key activities (Figure 14.9); each activity is described in detail (Figure 14.10); a precedence diagram and an associated time-scaled bar chart are constructed to show precedence and time relationships; all significant sources of risk associated with each activity are listed and described (Figure 14.11); all feasible responses to each source of risk are then listed and described (Figure 14.12). Where appropriate, secondary sources of risk associated with responses are similarly considered. The overall problem structure is then studied, distinguishing responses which are specific to single sources of risk from responses which apply to sets of risks, ordering responses, and studying associated decision rules.
This structure is then summarized in diagrammatic form (Figure 14.13). A distinction is made between minor sources of risk, major sources of risk whose non-realization will be a condition of the quantitative analysis, and major sources of risk to be modelled probabilistically. All project staff with relevant information are involved at appropriate stages, and the problem perception captured by this documentation is agreed with management before the qualitative phase ends.

Number	Title/Details
8	*Fabrication, erection and tests* This activity covers fabrication of the platform structure including offsite fabrication of nodes, erection of the structure, installation of all associated equipment cathodic protection, flooding, grouting, internal leg inspection, strain gauges, environmental monitoring and 'J' tubes risers, painting of the structure, inspection of welds and any resulting repairs and testing all equipment prior to float and tow-out. The award of the fabrication contract is planned for the beginning of November 1986. The subcontract(s) for rolling Jacket tubulars, rolling and fabricating nodes/ piles and assembly of the structure should be in accordance with the procedures summarized in Activity No.5 Details List. Fabrication of the jacket assembly has a planned duration of 29 months available and should be complete by the end of March 1989. The principal keydate associated with this activity is 1 April 1989 planned float-out in preparation for the tow to site.
Prepared by	Date

Fig. 14.10 Activity details list

Number	Title/Description and notes
	Fabrication, erection and tests
8.1	Yard availability: Contracted yard may not be available at the required fabrication start date (e.g. because of modifications necessary to accommodate the platform or work on a previous contract).
8.2	Mobilization: Yard may have been idle for some time leading to equipment and manpower mobilization problems.
8.3	Productivity: Contractor's ability to proceed at expected rate to the required specifications (speed and effectiveness of work and minor interruptions).
8.4	Industrial disputes: Significant interruptions not covered under productivity (8.3).
8.5	Equipment breakdowns: Significant breakdowns under productivity (8.3).
Prepared by	Date

Fig. 14.11 Risk/response list

Number	Title/Description and notes	Secondary risks
8.0	*Fabrication, erection and tests*	
8.1	*Yard availability:*	
8.1.1	Mobilize: Mobilize as much equipment/personnel as possible (preferred if delay if likely to be short).	No
8.1.2	Alternative yard: Search for an alternative yard if delay is likely to be very long – this will depend on activities in offshore structure fabrication yards.	Yes
	N.B. The costs involved in implementing the above and/or compensation for delays could be (partially) recovered from contractor, if appropriate contractual clauses are incorporated.	
8.1.3	Accept delay: In awaiting availability of contracted yard.	No
Prepared by	Date	

Fig. 14.12 Risk list

The second phase is quantitative. The probabilities that risks will occur and the probabilities of various consequences given assumed decision rules for responses are assessed, considering structural and statistical dependences. The cumulative effect of successive sources of risk is then computed, to provide a portrait of the relative role of each source of risk within an activity (Figure 14.8), and the relative contribution of each activity to total time risks (Figure 14.14). At intermediate stages in this accumulation process, feedback occurs. Feedback continues until everyone is satisfied with the assumed probabilities and decision rules in relation to the conditions identified earlier.

When the project team is satisfied with the schedule risk analysis, associated cost risk is considered. For example, activity duration distributions are combined with unit of resource per unit time distributions, and probabilistic treatment of exchange rates and inflation rates are used if appropriate. The relative impact of each source of cost risk is portrayed in the format of Figures 14.8 and 14.14. Only when management is satisfied that they have an understandable and justifiable case for all major decisions inherent in this overall result

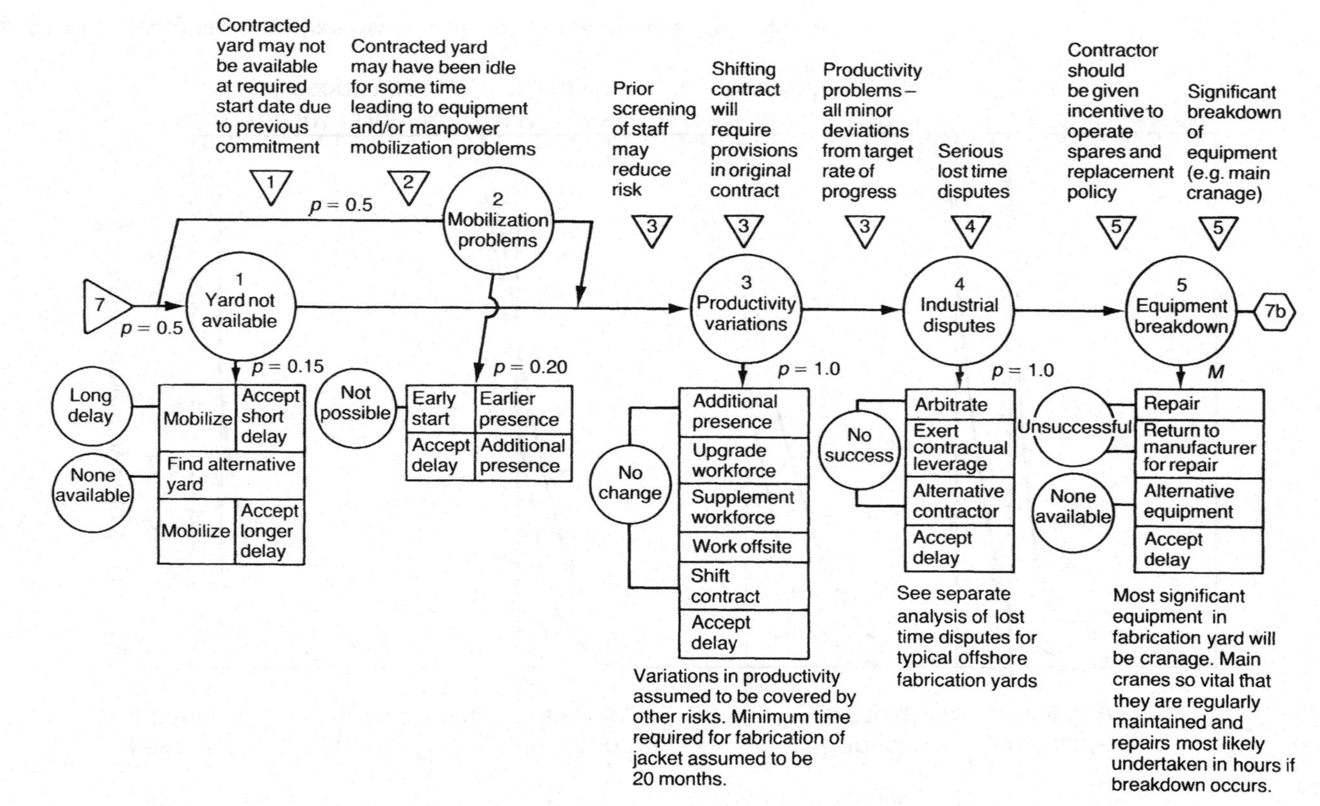

Fig. 14.13 Risk/response diagram for an offshore project

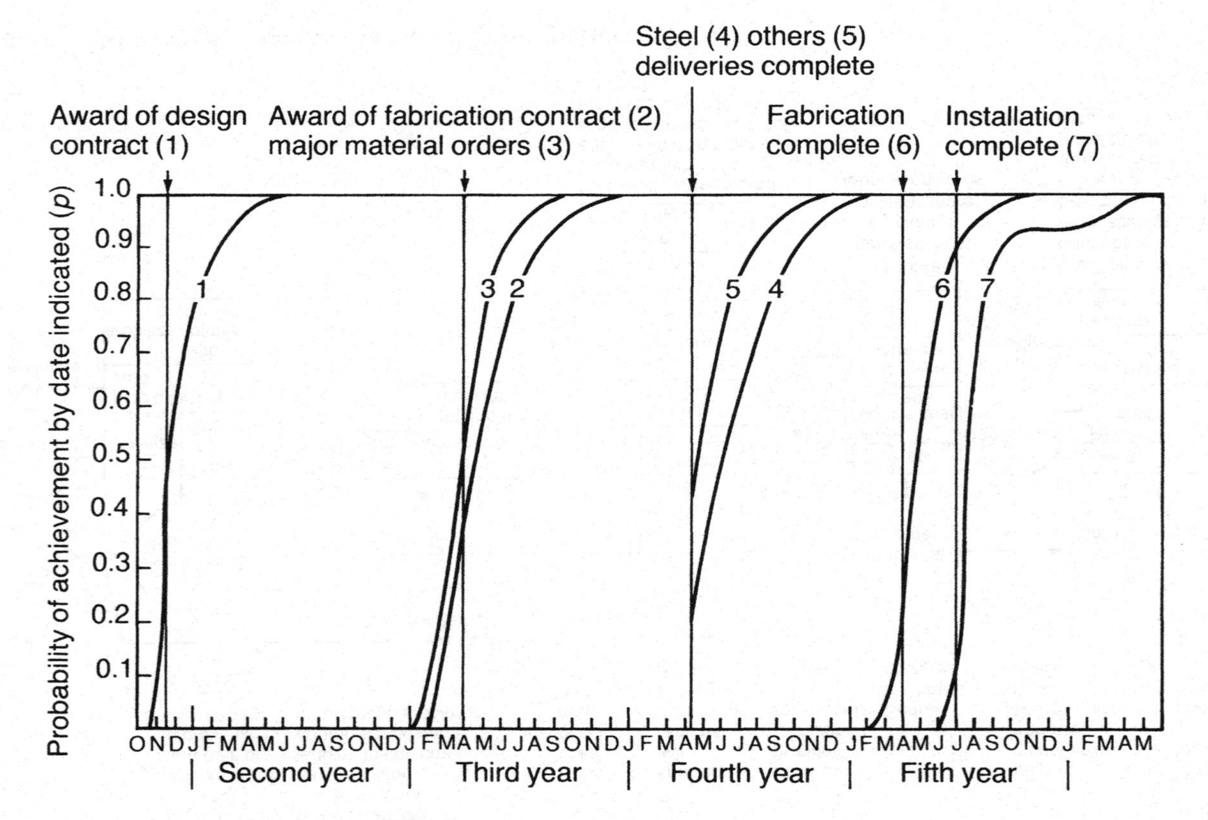

Fig. 14.14 Cumulative probability distributions for an activity sequence

are the associated plans (including contingency plans) and budgets released.

This method has provided an experience base for the design of a number of other methods explicitly recognising the model, method and method design distinctions. Some previous OR experience might be associated with these distinctions in retrospect, but explicit recognition is important in the authors' view.

14.4.1 Risk analysis computer software

Some standardized risk analysis software packages are available. However, appropriate computer software must reflect the choice of model and method, and there is no one best approach for all forms of risk analysis.

Families of related programs have been developed for associated families of models and methods. They are sophisticated, interactive program suites which provide a means of entering information in a wide range of different formats and synthesizing it in any required manner. They are designed to take advantage of the general characteristics and terminology of the application area being addressed.

14.4.2 Risk analysis methodology design

Methodology design involves choosing or developing an appropriate model/method/software combination for a particular kind of risk analysis in a particular context. The time and money available to perform the analysis, and the expected future use, are obviously important considerations, as well as the immediate task. Methodology design is a process which is necessarily dependent upon experience and intuition.

14.3.4 Risk analysis and specialist expertise

Risk analysis models, methods and software provide valuable tools for project planning and design, but obtaining the right answers still depends upon specialist expertise. Judgements must be made, in some cases based upon hard data, in others based on sound conventional guidelines. In other cases, creative innovation and well-schooled intuition based upon a wide range of relevant experience must be used. Expertise involving an effective and efficient blend of all these aspects is not made less important by adopting risk analysis methodology; it is simply made use of more effectively.

Risk analysis requires may forms of expertise. As well as detailed engineering and geological skills, skills in economics, finance, environmental issues, contractual issues, etc., are involved in most major underground projects. What is not always appreciated is that they may directly affect the way the design and planning is done, and they should be properly considered in a timely manner. Risk analysis helps to provide this integration.

14.5 Summary

Successful risk analysis requires:

(a) A flexible and general set of verbal, graphical and mathematical models, supported by sophisticated interactive computer software.
(b) A family of related methods, designed to suit the models, which link the models and the circumstances in which they are to be used.
(c) A wide range of relevant expertise and specialist skills.
(d) The experience and leadership to design and integrate models, methods and software for specific risk analysis tasks, to organize and manage risk analysis study teams, and to execute successfully large risk analyses for major projects.

The risks in large projects may have consequences which involve physical loss or damage, commercial loss, environmental damage, and third-party liability. The risk analysis methods discussed here identify the sources of risks, possible responses to them, and their consequences, including any secondary risks which arise as a result of the response mode chosen. This explicit identification of risk sources permits a reasoned assessment of who should appropriately bear each risk if it arises — the insurer, the owner or the contractors — and how much it might be worth to do so. The detailed examination of the project risk structure usually also leads to better contingency planning, and hence to an overall reduction in risk, to the benefit of all parties. The approach has been used successfully in a variety of applications, including seismic risk assessment, optimization and evaluation of onshore and offshore hydrocarbon developments, contract examination for a thermal power station, underground power storage, choice of river crossing methods for a gas pipeline, and reliability analysis of an Arctic LNG facility.

14.6 References and further reading

Chapman, C. B., 'Large engineering project risk analysis', *I.E.E.E. Transactions on Engineering Management*, EM-**26**(3), 78–86 (1979).

Chapman, C. B. and Cooper, D. F., 'Risk engineering: basic controlled interval and memory models', *J. Opl. Res. Soc.*, **34**(1), 51–60 1983(a)).

Chapman, C. B. and Cooper, D. F., 'Risk analysis: testing some prejudices', *European J. Opl. Res.*, **14**(3), 238–247 (1983(b)).

Chapman, C. B. and Cooper, D. F., 'Risk analysis' in *Further Developments in Operational Research* (G. K. Rand and R. W. Eglese, editors), Pergamon, Oxford (1985).

Chapman, C. B., Cooper, D. F. and Cammaert, A. B., 'Model and situation specific OR methods: risk engineering reliability analysis of an LNG facility', *J. Opl. Res. Soc.*, **35**(1) 27–35 (1984).

Chapman, C. B., Cooper, D. F., Debelius, C. A. and Pecora, A. G., 'Problem-solving methodology design on the run', *J. Opl. Res. Soc.*, **36**(9) 769–778 (1985).

Chapman, C. B., Phillips, E. D., Cooper, D. F. and Lightfoot, L., 'Selecting an approach to project planning', *International Journal of Project Management*, **3**, 19–26 (1985).

Clark, P. and Chapman, C. B., 'The development of computer software for risk analysis: a decision support system case study', *European J. Opl. Res.* **29**(3) 252–261 (1987).

Cooper, D. F. and Chapman, C. B., 'An overview of risk analysis for underground projects', *Underground Space*, **9**, 1, 35–40 (1985).

Cooper, D. F. and Chapman, C. B., *Risk Analysis for Large Projects: Models, Methods and Cases*, John Wiley, Chichester (1987).

Cooper, D. F., MacDonald, D. H. and Chapman, C. B., 'Risk analysis of a construction cost estimate', *Études et Dossiers Nr. 84*, Geneva Association for Risk and Insurance, 89–130 (1984).

Howard, R. A., *Dynamic Probability Systems: Vol. 1 Markov Models, Vol. 2 Semi-Markov and Decision Processes*, John Wiley, New York, (1971).

Raiffa, H., *Decision Analysis: Introductory Lectures on Choices under Uncertainty*, Addison-Wesley, Reading, Massachusetts (1970).

U. S. Nuclear Regulatory Commission Reactor Safety Study, *An Assessment of Accident Risks in US Commercial Nuclear Power Plants*, WASH-1400 NUREG 75/014 (1975).

14.7 Problems and practical exercises

1 You have been invited to attend a job interview in another city, and you would prefer to travel by public transport. It is important that you are not late for your appointment. Perform a simple risk analysis of your travel plans:

(a) Specify in detail the transport modes you will use (e.g. walk from home to bus stop, take bus to station, . . .).
(b) For each travel mode, try to identify the circumstances which might cause you to be delayed (the risks), and estimate the consequences of each risk if it arises.
(c) For each risk, write down the different ways you might cope with it (the responses), and select the most appropriate one. Are there any major secondary risks associated with your chosen response?
(d) Would you need to quantify any parts of your analysis? If so, say what parts, why quantification would be helpful, and how you might do it.
(e) Extend this approach to the problem of selecting the best way to travel to your interview.

2 A small construction job requires carpenters and timber, with uncertain costs as follows:

Skilled labour	about £450
Dressed timber	about £350
Undressed timber	about £300.

(a) Making some reasonable assumptions, specify the uncertainty about the costs in the form of histograms.
(b) Assuming the costs are independent, use a probability tree (cf. Figure 14.3) to calculate the distribution of total cost.
(c) The costs of dressed and undressed timber are likely to be positively related (e.g. if the cost of dressed timber is high, the cost of undressed timber is likely to be high as well). Modify the specification in part (a) to show positive dependence. Again use a probability tree to recalculate the distribution of total cost.
(d) Plot the distributions of total cost from parts (b) and (c) as histograms on the same graph. The distribution in part (b) is narrower than that from part (c). In practice, the assumption of

independence, when there should be positive relationships between items, leads to narrower distributions, and hence to a major under-estimation of risk.

3 Discuss the forms of risk analysis that might be appropriate in each of the following areas, commenting on the kinds of risks and responses involved, the relative levels of detail into which the task and the risks might be decomposed, and the need for quantification of uncertainty, the criteria for decision, and any other major consequences or results from the risk analysis process itself:

(a) Seismic risks for a nuclear reprocessing plant.
(b) Profitability for a new production line.
(c) Contract conditions for a turnkey project in another country.
(d) Design parameters for a hydroelectric power generation development.
(e) Capital cost for a new factory.
(f) Operating costs for a chemical plant.
(g) Laying pipelines to an offshore gas production platform.
(h) Insurance for a telecommunications satellite.
(i) Buying shares in a company about to begin mining gold.
(j) Lending money to a fishing co-operative to purchase a new trawler.

4 Constructing an office block for sale on completion has been associated with an uncertain cost of £700 000 and an uncertain sale price of £900 000. Most of the uncertainty associated with both arises because it is not known what rates of inflation will apply over the three-year life of the project. Making some reasonable assumptions, specify the uncertainty associated with the cost and revenue in the form of histograms. Use a probability tree to compute the gross profit (revenue less cost) distribution, and comment on the form and effect of the assumed dependence.

5 Constructing a road in an underdeveloped country has been associated with uncertain costs as follows:

Labour	about 10 (million pounds)
Equipment	about 15 (million pounds)
Materials	about 5 (million pounds).

Most of the uncertainly associated with labour and equipment arises because it is not known whether a labour-intensive or an equipment-intensive approach will be adopted, and most of the residual uncertainty associated with these items arises because it is not known whether or not another major project will start at the same time in the region, causing a shortage of appropriate labour and equipment. Making some reasonable assumptions, specify the uncertainty about the costs in the form of histograms. Use a probability tree to compute the distribution of total cost, and comment on the form and effect of the assumed dependence.

6 A friend cannot decide whether to buy a small two-year-old saloon car from his sister, or a six-year-old sports car from a local dealer. Perform a simple risk analysis which you could use to give him guidance on the key issues involved.

Appendix 1
Solutions to problems and practical examples

Chapter 2

1. (a) $\frac{1}{4}$ (b) $\frac{5}{24}$ (c) $\frac{13}{24}, \frac{16}{25}$.
2. Solution of $1 - (1 - \pi^3)^2 = 0.99$, $\pi = 0.965$.
3. (a) $\frac{2}{3}$ (b) $\frac{3}{4}$.
4. (a) 0.75 (b) 0.282.
5. (a) 0.08 (b) 0.64.
6. (a) 36.8% (b) 60.6% (increases, but does not double).
7. (a) 0.577
   (b) Overbook to a level where there is a high probability of 300 or less claimed seats. Balance costs of empty seats against compensation to overbooked passengers who turn up.
8. (a) 0.331 (b) 0.071.
9. Sample average = 117.0
   Sample standard deviation = 46.7
   Median = $\frac{1}{2}$ (118 + 119) = 118.5
10. (a) 0.02275 (b) 0.0668 (c) 0.8351.
11. $\sigma = 8/1.175 = 6.81$ (a) 1011.2 (b) 0.098.
12. $f(x|1) = 0.252xe^{-x/5}$ $15 \leq x \leq 25$ mean = 18.81.

Chapter 3

1.

	XYZ	XZY	YXZ	YZX	ZXY	ZYX
XYZ	18	14	10	6	14	14
YXZ	12	8	18	14	14	14

reducing by dominance arguments to

	XZY	YZX
XYZ	14	6
YXZ	8	14

Player I plays *XYZ* with probability $\frac{3}{7}$, *YXZ* with probability $\frac{4}{7}$
Player II plays *XYZ* with probability $\frac{4}{7}$, *YZX* with probability $\frac{3}{7}$
Value of game is $\frac{74}{7}$.

2. No. He would always visit customer Z on day 2.
3. Denote strategies by pairs of initial letters of salesman's surname, the first being allocated to the £10 000 contract.

		Company B	
		(F, H)	(H, F)
Company A	(R, L)	15	10
	(L, R)	5	15
		(units of £000)	

Optimum strategy for company A is (R, L) with probability $\frac{2}{3}$, (L, R) with probability $\frac{1}{3}$; for company B (F, H) with probability $\frac{1}{3}$, (H, F) with probability $\frac{2}{3}$.

4. Pay-off matrix is now

	(F, H)	(H, F)
(R, L)	10	10
(L, R)	5	5

5. Company adopts strategy M for maximum expected return.

Chapter 4

1. Player I: Maximin YXZ or YZX
   Minimax regret YXZ or YZX
   Laplace YXZ
   Player II: Laplace YXZ
   All profit goes to Player I.
2. $\pi_1 = \frac{1}{2}$.
3. Maximin 10 000 Minimax regret 30 000 EMV 30 000.
4. $\pi < \frac{1}{4}$ £1 million.
5. Smallest probability is 0.48.
   If x is the return and π the probability of the development, buy if $\pi > 12/(x + 5)$.

Chapter 5

1. £1 million.
   2 × £355 000 = £710 000
   Second firm would require £364 000.
   Greater assets, hence needs more incentive to take a large risk.

2. EMV criterion Project A 2.6 Project B 2.8
   Utility criterion Project A 0.7416 Project B 0.7390
3. Approximately 27.
4. $u(45) = 0.6$ $u(25) = 0.4$ $u(34) = 0.5$ $u(38) = 0.525$ $u(43) = 0.585$
   $u(44) = 0.5925$ $u(41) = 0.5625$
   $u(38)$ is out of line when a smooth curve is fitted through these points, suggesting that the premium is high for building B. Utilities are linear functions of the base choices. The relative positions of the above points are unaltered, i.e. the shape of the curve remains the same.

Chapter 6

1. Bid high for the airport contract. If this fails bid medium for the building society contract.
2. Make in regular time and also subcontract.
3. *A:* 200 now, further 400 if contract awarded
   B: 400 now, further 200 if contract awarded
   C: 600 now.
   Maximin: *C*, but this fails to take profit into account.
   Minimax regret: *B*, more plausible.
   Max. EMV: A if $\pi < 0.5$, C if $\pi > 0.5$, never B.
4. Optimum EMV strategy: Do not support Luftkopf. Build small, but if Shepshed win, extend.
   The research is worth no more than £350 000 to you.
5. Use market research. If they predict 'high', market. If they predict 'low', sell the design. With a cost of £800, the market research firm would not be engaged and he would market the product himself.

Chapter 7

1. Condition depends on taking the same launch/do not launch action whatever the outcome of the market research.
2. EVSI is now 0.51. Market research probabilities indicate a stronger relationship between the verdict and the market share. It is therefore more valuable.
3. They would still bid high.
4. EMVUC $= 2000 - 1600 = 400$
   EVSI $= 2000 - 1660 = 340$.
5. (a) £600 (b) £12 000
   Difference owing to poor record or prediction.

Chapter 8

1. 80 singles, 30 doubles.
   (a) shadow price is £2 per bedroom added. Five bedrooms is within the dining room capacity.
   (b) no benefit. Optimum solution is below capacity.
   (c) shadow price is £24.
   Limit is $1\frac{2}{3}$ hours.
2. 40 cars, 30 cars plus caravans. Profit of £1700 per trip. Limit of £40.
   (a) Change strategy to 48 cars, 26 cars plus caravans. Profit increases to £1740 per trip.
   (b) No change.
3. 110 noisy evenings, 220 quiet evenings.
   Optimum if charge increased by less than £2.50.
   Between £2 and £15 the optimal would be 92 noisy and 256 quiet evenings.
   Beyond £15 there would be 348 quiet evenings and no noisy evenings.
4. $X = \frac{1}{3}$, $Y = \frac{2}{3}$, $Z = 0$. Optimal if cost of $Z > 31\frac{2}{3}$.
5.

Earbrook →	Jandon	$5 + x$	Tapton	→	Jandon	$9 - x$
Earbrook →	Teiby	4	Tapton	→	Veeton	11
Earbrook →	Kenton	$21 - x$	Tapton	→	Kenton	x
Porcston →	Lipton	1	Sandiland	→	Lipton	17
Porcston →	Teiby	9				
Porcston →	Lumham	15	$0 \leqslant x \leqslant 9$			

   Increase in cost is £269 per day.
6.

Weydon-Priors	→	Melchester	3
Bulbarrow	→	Wintoncester	2
Bulbarrow	→	Casterbridge	1
Pilsdon	→	Melchester	$6 - x$
Pilsdon	→	Emminster	$2 + x$
Sherton Abbas	→	Casterbridge	x
Sherton Abbas	→	Emminster	$3 - x$
Stourcastle	→	Melchester	$4 + x$
Stourcastle	→	Casterbridge	$4 - x$
Shottsford	→	Casterbridge	8

   where $x = 0, 1, 2$ or 3.
7. Exchange V by firm F
   Exchange W by firm B
   Exchange X by firm B
   Exchange Y by firm A
   Exchange Z by firm C
   Reduce bid for Y by at least £100 000.
8. Total penalty is £1400.

9.

Bath	→	Southampton	5
Birmingham	→	Folkestone	$3 - x$
Birmingham	→	Gatwick	$16 + x$
Birmingham	→	Southampton	3
Canterbury	→	Folkestone	5
London	→	Folkestone	x
London	→	Gatwick	$15 - x$
Oxford	→	Southampton	2
Nottingham	→	Harwich	3

Spare cars at Birmingham (2) and Nottingham (1)

$$x = 0, 1, 2, 3$$

Reduction is £42.20.

Chapter 9

1. 40 single rooms, 60 double rooms.
   Any non-inferior strategy is of the form $X = 40 + 40a$, $Y = 60 - 3a$, where X = number of single rooms, Y = number of double rooms and $0 \leqslant a \leqslant 1$.
   Best integer solution is $X = 56$, $Y = 48$.
2. Optimum profit is £1700/trip. Loading time 130 minutes. Every minute saved costs £2 up to loading time of 105 minutes. Between 105 minutes and 75 minutes loading time, every minute saved costs £5. Below 75 minutes loading time, every minute saved costs £20.
3. X = number of 'noisy' evenings
   Y = number of 'quiet' evenings
   Non-inferior activities
   For $0 \leqslant X \leqslant 92$, $Y = 348 - X$, Profit $= 3480 + 15X$,
   Attendance $= 27\,840 - 30X$
   For $92 \leqslant X \leqslant 110$, $Y = 440 - 2X$, Profit $= 4400 + 5X$,
   Attendance $= 35\,200 - 110X$
   $X = 92$, $Y = 256$ is the new optimum.

Chapter 10

1.

Firm	A	B	C	D	E
Contract	Y	X	Z	W	V

2. $A \rightarrow C \rightarrow D \rightarrow B \rightarrow A$ (or the reverse).
3. 87 paying guests, 63 conference delegates.

Chapter 11

1. Manufacture 5 in January, 0 in February, 6 in March, 5 in April.
2. The previous policy shifts forward a day. On day 1 they also purchase only if the exchange rate is at its maximum.
3. Day 6: sell if price is 100, 105, 110
   Day 2–Day 5: sell if price is 105, 110
   Day 1: sell if price is 110.
4. Make the decision in three stages, each one corresponding to an allocation to one of the media.
   Optimum strategy is £20 000 to roadside posters, £30 000 to newspapers and zero to television, producing a return of 45.

Chapter 12

1. Reject remainders 101, 102. Take the remainder 100 as 00. The method cycles after 28 numbers.
2. Balance loss through being out of stock against cost of carrying inventory. Simulate various re-order levels.
3. Simulation should produce results which are a compromise between 'all 15 minutes' and 'all 20 minutes'.
   It would generally be to the physiotherapist's advantage to have the last appointment at 11.30 a.m., but it would tend to increase the customer waiting time towards the end of the morning.

Chapter 13

1. NPV = £27 430 IRR = 25.7%.
2. NPV = 893 IRR = 104 per cent.
3. NPV (£000) 3.06 1.40 0.33 − 1.33
   Probability 0.35 0.15 0.35 0.15
   Expected value is £1200.
4. NPV (£000) 2.15 1.32 1.24 0.41
   Probability 0.16 0.24 0.24 0.36
   Expected value is £1110.
5. Project *A*, IRR = 25.7% Project *B*, IRR = 24.7%
   If the NPV is used, Project *B* is better if the cost of capital is 16 per cent or less. Otherwise Project *A* would be selected.

Appendix 2
Areas in tail of the normal distribution

The function tabulated is $\Pr(Y \geqslant a)$ where Y is a standardized normal random variable.

$\Pr\ (Y \geqslant a) \qquad 0 \leqslant \qquad a \leqslant 3.29$

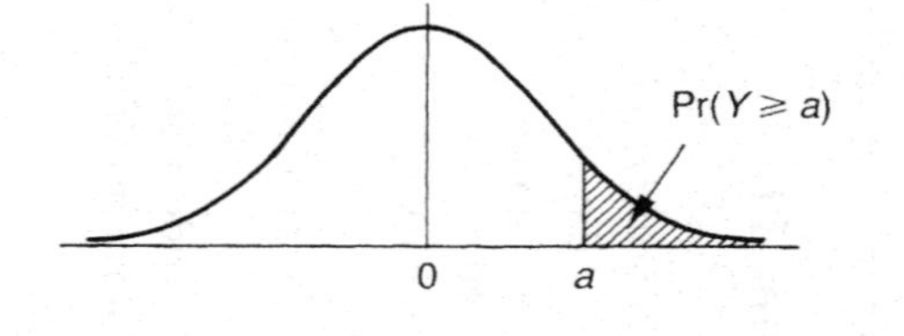

a	.00	.01	.02	.03	.04	.05	.06	.07	.08	.09
0.0	.5000	.4960	.4920	.4880	.4840	.4801	.4761	.4721	.4681	.4641
0.1	.4602	.4562	.4522	.4483	.4443	.4404	.4364	.4325	.4286	.4247
0.2	.4207	.4168	.4129	.4090	.4052	.4013	.3974	.3936	.3897	.3859
0.3	.3821	.3783	.3745	.3707	.3669	.3632	.3594	.3557	.3520	.3483
0.4	.3446	.3409	.3372	.3336	.3300	.3264	.3228	.3192	.3156	.3121
0.5	.3085	.3050	.3015	.2981	.2946	.2912	.2877	.2843	.2810	.2776
0.6	.2743	.2709	.2676	.2643	.2611	.2578	.2546	.2514	.2483	.2451
0.7	.2420	.2389	.2358	.2327	.2296	.2266	.2236	.2206	.2177	.2148
0.8	.2119	.2090	.2061	.2033	.2005	.1977	.1949	.1922	.1894	.1867
0.9	.1841	.1814	.1788	.1762	.1736	.1711	.1685	.1660	.1635	.1611

1.0	.1587	.1562	.1539	.1515	.1492	.1469	.1446	.1423	.1401	.1379
1.1	.1357	.1335	.1314	.1292	.1271	.1251	.1230	.1210	.1190	.1170
1.2	.1151	.1131	.1112	.1093	.1075	.1056	.1038	.1020	.1003	.0985
1.3	.0968	.0951	.0934	.0918	.0901	.0885	.0869	.0853	.0838	.0823
1.4	.0808	.0793	.0778	.0764	.0749	.0735	.0721	.0708	.0694	.0681
1.5	.0668	.0655	.0643	.0630	.0618	.0606	.0594	.0582	.0571	.0559
1.6	.0548	.0537	.0526	.0516	.0505	.0495	.0485	.0475	.0465	.0455
1.7	.0446	.0436	.0427	.0418	.0409	.0401	.0392	.0384	.0375	.0367
1.8	.0359	.0351	.0344	.0336	.0329	.0322	.0314	.0307	.0301	.0294
1.9	.0287	.0281	.0274	.0268	.0262	.0256	.0250	.0244	.0239	.0233
2.0	.02275	.02222	.02169	.02118	.02068	.02018	.01970	.01923	.01876	.01831
2.1	.01786	.01743	.01700	.01659	.01618	.01578	.01539	.01500	.01463	.01426
2.2	.01390	.01355	.01321	.01287	.01255	.01222	.01191	.01160	.01130	.01101
2.3	.01072	.01044	.01017	.00990	.00964	.00939	.00914	.00889	.00866	.00842
2.4	.00820	.00798	.00776	.00755	.00734	.00714	.00695	.00676	.00657	.00639
2.5	.00621	.00604	.00587	.00570	.00554	.00539	.00523	.00508	.00494	.00480
2.6	.00466	.00453	.00440	.00427	.00415	.00402	.00391	.00379	.00368	.00357
2.7	.00347	.00336	.00326	.00317	.00307	.00298	.00289	.00280	.00272	.00264
2.8	.00256	.00248	.00240	.00233	.00226	.00219	.00212	.00205	.00199	.00193
2.9	.00187	.00181	.00175	.00169	.00164	.00159	.00154	.00149	.00144	.00139
3.0	.00135	.00131	.00126	.00122	.00118	.00114	.00111	.00107	.00104	.00100
3.1	.00097	.00094	.00090	.00087	.00084	.00082	.00079	.00076	.00074	.00071
3.2	.00069	.00066	.00064	.00062	.00060	.00058	.00056	.00054	.00052	.00050

Appendix 3

Table of random numbers

74	79	25	87	00	36	24	63	44	29	37	00	73	10	81	22	27	66	34	99
38	35	66	92	57	61	08	86	10	96	76	90	92	36	81	66	55	30	22	23
47	07	47	82	31	89	93	84	08	01	95	36	36	23	99	23	80	90	06	21
62	95	54	27	15	09	94	99	66	61	66	95	52	66	84	65	08	48	49	05
55	27	94	52	22	77	00	71	75	88	87	23	96	50	53	56	55	19	07	59
37	87	74	20	51	84	14	25	98	97	57	56	15	68	79	28	51	60	24	32
60	01	54	56	68	73	20	30	58	90	82	51	33	38	70	93	30	25	45	54
61	92	62	47	21	58	41	20	48	92	81	59	49	42	70	50	06	50	61	38
21	80	86	65	09	76	69	74	44	91	29	29	97	83	17	02	74	22	69	95
63	52	14	82	66	32	81	16	01	69	57	99	13	67	50	16	35	26	80	30
67	88	99	90	59	44	39	11	13	02	44	80	01	57	08	07	92	08	83	57
60	76	82	00	30	91	46	15	37	75	76	70	75	73	29	31	31	07	98	39
29	97	32	40	22	94	96	98	17	50	58	28	73	50	44	50	03	53	74	85
88	62	96	97	87	82	38	51	80	62	70	85	90	22	18	02	42	44	39	25
30	15	76	71	22	43	69	50	62	35	03	97	49	16	13	67	28	91	54	05
82	75	78	81	81	40	38	51	65	60	73	14	24	91	17	77	17	01	13	71
24	75	78	30	58	59	48	30	30	33	23	34	45	85	30	17	28	93	41	39
16	25	13	08	43	27	99	28	18	39	55	67	95	22	97	82	83	69	93	71

51	40	86	70	83	42	71	25	72	81	36	71	37	48	69	67	00	40	97	55
32	19	21	68	74	68	32	65	79	72	05	76	39	67	32	59	56	01	10	60
29	92	72	80	03	92	30	76	00	26	71	23	48	80	83	54	19	18	36	28
86	32	53	98	71	93	09	81	41	60	07	37	46	14	93	49	87	58	57	22
97	91	30	18	71	79	83	02	73	05	42	43	12	88	83	43	18	62	82	61
19	21	09	08	86	98	70	43	94	63	20	23	31	78	70	45	96	67	91	24
94	70	89	84	75	35	47	63	28	88	45	52	78	99	23	32	01	70	82	07
81	80	12	65	26	82	32	12	40	81	39	14	01	18	05	88	93	77	96	64
41	12	81	00	81	13	72	76	47	54	18	79	47	33	46	68	80	88	99	52
70	68	96	10	93	80	60	60	27	60	65	90	24	01	18	70	06	95	23	22
67	56	99	52	07	84	91	23	28	17	42	28	18	04	34	96	99	83	33	51
12	43	71	22	90	14	04	49	59	75	15	55	32	54	22	76	20	52	69	22
47	59	79	89	78	66	63	76	53	67	08	02	24	86	30	40	07	41	15	16
89	08	29	44	18	73	32	13	34	84	29	23	61	80	08	41	54	90	32	58
46	32	86	67	82	68	88	43	08	58	70	03	81	92	80	61	01	40	70	54
14	55	51	16	90	79	87	78	51	76	94	02	86	28	61	99	84	19	94	78
03	17	24	13	66	48	70	53	43	90	27	17	72	42	63	87	08	79	93	92
81	40	88	62	71	55	71	64	04	72	64	42	69	66	27	13	38	32	30	89
46	59	36	95	28	94	11	62	63	02	10	59	30	85	29	20	79	64	95	41
76	06	09	53	77	75	72	70	59	55	11	21	97	17	37	26	91	05	10	30
48	11	80	88	99	50	60	52	73	62	12	56	48	00	00	43	50	01	64	60
66	85	42	15	40	02	76	66	41	99	15	77	11	67	08	61	51	30	53	31

Author Index

Index

9 780306 428548